Local Fractional Integral Transforms and Their Applications

Local Fractional Integral Transforms and Their Applications

Xiao-Jun Yang
Department of Mathematics and Mechanics, China University of Mining and Technology, Xuzhou, China

Dumitru Baleanu
Department of Mathematics and Computer Sciences, Faculty of Arts and Sciences, Cankaya University, Ankara, Turkey
and
Institute of Space Sciences, Magurele-Bucharest, Romania

H. M. Srivastava
Department of Mathematics and Statistics, University of Victoria, Victoria, British Columbia V8W 3R4, Canada

Academic Press is an imprint of Elsevier
32 Jamestown Road, London NW1 7BY, UK
The Boulevard, Langford Lane, Kidlington, Oxford OX5 1GB, UK
225 Wyman Street, Waltham, MA 02451, USA
525 B Street, Suite 1800, San Diego, CA 92101-4495, USA

ISBN: 978-0-12-804002-7

Library of Congress Cataloging-in-Publication Data
A catalog record for this book is available from the Library of Congress

British Library Cataloguing in Publication Data
A catalogue record for this book is available from the British Library

For information on all Academic Press publications
visit our website at http://store.elsevier.com/

Contents

List of figures

List of tables

Preface

The purpose of this book is to give a detailed introduction to the local fractional integral transforms and their applications in various fields of science and engineering. The local fractional calculus is utilized to handle various nondifferentiable problems that appear in complex systems of the real-world phenomena. Especially, the nondifferentiability occurring in science and engineering was modeled by the local fractional ordinary or partial differential equations. Thus, these topics are important and interesting for researchers working in such fields as mathematical physics and applied sciences.

In light of the above-mentioned avenues of their potential applications, we systematically present the recent theory of local fractional calculus and its new challenges to describe various phenomena arising in real-world systems. We describe the basic concepts for fractional derivatives and fractional integrals. We then illustrate the new results for local fractional calculus. Specifically, we have clearly stated the basic ideas of local fractional integral transforms and their applications.

The book is divided into five chapters with six appendices.

Chapter 1 points out the recent concepts involving fractional derivatives. We give the properties and theorems associated with the local fractional derivatives and the local fractional integrals. Some of the local fractional differential equations occurring in mathematical physics are discussed. With the help of the Cantor-type circular coordinate system, Cantor-type cylindrical coordinate system, and Cantor-type spherical coordinate system, we also present the local fractional partial differential equations in fractal dimensional space and their forms in the Cantor-type cylindrical symmetry form and in the Cantor-type spherical symmetry form.

In Chapter 2, we address the basic idea of local fractional Fourier series via the analogous trigonometric functions, which is derived from the complex Mittag–Leffler function defined on the fractal set. The properties and theorems of the local fractional Fourier series are discussed in detail. We mainly focus on the Bessel inequality for local fractional Fourier series, the Riemann–Lebesgue theorem for local fractional Fourier series, and convergence theorem for local fractional Fourier series. Some applications to signal analysis, ODEs and PDEs are also presented. We specially discuss the local fractional Fourier solutions of the homogeneous and nonhomogeneous local fractional heat equations in the nondimensional case and the local fractional Laplace equation and the local fractional wave equation in the nondimensional case.

Chapter 3 is devoted to an introduction of the local fractional Fourier transform operator via the Mittag–Leffler function defined on the fractal set, which is derived by approximating the local fractional integral operator of the local fractional Fourier series. The properties and theorems of the local fractional Fourier transform operator are discussed. A particular attention is paid to the logical explanation for the theorems for the local fractional Fourier transform operator and for another version of the local fractional Fourier transform operator (which is called the generalized local fractional Fourier transform operator). Meanwhile, we consider some application of the local fractional Fourier transform operator to signal processing, ODEs, and PDEs with the help of the local fractional differential operator.

Chapter 4 addresses the study of the local fractional Laplace transform operator based on the local fractional calculus. Our attentions are focused on the basic properties and theorems of the local fractional Laplace transform operator and its potential applications, such as those in signal analysis, ODEs, and PDEs involving the local fractional derivative operators. Some typical examples for the PDEs in mathematical physics are also discussed.

Chapter 5 treats the variational iteration and decomposition methods and the coupling methods of the Laplace transform with them involved in the local fractional operators. These techniques are then utilized to solve the local fractional partial differential equations. Their nondifferentiable solutions with graphs are also discussed.

We take this opportunity to thank many friends and colleagues who helped us in our writing of this book. We would also like to express our appreciation to several staff members of Elsevier for their cooperation in the production process of this book.

Xiao-Jun Yang
Dumitru Baleanu
H.M. Srivastava

Introduction to local fractional derivative and integral operators

1

1.1 Introduction

1.1.1 Definitions of local fractional derivatives

The concept of local fractional calculus (also called fractal calculus), which was first proposed by Kolwankar and Gangal [1, 2] based on the Riemann–Liouville fractional derivative [3–6], was applied to deal with nondifferentiable problems from science and engineering [7–16]. Several other points of fractal calculus were presented, such as the fractal derivative via Hausdorff measure [1, 17, 18], fractal derivative using fractal geometry [1, 19, 20], and local fractional derivative using the fractal geometry [1, 21–25]. Here, in this chapter, we present the logical extensions of the definitions to the subject of local derivative on fractals.

Let us recall the basic definitions as follows.

Local fractional derivative of $\Phi(\mu)$ of order ε $(0 < \varepsilon \leq 1)$ defined in [1, 2, 7–16] is given by

$$D^{(\varepsilon)}\Phi(\mu) = \left.\frac{\mathrm{d}^{\varepsilon}\Phi(\mu)}{\mathrm{d}\mu^{\varepsilon}}\right|_{\mu=\mu_0} = \lim_{\mu\to\mu_0}\frac{\mathrm{d}^{\varepsilon}\left[\Phi(\mu)-\Phi(\mu_0)\right]}{\left[\mathrm{d}(\mu-\mu_0)\right]^{\varepsilon}}, \tag{1.1}$$

where the term $\mathrm{d}^{\varepsilon}\left[\Phi(\mu)\right]/\left[\mathrm{d}(\mu-\mu_0)\right]^{\varepsilon}$ is the Riemann–Liouville fractional derivative of order ε of $\Phi(\mu)$.

Local fractional (fractal) derivative of $\Phi(\mu)$ of order ε $(0 < \varepsilon \leq 1)$ via Hausdorff measure μ^{ε} defined in [1, 17, 18] is given by

$$D^{(\varepsilon)}\Phi(\mu) = \left.\frac{\mathrm{d}^{\varepsilon}\Phi(\mu)}{\mathrm{d}\mu^{\varepsilon}}\right|_{\mu=\mu_0} = \lim_{\mu\to\mu_0}\frac{\Phi(\mu)-\Phi(\mu_0)}{\mu^{\varepsilon}-\mu_0^{\varepsilon}}, \tag{1.2}$$

where μ^{ε} is a fractal measure.

Local fractional (fractal) derivative using fractal geometry of $\Phi(\mu)$ of order ε $(0 < \varepsilon \leq 1)$ defined in [1, 19, 20] is written as

$$D^{(\varepsilon)}\Phi(\mu) = \left.\frac{\mathrm{d}\Phi(\mu)}{\mathrm{d}\mu^{\varepsilon}}\right|_{\mu=\mu_0} = \frac{\mathrm{d}\Phi(\mu)}{\mathrm{d}\sigma} = \lim_{\Delta\mu\to\mu_0}\frac{\Phi(\mu_B)-\Phi(\mu_A)}{\Upsilon\eta_0^{\varepsilon}}, \tag{1.3}$$

where $\mathrm{d}\sigma = \Upsilon\eta_0^{\varepsilon}$ with geometric parameter Υ and measure scale η_0 is shown in Figure 1.1.

Local Fractional Integral Transforms and Their Applications. http://dx.doi.org/10.1016/B978-0-12-804002-7.00001-2

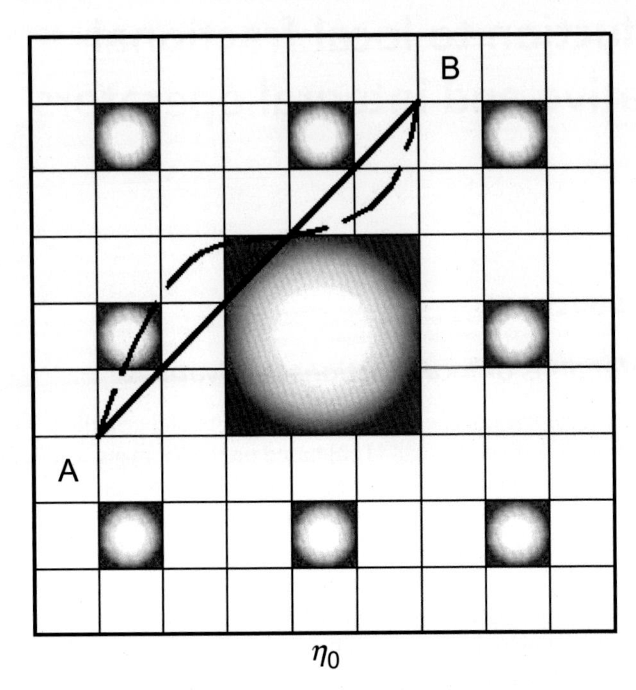

Figure 1.1 The distance between two points of A and B in a discontinuous space-time.

The local fractional derivative using the fractal geometry $\Phi(\mu)$ of order ε $(0 < \varepsilon \leq 1)$ defined in [1, 21–25] has the following form:

$$D^{(\varepsilon)}\Phi(\mu) = \left.\frac{d^{\varepsilon}\Phi(\mu)}{d\mu^{\varepsilon}}\right|_{\mu=\mu_0} = \lim_{\mu\to\mu_0}\frac{\Delta^{\varepsilon}[\Phi(\mu)-\Phi(\mu_0)]}{(\mu-\mu_0)^{\varepsilon}}, \tag{1.4}$$

where $\Delta^{\varepsilon}[\Phi(\mu)-\Phi(\mu_0)] \cong \Gamma(1+\varepsilon)[\Phi(\mu)-\Phi(\mu_0)]$ with the Euler's Gamma function $\Gamma(1+\varepsilon) =: \int_0^{\infty}\mu^{\varepsilon-1}\exp(-\mu)\,d\mu$.

Following (1.4), we define ε $(0 < \varepsilon \leq 1)$-dimensional Hausdorff measure given by [1–25]

$$H^{\varepsilon}[\Omega\cap(\mu_0,\mu)] = (\mu-\mu_0)^{\varepsilon}, \tag{1.5}$$

and its plot when $\varepsilon = \ln 2/\ln 3$ is the dimension of the fractal set Ω and $\mu_0 = 0$ is shown in Figure 1.2.

1.1.2 Comparisons of fractal relaxation equation in fractal kernel functions

The fractal relaxation equation with the help of (1.1) is given as

$$D^{(\varepsilon)}\Phi(\mu) + \omega\Phi(\mu) = 0, \tag{1.6}$$

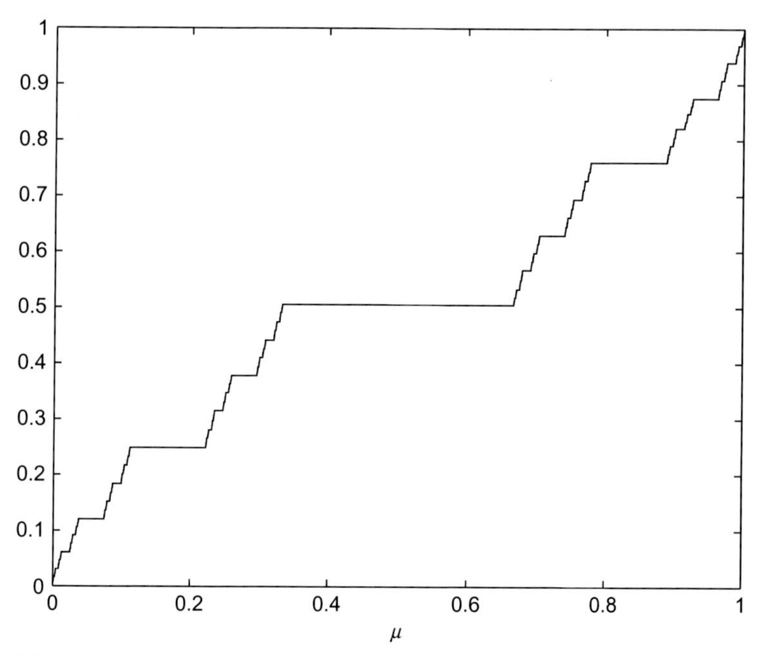

Figure 1.2 The curve of ε-dimensional Hausdorff measure with $\varepsilon = \ln 2/\ln 3$.

where $\Phi(0) = 1$. Its solution is written as follows:

$$\Phi(\mu) = \exp(-\omega F_c(\mu)), \tag{1.7}$$

where $F_c(\mu)$ is a Lebesgue–Cantor function and $F_c(\mu) \sim \mu^{\varepsilon}$.

The fractal relaxation equation with the help of (1.2) is given as follows [26]:

$$D^{(\varepsilon)}\Phi(\mu) + \omega\Phi(\mu) = 0, \tag{1.8}$$

where $\Phi(0) = 1$, and its solution is given by

$$\Phi(\mu) = \exp\left(-\omega\mu^{\varepsilon}\right). \tag{1.9}$$

The fractal relaxation equation by using (1.3) (see [19]):

$$D^{(\varepsilon)}\Phi(\mu) + \omega\Phi(\mu) = 0, \tag{1.10}$$

with $\Phi(0) = 1$ that has the solution

$$\Phi(\mu) = \exp\left(-\omega\Upsilon\eta_0^{\varepsilon-1}\mu\right), \tag{1.11}$$

where $\sigma = \Upsilon\eta_0^{\varepsilon-1}\mu$.

The fractal relaxation equation based on (1.4) is given as follows (see [26]):

$$D^{(\varepsilon)}\Phi(\mu) + \omega\Phi(\mu) = 0, \tag{1.12}$$

and its solution is presented as

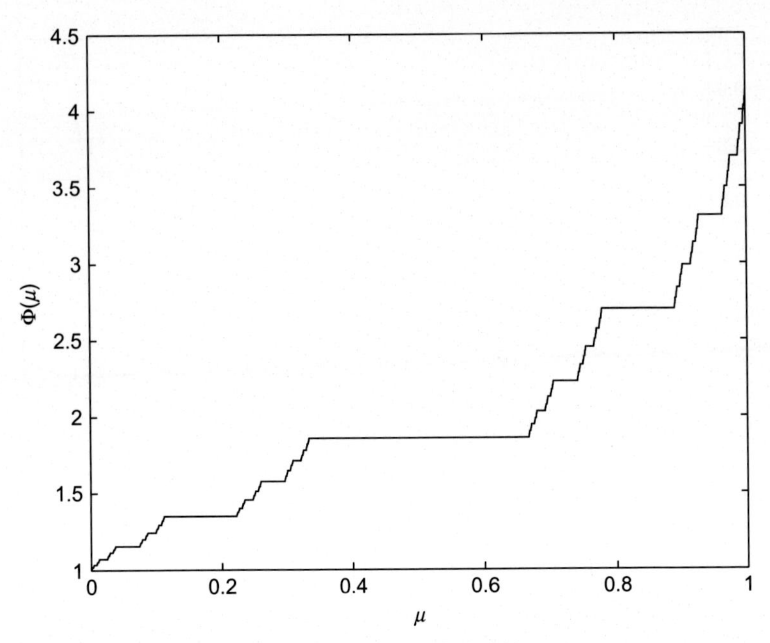

Figure 1.3 The chart of $\Phi(\mu)$ when $\omega = 1$ and $\varepsilon = \ln 2/\ln 3$.

$$\Phi(\mu) = E_{\varepsilon}\left(-\omega\mu^{\varepsilon}\right), \tag{1.13}$$

where $E_{\varepsilon}(-\omega\mu^{\varepsilon}) = \sum_{i=0}^{\infty} \frac{(-1)\omega^{i}\mu^{\varepsilon i}}{\Gamma(1+\varepsilon i)}$ is defined on the Cantor sets. The corresponding graph for $\omega = 1$ and $\varepsilon = \ln 2/\ln 3$ is shown in Figure 1.3.

1.1.3 Comparisons of fractal diffusion equation in fractal kernel functions

The fractal diffusion equation based on (1.1) is presented as follows (see [8]):

$$\frac{\partial^{\varepsilon}\Phi(\sigma,\mu)}{\partial\mu^{\varepsilon}} = \Lambda\frac{\partial^{2}\Phi(\sigma,\mu)}{\partial\sigma^{2}}, \tag{1.14}$$

where $\Lambda = \Gamma(1+\varepsilon)\chi_{c}(\mu)/4$, and its solution is given by

$$\Phi(\sigma,\mu) = \frac{1}{\sqrt{\pi F_{c}(\mu)}}\exp\left(-\frac{\sigma^{2}}{F_{c}(\mu)}\right), \tag{1.15}$$

where $F_{c}(\mu)$ is a Lebesgue–Cantor function and $\chi_{c}(\mu)$ is the membership function of a Cantor set.

We mention that the fractal diffusion equation within (1.2) has the form [17]:

$$\frac{\partial^{\varepsilon}\Phi(\sigma,\mu)}{\partial\mu^{\varepsilon}} = \Lambda\frac{\partial^{2\varsigma}\Phi(\sigma,\mu)}{\partial\sigma^{2\varsigma}}, \tag{1.16}$$

where $0 < \varsigma \leq 1$ and Λ is a contact, and its solution is

$$\Phi(\sigma,\mu) = \frac{1}{\sqrt{4\pi\Lambda\mu^{\varepsilon}}} \exp\left(-\frac{\sigma^{2\sigma}}{4\Lambda\mu^{\varepsilon}}\right). \tag{1.17}$$

The fractal diffusion equation based on (1.3) has the form

$$\frac{\partial^{\varepsilon}\Phi(\sigma,\mu)}{\partial\mu^{\varepsilon}} = \Lambda\frac{\partial^{2\varsigma}\Phi(\sigma,\mu)}{\partial\sigma^{2\varsigma}}, \tag{1.18}$$

where $0 < \varsigma \leq 1$ and Λ is a contact, and its solution is given by

$$\Phi(\sigma,\mu) = \frac{1}{\sqrt{4\pi\Lambda\Upsilon\eta_0^{1-\varepsilon}\mu}} \exp\left(-\frac{\left(\iota\xi_0^{1-\varsigma}\sigma\right)^2}{4\Lambda\Upsilon\eta_0^{1-\varepsilon}\mu}\right). \tag{1.19}$$

We mention below the fractal diffusion equation based on (1.4) [23]

$$\frac{\partial^{\varepsilon}\Phi(\sigma,\mu)}{\partial\mu^{\varepsilon}} = \Lambda\frac{\partial^{2\varepsilon}\Phi(\sigma,\mu)}{\partial\sigma^{2\varepsilon}}, \tag{1.20}$$

where Λ is a contact. The solution is given by

$$\Phi(\sigma,\mu) = \Phi_0\mu^{\beta\varepsilon}E_{\varepsilon}\left(-\frac{\sigma^{2\varepsilon}}{(4\Lambda\mu)^{\varepsilon}}\right), \tag{1.21}$$

where $E_{\varepsilon}\left(-\omega\mu^{2\varepsilon}\right) = \sum_{i=0}^{\infty}\frac{2^{\varepsilon}}{2}\frac{(-1)^i\omega^i\mu^{2\varepsilon i}}{\Gamma(1+\varepsilon i)}$ is defined on the Cantor sets and its graph, when $\omega = 1$ and $\varepsilon = \ln 2/\ln 3$ [23], is shown in Figure 1.4.

When $\Lambda = 1$, we conclude that

$$\Phi(\sigma,\mu) = \Phi_{0,0}\mu^{\beta\varepsilon}E_{\varepsilon}\left(-\frac{\sigma^{2\varepsilon}}{(4\mu)^{\varepsilon}}\right), \tag{1.22}$$

such that [21]

$$\Phi(\sigma,0) = \delta_{\varepsilon}(\sigma). \tag{1.23}$$

Below, we present a new definition of the local fractional Dirac function, namely,

$$\delta_{\varepsilon}(\sigma) = \lim_{\mu\to 0}\Phi_{0,0}\mu^{\beta\varepsilon}E_{\varepsilon}\left(-\frac{\sigma^{2\varepsilon}}{(4\mu)^{\varepsilon}}\right). \tag{1.24}$$

Using the reference [27], we have

$$\frac{1}{\Gamma(1+\varepsilon)}\int_{-\infty}^{\infty}\frac{1}{\frac{(4\pi\mu)^{\frac{\varepsilon}{2}}}{\Gamma(1+\varepsilon)}}E_{\varepsilon}\left(-\frac{\sigma^{2\varepsilon}}{(4\mu)^{\varepsilon}}\right)(d\sigma)^{\varepsilon}, \tag{1.25}$$

so that

$$\delta_{\varepsilon}(\sigma) = \lim_{\mu\to 0}\frac{1}{\frac{(4\pi\mu)^{\frac{\varepsilon}{2}}}{\Gamma(1+\varepsilon)}}E_{\varepsilon}\left(-\frac{\sigma^{2\varepsilon}}{(4\mu)^{\varepsilon}}\right). \tag{1.26}$$

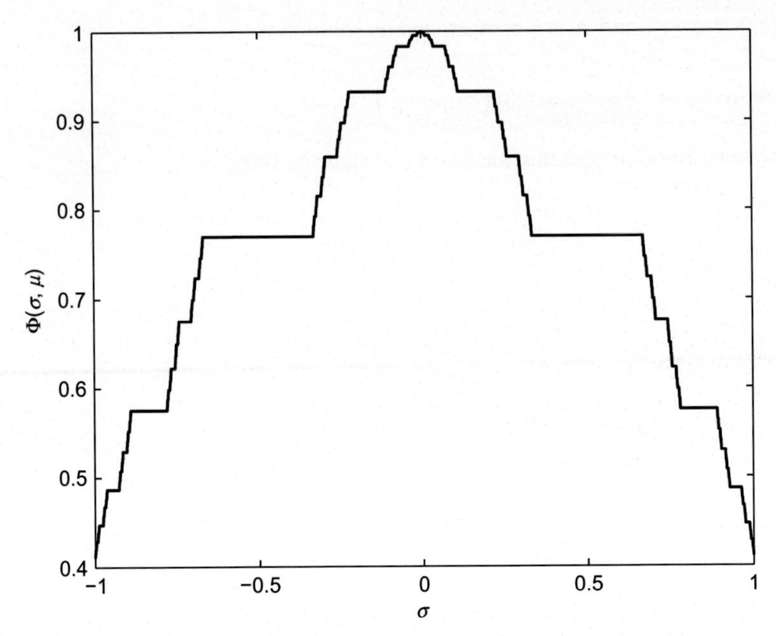

Figure 1.4 The concentration-distance curves for nondifferentiable source (see [23]).

Hence, with the help of (1.24) and (1.26), we get

$$\Phi_{0,0} = \frac{1}{\frac{(4\pi)^{\frac{\varepsilon}{2}}}{\Gamma(1+\varepsilon)}} \tag{1.27}$$

and

$$\beta = -\frac{\varepsilon}{2}. \tag{1.28}$$

In a similar manner, we obtain

$$\frac{1}{\Gamma(1+\varepsilon)} \int_{-\infty}^{\infty} \frac{1}{\frac{(4\pi\Lambda\mu)^{\frac{\varepsilon}{2}}}{\Gamma(1+\varepsilon)}} E_\varepsilon\left(-\frac{\sigma^{2\varepsilon}}{(4\Lambda\mu)^\varepsilon}\right)(d\sigma)^\varepsilon. \tag{1.29}$$

Therefore, there is a local fractional Dirac function defined by

$$\delta_\varepsilon(\sigma) = \lim_{\mu\to 0} \frac{1}{\frac{(4\pi\Lambda\mu)^{\frac{\varepsilon}{2}}}{\Gamma(1+\varepsilon)}} E_\varepsilon\left(-\frac{\sigma^{2\varepsilon}}{(4\Lambda\mu)^\varepsilon}\right), \tag{1.30}$$

so that

$$\Phi_0 = \frac{1}{\frac{(4\pi\mu)^{\frac{\varepsilon}{2}}}{\Gamma(1+\varepsilon)}}. \tag{1.31}$$

1.1.4 *Fractional derivatives via fractional differences*

Fractional derivatives via fractional differences were applied to solve the numerical problems for fractional differential equations in mathematical physics. We present the basic definitions of them given below:

The Grünwald–Letnikov derivative of the function $\Phi(\mu)$ of fractional order $\varepsilon\ (0 < \varepsilon \leq 1)$ [6, 28–34] is a fractional derivative via fractional difference, given by

$$D^{(\varepsilon)}\Phi(\mu) = \left.\frac{d^{\varepsilon}\Phi(\mu)}{d\mu^{\varepsilon}}\right|_{\mu=\mu_0} = \lim_{\rho\to 0}\frac{\Delta^{\varepsilon}\Phi(\mu)}{\rho^{\varepsilon}}, \tag{1.32}$$

where the fractional difference term is

$$\Delta^{\varepsilon}\Phi(\mu) = \sum_{i=0}^{\infty}(-1)^{i}\binom{\varepsilon}{i}\Phi(\mu - i\rho), \tag{1.33}$$

with $\binom{\varepsilon}{i} = \frac{\Gamma(1+\varepsilon)}{\Gamma(1+i)\Gamma(1+\varepsilon-i)}$.

The fractional derivative of the function $\Phi(\mu)$ of fractional order $\varepsilon\ (0 < \varepsilon \leq 1)$ [35–37] is a fractional derivative via fractional difference, given by

$$D^{(\varepsilon)}\Phi(\mu) = \left.\frac{d^{\varepsilon}\Phi(\mu)}{d\mu^{\varepsilon}}\right|_{\mu=\mu_0} = \lim_{\rho\to 0}\frac{\Delta^{\varepsilon}\Phi(\mu)}{\rho^{\varepsilon}}. \tag{1.34}$$

Here the fractional difference term is given by

$$\Delta^{\varepsilon}\Phi(\mu) = \sum_{i=0}^{\infty}(-1)^{i}\binom{\varepsilon}{i}\Phi(\mu - (\varepsilon - i)\rho). \tag{1.35}$$

The fractional derivative of the function $\Phi(\mu)$ of fractional order $\varepsilon\ (0 < \varepsilon \leq 1)$ introduced in [38] is a fractional derivative via fractional difference, given by

$$D^{(\varepsilon)}\Phi(\mu) = \left.\frac{d^{\varepsilon}\Phi(\mu)}{d\mu^{\varepsilon}}\right|_{\mu=\mu_0} = \lim_{\rho\to 0}\frac{\Delta^{\varepsilon}\left[\Phi(\mu) - \Phi(\mu_0)\right]}{\rho^{\varepsilon}}, \tag{1.36}$$

where the fractional difference term is

$$\Delta^{\varepsilon}\Phi(\mu) = \sum_{i=0}^{\infty}(-1)^{i}\binom{\varepsilon}{i}\Phi(\mu - (\varepsilon - i)\rho). \tag{1.37}$$

The fractional derivative of the function $\Phi(\mu)$ of variational order $\varepsilon(\mu)\ (0 < \varepsilon(\mu) \leq 1)$ [24] is defined as

$$D^{(\varepsilon(\mu))}\Phi(\mu) = \left.\frac{d^{\varepsilon(\mu)}\Phi(\mu)}{d\mu^{\varepsilon(\mu)}}\right|_{\mu=\mu_0} = \lim_{\rho\to 0}\frac{\Delta^{\varepsilon(\mu)}\left[\Phi(\mu) - \Phi(0)\right]}{\rho^{\varepsilon(\mu)}}, \tag{1.38}$$

where the fractional difference term is given by

$$\Delta^{\varepsilon(\mu)}\Phi(\mu) = \sum_{i=0}^{\infty}(-1)^{i}\frac{1}{\Gamma(i - \varepsilon(\mu))}\Phi(\mu \quad (\varepsilon(\mu) - i)\rho). \tag{1.39}$$

The Grünwald–Letnikov–Riesz derivative of the function $\Phi(\mu)$ of fractional order ε $(0 < \varepsilon \leq 1)$ via Grünwald–Letnikov derivative introduced in [39] is defined as

$$D^{(\varepsilon)}\Phi(\mu) = \left.\frac{d^{\varepsilon}\Phi(\mu)}{d\mu^{\varepsilon}}\right|_{\mu=\mu_0} = c_{\varepsilon}\lim_{\rho\to 0}\frac{\left[\Delta_{+}^{\varepsilon}\Phi(\mu) + \Delta_{-}^{\varepsilon}\Phi(\mu)\right]}{\rho^{\varepsilon}}, \tag{1.40}$$

where

$$c_{\varepsilon} = \frac{1}{2\cos\left(\frac{\pi\varepsilon}{2}\right)}, \tag{1.41}$$

and the fractional difference terms are for $\rho > 0$ and $\rho < 0$,

$$\Delta_{+}^{\varepsilon}\Phi(\mu) = \sum_{i=0}^{\infty}(-1)^{|i|}\binom{\varepsilon}{i}\Phi(\mu - i\rho), \tag{1.42}$$

$$\Delta_{-}^{\varepsilon}\Phi(\mu) = \sum_{i=0}^{\infty}(-1)^{|i|}\binom{\varepsilon}{i}\Phi(\mu + i\rho), \tag{1.43}$$

respectively.

1.1.5 Fractional derivatives with and without singular kernels and other versions of fractional derivatives

Fractional derivatives with singular kernel [28–69] have found popular applications in the fields of science and engineering. We mention some of them, for example, Liouville, Riemann–Liouville, Caputo, Weyl, Marchaud, Hadamard, Chen, Canavati, Riesz, and Cossar. The details on the conformable fractional derivatives were discussed recently in [40, 41]. A tempered fractional derivative was proposed in [42]. Generalized Riemann and Caputo versions of fractional derivatives were proposed in [43]. A fractional derivative without singular kernel and some of its properties were discussed very recently in [44, 45]. Below, we present the definitions of fractional derivatives with and without singular kernels as well as the conformable and tempered fractional derivatives.

Liouville fractional derivative of the function $\Phi(\mu)$ of fractional order ε is defined as

$$D^{(\varepsilon)}\Phi(\mu) = \frac{1}{\Gamma(1-\varepsilon)}\frac{d}{d\mu}\int_{-\infty}^{\mu}\frac{\Phi(\lambda)}{(\mu-\lambda)^{\varepsilon}}d\lambda, \tag{1.44}$$

where $-\infty < \mu < \infty$ and ε is a real number.

Liouville left-sided fractional derivative of the function $\Phi(\mu)$ of fractional order ε is defined by

$$D_{+}^{(\varepsilon)}\Phi(\mu) = \frac{1}{\Gamma(n-\varepsilon)}\frac{d^{n}}{d\mu^{n}}\int_{0}^{\mu}\frac{\Phi(\lambda)}{(\mu-\lambda)^{\varepsilon+1-n}}d\lambda, \tag{1.45}$$

where $0 < \mu$, n is integer, and ε denotes a real number.

Liouville right-sided fractional derivative of the function $\Phi(\mu)$ of fractional order ε is given by

$$D_{-}^{(\varepsilon)}\Phi(\mu) = \frac{(-1)^n}{\Gamma(n-\varepsilon)}\frac{d^n}{d\mu^n}\int_{-\infty}^{\mu}\frac{\Phi(\lambda)}{(\mu-\lambda)^{\varepsilon+1-n}}d\lambda, \tag{1.46}$$

where $\mu < \infty$, n is integer, and ε is real number.

Riemann–Liouville left-sided fractional derivative of a function $\Phi(\mu)$ of fractional order ε is

$$D_{a+}^{(\varepsilon)}\Phi(\mu) = \frac{1}{\Gamma(n-\varepsilon)}\frac{d^n}{d\mu^n}\int_{a}^{\mu}\frac{\Phi(\lambda)}{(\mu-\lambda)^{\varepsilon+1-n}}d\lambda, \tag{1.47}$$

where $a \le \mu$, n is integer, and ε is real number.

Riemann–Liouville right-sided fractional derivative of the function $\Phi(\mu)$ of fractional order ε is defined as

$$D_{a+}^{(\varepsilon)}\Phi(\mu) = \frac{(-1)^n}{\Gamma(n-\varepsilon)}\frac{d^n}{d\mu^n}\int_{\mu}^{b}\frac{\Phi(\lambda)}{(\mu-\lambda)^{\varepsilon+1-n}}d\lambda, \tag{1.48}$$

where $\mu \le b$, n is integer, and ε denotes a real number.

Caputo left-sided fractional derivative of the function $\Phi(\mu)$ of fractional order ε is defined as

$$D_{a+}^{(\varepsilon)}\Phi(\mu) = \frac{1}{\Gamma(n-\varepsilon)}\int_{a}^{\mu}\frac{1}{(\mu-\lambda)^{\varepsilon+1-n}}\left[\frac{d^n}{d\lambda^n}\Phi(\lambda)\right]d\lambda. \tag{1.49}$$

Here $a \le \mu$, n denotes an integer, and ε is real number.

Caputo right-sided fractional derivative of the function $\Phi(\mu)$ of fractional order ε is defined by

$$D_{a+}^{(\varepsilon)}\Phi(\mu) = \frac{(-1)^n}{\Gamma(n-\varepsilon)}\int_{\mu}^{b}\frac{1}{(\mu-\lambda)^{\varepsilon+1-n}}\left[\frac{d^n}{d\lambda^n}\Phi(\lambda)\right]d\lambda, \tag{1.50}$$

where $\mu \le b$, n is integer, and ε is real number.

Weyl fractional derivative of the function $\Phi(\mu)$ of fractional order ε (alternative definition; see [24]) is defined as

$$D_{\mu}^{(\varepsilon)}\Phi(\mu) = \frac{1}{\Gamma(n-\varepsilon)}\frac{d^n}{d\mu^n}\int_{\mu}^{\infty}\frac{\Phi(\lambda)}{(\mu-\lambda)^{\varepsilon+1-n}}d\lambda. \tag{1.51}$$

Here n is an integer and ε denotes a real number.

Marchaud fractional derivative of the function $\Phi(\mu)$ of fractional order ε is defined as

$$D_{+}^{(\varepsilon)}\Phi(\mu) = \frac{\{\varepsilon\}}{\Gamma(1-\{\varepsilon\})}\int_{\mu}^{\infty}\frac{[\Phi(\mu)-\Phi(\lambda)]}{(\mu-\lambda)^{\{\varepsilon\}+1}}d\lambda, \tag{1.52}$$

where $\varepsilon = [\varepsilon] + \{\varepsilon\}$.

Marchaud left-sided fractional derivative of the function $\Phi(\mu)$ of fractional order ε is written as

$$D_{+}^{(\varepsilon)}\Phi(\mu)=\frac{\{\varepsilon\}}{\Gamma(1-\{\varepsilon\})}\int_{\mu}^{\infty}\frac{\left[\Phi^{([\varepsilon])}(\mu)-\Phi^{([\varepsilon])}(\mu-\lambda)\right]}{\lambda^{\{\varepsilon\}+1}}\mathrm{d}\lambda, \tag{1.53}$$

for $\varepsilon=[\varepsilon]+\{\varepsilon\}$.

Marchaud right-sided fractional derivative of a function $\Phi(\mu)$ of fractional order ε has the form

$$D_{-}^{(\varepsilon)}\Phi(\mu)=\frac{\{\varepsilon\}}{\Gamma(1-\{\varepsilon\})}\int_{0}^{\mu}\frac{\left[\Phi^{([\varepsilon])}(\mu)-\Phi^{([\varepsilon])}(\mu+\lambda)\right]}{\lambda^{\{\varepsilon\}+1}}\mathrm{d}\lambda, \tag{1.54}$$

where $\varepsilon=[\varepsilon]+\{\varepsilon\}$.

Below, the Hadamard fractional derivative of a function $\Phi(\mu)$ of fractional order ε is defined as

$$D_{+}^{(\varepsilon)}\Phi(\mu)=\frac{\varepsilon}{\Gamma(1-\varepsilon)}\int_{0}^{\mu}\frac{[\Phi(\mu)-\Phi(\lambda)]}{[\ln(\mu/\lambda)]^{\varepsilon+1}}\frac{\mathrm{d}\lambda}{\lambda}, \tag{1.55}$$

where ε is real number.

Now, we define the Chen left-sided fractional derivative of the function $\Phi(\mu)$ of fractional order ε has the form

$$D_{a}^{(\varepsilon)}\Phi(\mu)=\frac{1}{\Gamma(1-\varepsilon)}\frac{\mathrm{d}}{\mathrm{d}\mu}\int_{a}^{\mu}\frac{\Phi(\lambda)}{(\mu-\lambda)^{\varepsilon}}\mathrm{d}\lambda, \tag{1.56}$$

where $a\leq\mu$ and ε is real number.

Chen right-sided fractional derivative of the function $\Phi(\mu)$ of fractional order ε is defined as

$$D_{a}^{(\varepsilon)}\Phi(\mu)=-\frac{1}{\Gamma(1-\varepsilon)}\frac{\mathrm{d}}{\mathrm{d}\mu}\int_{\mu}^{b}\frac{\Phi(\lambda)}{(\lambda-\mu)^{\varepsilon}}\mathrm{d}\lambda, \tag{1.57}$$

where $\mu\leq b$ and ε is real number.

Canavati fractional derivative of the function $\Phi(\mu)$ of fractional order ε is given by

$$D^{(\varepsilon)}\Phi(\mu)=\frac{1}{\Gamma(1-\varepsilon)}\frac{\mathrm{d}}{\mathrm{d}\mu}\int_{0}^{\mu}\frac{1}{(\mu-\lambda)^{\varepsilon-n}}\left[\frac{\partial^{n}}{\partial\lambda^{n}}\Phi(\lambda)\right]\mathrm{d}\lambda, \tag{1.58}$$

where $0\leq\mu$, ε is real number, and $[\varepsilon]=n$ is integral.

Riesz fractional derivative of the function $\Phi(\mu)$ of fractional order ε has the form

$$D^{(\varepsilon)}\Phi(\mu)=\frac{-c_{\varepsilon}}{\Gamma(\varepsilon)}\frac{\partial^{n}}{\partial\mu^{n}}\left[\int_{-\infty}^{\mu}\frac{\Phi(\lambda)}{(\mu-\lambda)^{\varepsilon+1-n}}\mathrm{d}\lambda+\int_{\mu}^{\infty}\frac{\Phi(\lambda)}{(\lambda-\mu)^{\varepsilon+1-n}}\mathrm{d}\lambda\right], \tag{1.59}$$

where $c_{\varepsilon}=1/\left[2\cos\left(\frac{\pi\varepsilon}{2}\right)\right]$, ε is real number, and n is integer.

Cossar fractional derivative of the function $\Phi(\mu)$ of fractional order ε is defined as

$$D^{(\varepsilon)}\Phi(\mu) = \frac{-1}{\Gamma(1-\varepsilon)} \lim_{N\to\infty} \frac{\partial}{\partial\mu}\left[\int_{\mu}^{N} \frac{\Phi(\lambda)}{(\lambda-\mu)^{\varepsilon}} d\lambda\right], \tag{1.60}$$

where ε is real number.

Modified Riemann–Liouville fractional derivative of the function $\Phi(\mu)$ of fractional order ε is defined as

$$D^{(\varepsilon)}\Phi(\mu) = \frac{1}{\Gamma(1-\varepsilon)} \frac{\partial}{\partial\mu} \int_{0}^{\mu} \frac{\Phi(\lambda)-\Phi(0)}{(\lambda-\mu)^{\varepsilon}} d\lambda, \tag{1.61}$$

where ε is real number.

The conformable fractional derivative of the function $\Phi(\mu)$ of fractional order ε [40] is defined as

$$D^{(\varepsilon)}\Phi(\mu) = \lim_{\kappa\to 0} \frac{\Phi\left(\mu+\kappa\mu^{1-\varepsilon}\right)-\Phi(\mu)}{\kappa}, \tag{1.62}$$

where $\varepsilon(0<\varepsilon\le 1)$ is real number.

The modified conformable left-sided fractional derivative of the function $\Phi(\mu)$ of fractional order ε [41] is defined as

$$D^{(\varepsilon)}\Phi(\mu) = \lim_{\kappa\to 0} \frac{\Phi\left(\mu+\kappa(\mu-a)^{1-\varepsilon}\right)-\Phi(\mu)}{\kappa}, \tag{1.63}$$

where $\varepsilon(0<\varepsilon\le 1)$ is real number.

The modified conformable right-sided fractional derivative of the function $\Phi(\mu)$ of fractional order ε [41] is defined as

$$D^{(\varepsilon)}\Phi(\mu) = -\lim_{\kappa\to 0} \frac{\Phi\left(\mu+\kappa(\mu-a)^{1-\varepsilon}\right)-\Phi(\mu)}{\kappa}, \tag{1.64}$$

where $\varepsilon(0<\varepsilon\le 1)$ is real number.

Tempered left-sided fractional derivative of the function $\Phi(\mu)$ of fractional order ε introduced in [42] is defined as

$$D_a^{(\varepsilon)}\Phi(\mu) = \frac{\varepsilon}{\Gamma(1-\varepsilon)} \int_{0}^{\infty} \frac{\Phi(\mu)-\Phi(\mu-\lambda)}{\lambda^{\varepsilon+1}} \exp(-\iota\lambda)\, d\lambda, \tag{1.65}$$

where ε is real number.

Tempered left-sided fractional derivative of the function $\Phi(\mu)$ of fractional order ε introduced in [42] is defined as

$$D_a^{(\varepsilon)}\Phi(\mu) = \frac{\varepsilon}{\Gamma(1-\varepsilon)} \int_{0}^{\infty} \frac{\Phi(\mu)-\Phi(\mu+\lambda)}{\lambda^{\varepsilon+1}} \exp(-\iota\lambda)\, d\lambda, \tag{1.66}$$

where ε is real number.

Generalized Riemann fractional derivative of the function $\Phi(\mu)$ of fractional order ε introduced in [43] is defined as

$${}^{\gamma}D_a^{(\varepsilon)}\Phi(\mu) = \frac{(1+\gamma)\varepsilon}{\Gamma(1-\varepsilon)} \frac{d}{d\mu} \int_{a}^{\mu} \frac{\lambda^{\gamma}\Phi(\mu)}{\left(\mu^{\gamma+1}-\lambda^{\gamma+1}\right)^{\varepsilon}} d\lambda, \tag{1.67}$$

where $a \leq \mu$ and ε is real number.

Generalized Caputo fractional derivative of the function $\Phi(\mu)$ of fractional order ε introduced in [43] is defined as

$$ {}^{\gamma}D_{0}^{(\varepsilon)}\Phi(\mu) = \frac{(1+\gamma)\,\varepsilon}{\Gamma(1-\varepsilon)}\frac{\mathrm{d}}{\mathrm{d}\mu}\int_{a}^{\mu}\frac{\lambda^{\gamma}\Phi(\lambda)}{\left(\mu^{\gamma+1}-\lambda^{\gamma+1}\right)^{\varepsilon}}\mathrm{d}\lambda, \tag{1.68}$$

where $0 \leq \mu$ and ε is real number.

Erdelyi–Kober fractional derivative of the function $\Phi(\mu)$ of fractional order ε is defined as

$$D_{0,\xi,\zeta}^{(\varepsilon)}\Phi(\mu) = \mu^{-n\zeta}\left(\frac{1}{\xi\mu^{\zeta-1}}\frac{\mathrm{d}}{\mathrm{d}\mu}\right)^{n}\mu^{-\zeta(n+\zeta)}I_{0,\xi,\xi+\zeta}^{n-\varepsilon}\Phi(\mu), \tag{1.69}$$

where

$$I_{0,\xi,\xi+\zeta}^{n-\varepsilon}\Phi(\mu) = \frac{\xi\mu^{-\zeta(\zeta+\varepsilon)}}{\Gamma(\varepsilon)}\int_{0}^{\mu}\frac{\lambda^{\xi\zeta+\xi-1}\Phi(\mu)}{\left(\mu^{\zeta}-\lambda^{\zeta}\right)^{1-\varepsilon}}\mathrm{d}\lambda, \tag{1.70}$$

with real number ε.

Caputo–Fabrizio fractional derivative of the function $\Phi(\mu)$ of fractional order ε introduced in [44, 45] is defined as

$$D^{(\varepsilon)}\Phi(\mu) = \frac{1}{1-\varepsilon}\int_{0}^{\mu}\exp\left(-\frac{\varepsilon}{1-\varepsilon}(\mu-\lambda)\right)\Phi^{(1)}(\mu)\,\mathrm{d}\lambda, \tag{1.71}$$

where $0 < \mu$ and ε is real number.

Coimbra fractional derivative of the function $\Phi(\mu)$ of fractional order $\varepsilon(\mu)$ is defined as

$$D^{(\varepsilon(\mu))}\Phi(\mu) = \frac{1}{\Gamma(1-\varepsilon(\mu))}\left\{\int_{a}^{\mu}\frac{1}{(\mu-\lambda)^{\varepsilon(\mu)}}\left[\frac{\partial\Phi(\lambda)}{\partial\lambda}\right]\mathrm{d}\lambda + \Phi(0)\,\mu^{-\varepsilon(\mu)}\right\}, \tag{1.72}$$

where $a < \mu$ and $\varepsilon(\mu)$ $(0 < \varepsilon(\mu) < 1)$ is real number related to μ.

Left-sided Riemann–Liouville fractional derivative of the function $\Phi(\mu)$ of variable fractional order $\varepsilon(\lambda,\mu)$ is defined as

$$D_{a+}^{(\varepsilon(\lambda,\mu))}\Phi(\mu) = \frac{\mathrm{d}}{\mathrm{d}\mu}\int_{a}^{\mu}\frac{\Phi(\lambda)}{(\mu-\lambda)^{\varepsilon(\lambda,\mu)}}\frac{\mathrm{d}\lambda}{\Gamma[1-\varepsilon(\lambda,\mu)]}, \tag{1.73}$$

where $a < \mu$ and $\varepsilon(\lambda,\mu)$ $(0 < \varepsilon(\lambda,\mu) < 1)$ is real number related to μ.

Right-sided Riemann–Liouville fractional derivative of the function $\Phi(\mu)$ of variable fractional order $\varepsilon(\lambda,\mu)$ is defined as

$$D_{b-}^{(\varepsilon(\lambda,\mu))}\Phi(\mu) = \frac{\mathrm{d}}{\mathrm{d}\mu}\int_{\mu}^{b}\frac{\Phi(\lambda)}{(\lambda-\mu)^{\varepsilon(\lambda,\mu)}}\frac{\mathrm{d}\lambda}{\Gamma[1-\varepsilon(\lambda,\mu)]}, \tag{1.74}$$

where $\mu < b$ and $\varepsilon(\lambda,\mu)$ $(0 < \varepsilon(\lambda,\mu) < 1)$ is real number related to μ.

Left-sided Caputo fractional derivative of the function $\Phi(\mu)$ of variable fractional order $\varepsilon(\lambda,\mu)$ is defined as

$$D_{a^+}^{(\varepsilon(\lambda,\mu))}\Phi(\mu) = \int_a^{\mu} \frac{1}{(\mu-\lambda)^{\varepsilon(\lambda,\mu)}}\left[\frac{d}{d\mu}\Phi(\lambda)\right]\frac{d\lambda}{\Gamma[1-\varepsilon(\lambda,\mu)]}, \tag{1.75}$$

where $a < \mu$ and $\varepsilon(\lambda,\mu)$ $(0 < \varepsilon(\lambda,\mu) < 1)$ is real number related to μ.

Right-sided Caputo fractional derivative of the function $\Phi(\mu)$ of variable fractional order $\varepsilon(\lambda,\mu)$ is defined as

$$D_{b^-}^{(\varepsilon(\lambda,\mu))}\Phi(\mu) = \int_{\mu}^{b} \frac{1}{(\lambda-\mu)^{\varepsilon(\lambda,\mu)}}\left[\frac{d}{d\mu}\Phi(\lambda)\right]\frac{d\lambda}{\Gamma[1-\varepsilon(\lambda,\mu)]}, \tag{1.76}$$

where $\mu < b$ and $\varepsilon(\lambda,\mu)$ $(0 < \varepsilon(\lambda,\mu) < 1)$ is real number related to μ.

Caputo fractional derivative of variable fractional order is defined as

$$D_{a^+}^{(\varepsilon(\mu))}\Phi(\mu) = \frac{1}{\Gamma[1-\varepsilon(\mu)]}\int_a^{\mu} \frac{1}{(\mu-\lambda)^{\varepsilon(\mu)}}\left[\frac{d}{d\mu}\Phi(\lambda)\right]d\lambda, \tag{1.77}$$

where $\mu < b$ and $\varepsilon(\lambda,\mu)$ $(0 < \varepsilon(\lambda,\mu) < 1)$ is real number related to μ.

1.2 Definitions and properties of local fractional continuity

1.2.1 Definitions and properties

Let $\wp$ be a fractal set and let d_1 and d_0 be two metric spaces. Suppose Φ: $(\wp, d_0) \to (\aleph, d_1)$ is a bi-Lipschitz mapping, then, we have

$$\omega_1\Im^{\varepsilon}(\wp) \le \Im^{\varepsilon}(\Phi(\wp)) \le \omega_1\Im^{\varepsilon}(\wp) \tag{1.78}$$

such that

$$\omega_1|\mu_1-\mu_2| \le |\Phi(\mu_1)-\Phi(\mu_2)| \le \omega_1|\mu_1-\mu_2|, \tag{1.79}$$

where $\mu_1,\mu_2 \in \wp$, $\wp \subset R$, and $\omega_1,\omega_2 > 0$.

Using (1.79), for $\forall\rho > 0$ and $0 < \varepsilon < 1$, we have

$$|\Phi(\mu_1)-\Phi(\mu_2)| < \rho^{\varepsilon}, \tag{1.80}$$

where ε is fractal dimension of the fractal set $\wp$. This form is analogues of Lipschitz mapping.

Definition 1.1. Let Φ: $\wp \to \aleph$ be a function defined on a fractal set $\wp$ of fractal dimension $\varepsilon(0 < \varepsilon < 1)$. A real number χ is called a generalized limit of $\Phi(\mu)$ as μ tends to a, or the limit of $\Phi(\mu)$ at a, if to each $\tau > 0$ there corresponds $\delta > 0$ such that

$$|\Phi(\mu)-\chi| < \tau^{\varepsilon}, \tag{1.81}$$

whenever

$$0 < |\mu - a| < \delta. \tag{1.82}$$

The above statement is expressible in terms of inequalities as follows.

Suppose $\tau > 0$. Then, there is each $\delta > 0$ such that $|\Phi(\mu) - \chi| < \tau^{\varepsilon}$ if $0 < |\mu - a| < \delta$.

Thus, we write

$$\Phi(\mu) \to \chi \tag{1.83}$$

as $\mu \to a$, or

$$\lim_{\mu \to a} \Phi(\mu) = \chi. \tag{1.84}$$

We say that $\Phi(\mu)$ tends to χ as μ tends to a.

Definition 1.2. A function $\Phi(\mu)$ is said to be local fractional continuous at $\mu = \mu_0$ if for each $\tau > 0$, there exists for $\delta > 0$ such that

$$|\Phi(\mu) - \Phi(\mu_0)| < \tau^{\varepsilon}, \tag{1.85}$$

whenever $0 < |\mu - \mu_0| < \delta$.

It is written as

$$\lim_{\mu \to \mu_0} \Phi(\mu) = \Phi(\mu_0). \tag{1.86}$$

A function $\Phi(\mu)$ is said to be local fractional continuous at $\mu = \mu_0$ from the right if for each $\tau > 0$, there exists for $\delta > 0$ such that (1.82) holds whenever $\mu_0 < \mu < \delta + \mu_0$.

A function $\Phi(\mu)$ is said to be local fractional continuous at $\mu = \mu_0$ from the left if for each $\tau > 0$, there exists for $\delta > 0$ such that (1.82) holds whenever $\delta - \mu_0 < \mu < \mu_0$.

If $\lim_{\mu \to \mu_0^+} \Phi(\mu) = \Phi\left(\mu_0^+\right)$, $\lim_{\mu \to \mu_0^-} \Phi(\mu) = \Phi\left(\mu_0^-\right)$, and $\Phi\left(\mu_0^+\right) = \Phi\left(\mu_0^-\right)$ exist, then, we have

$$\lim_{\mu \to \mu_0} \Phi(\mu) = \lim_{\mu \to \mu_0^+} \Phi(\mu) = \lim_{\mu \to \mu_0^-} \Phi(\mu). \tag{1.87}$$

Suppose a function $\Phi(\mu)$ is local fractional continuous in the domain $I = (a, b)$, then, we write it as

$$\Phi(\mu) \in C_{\varepsilon}(a, b). \tag{1.88}$$

Theorem 1.1. *Suppose that* $\lim_{\mu \to \mu_0} \Phi(\mu) = \Phi(\mu_0)$ *and* $\lim_{\mu \to \mu_0} \Theta(\mu) = \Theta(\mu_0)$. *Then*

(a) $\lim_{\mu \to \mu_0} [\Phi(\mu) \pm \Theta(\mu)] = \Phi(\mu_0) \pm \Theta(\mu_0)$;
(b) $\lim_{\mu \to \mu_0} |\Phi(\mu)| = |\Phi(\mu_0)|$;
(c) $\lim_{\mu \to \mu_0} [\Phi(\mu) \Theta(\mu)] = \Phi(\mu_0) \Theta(\mu_0)$; *and*
(d) $\lim_{\mu \to \mu_0} [\Phi(\mu) / \Theta(\mu)] = \Phi(\mu_0) / \Theta(\mu_0)$, *provided* $\Theta(\mu_0) \neq 0$.

For the details of formal proofs of the validity of these four rules, see [1, 16, 21]. Theorem 1.1 is a natural generalized result of those known when the order is a positive integer.

1.2.2 Functions defined on fractal sets

Following the definition of ε-dimensional Hausdorff measure, we define the functions defined on fractal sets as follows:

Let $\Phi: \wp \to \aleph$ be a function defined on a fractal set $\wp$ of fractal dimension $\varepsilon (0 < \varepsilon < 1)$. A real-valued function $\Phi(\mu)$ defined on the fractal set $\wp$ is given by

$$\Phi(\mu) = \mu^{c}, \tag{1.89}$$

where $\mu^{\varepsilon} \in \wp$ and $0 < \varepsilon < 1$.

We now notice that (1.89) is a Lebesgue–Cantor function and $\lim_{\varepsilon \to 1} \Phi(\mu) = \mu \in R$ with real number set R.

The Mittag–Leffler function defined on the fractal set $\wp$ is given by

$$E_{\varepsilon}\left(\mu^{\varepsilon}\right) = \sum_{k=0}^{\infty} \frac{\mu^{k\varepsilon}}{\Gamma(1 + k\varepsilon)}, \tag{1.90}$$

where $\mu \in R$ and $0 < \varepsilon < 1$.

An extended version of (1.90) defined on the fractal set $\wp$ is given as

$$E_{\varepsilon}\left(\beta, \mu^{\varepsilon}\right) = \sum_{k=0}^{\infty} \frac{\mu^{k\varepsilon}}{\Gamma(\beta + k\varepsilon)}, \tag{1.91}$$

where β is real number, $\mu \in R$, and $0 < \varepsilon < 1$.

The following rules via Mittag–Leffler functions defined on the fractal set $\wp$ hold:

(a) $E_{\varepsilon}(\mu^{\varepsilon}) E_{\varepsilon}(\nu^{\varepsilon}) = E_{\varepsilon}(\mu^{\varepsilon} + \nu^{\varepsilon})$;
(b) $E_{\varepsilon}(\mu^{\varepsilon}) E_{\varepsilon}(-\nu^{\varepsilon}) = E_{\varepsilon}(\mu^{\varepsilon} - \nu^{\varepsilon})$;
(c) $E_{\varepsilon}(\mu^{\varepsilon}) E_{\varepsilon}(i^{\varepsilon}\nu^{\varepsilon}) = E_{\varepsilon}(\mu^{\varepsilon} + i^{\varepsilon}\nu^{\varepsilon})$;
(d) $E_{\varepsilon}(i^{\varepsilon}\mu^{\varepsilon}) E_{\varepsilon}(i^{\varepsilon}\nu^{\varepsilon}) = E_{\varepsilon}(i^{\varepsilon}\mu^{\varepsilon} + i^{\varepsilon}\nu^{\varepsilon})$; and
(e) $[E_{\varepsilon}(\mu^{\varepsilon} + i^{\varepsilon}\mu^{\varepsilon})]^{n} = E_{\varepsilon}(n^{\varepsilon}\mu^{\varepsilon} + n^{\varepsilon}i^{\varepsilon}\nu^{\varepsilon})$, where n is integer and i^{ε} is a imaginary unit of a fractal set $\wp$.

The sine function defined on the fractal set $\wp$ is given by

$$\sin_{\varepsilon}\left(\mu^{\varepsilon}\right) = \sum_{k=0}^{\infty} \frac{(-1)^{k} \mu^{(2k+1)\varepsilon}}{\Gamma(1 + (2k+1)\varepsilon)}, \tag{1.92}$$

where $\mu \in R$ and $0 < \varepsilon < 1$.

The cosine function defined on the fractal set $\wp$ is given by

$$\cos_{\varepsilon}\left(\mu^{\varepsilon}\right) = \sum_{k-0}^{\infty} \frac{(-1)^{k} \mu^{2k\varepsilon}}{\Gamma(1 + 2k\varepsilon)}, \tag{1.93}$$

where $\mu \in R$ and $0 < \varepsilon < 1$.

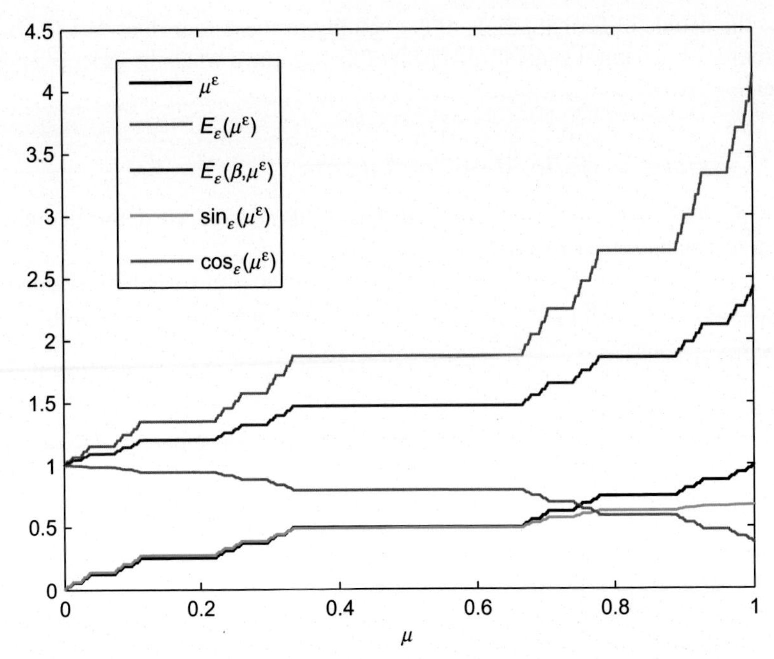

Figure 1.5 The comparisons of the nondifferentiable functions (1.89)–(1.93) when $\beta = 2$ and $\varepsilon = \ln 2/\ln 3$.

Their graphs corresponding to the fractal dimension $\varepsilon = \ln 2/\ln 3$ are shown in Figure 1.5.

The following rules via Mittag–Leffler, sine, and cosine functions defined on the fractal set $\wp$ hold:

(a) $E_{\varepsilon}(i^{\varepsilon}\mu^{\varepsilon}) = \cos_{\varepsilon}(\mu^{\varepsilon}) + i^{\varepsilon}\sin_{\varepsilon}(\mu^{\varepsilon})$;
(b) $\sin_{\varepsilon}(\mu^{\varepsilon}) = \frac{E_{\varepsilon}(i^{\varepsilon}\mu^{\varepsilon})-E_{\varepsilon}(-i^{\varepsilon}\mu^{\varepsilon})}{2i^{\varepsilon}}$;
(c) $\cos_{\varepsilon}(\mu^{\varepsilon}) = \frac{E_{\varepsilon}(i^{\varepsilon}\mu^{\varepsilon})+E_{\varepsilon}(-i^{\varepsilon}\mu^{\varepsilon})}{2}$;
(d) $\cos_{\varepsilon}(-\mu^{\varepsilon}) = \cos_{\varepsilon}(\mu^{\varepsilon})$;
(e) $\sin_{\varepsilon}(-\mu^{\varepsilon}) = -\sin_{\varepsilon}(\mu^{\varepsilon})$;
(f) $\sin^2_{\varepsilon}(\mu^{\varepsilon}) + \cos^2_{\varepsilon}(\mu^{\varepsilon}) = 1$; and
(g) $\frac{1}{2} + \sum_{k=1}^{n}\cos_{\varepsilon}(k\mu^{\varepsilon}) = \frac{\sin_{\varepsilon}((2n+1)\mu^{\varepsilon}/2)}{2\sin_{\varepsilon}(\mu^{\varepsilon}/2)}$, provided $\sin_{\varepsilon}(\mu^{\varepsilon}/2) \neq 0$.

Other properties are listed in Appendix A.

The hyperbolic functions via Mittag–Leffler function defined on the fractal set $\wp$ are given by

$$\sinh_{\varepsilon}\left(\mu^{\varepsilon}\right) = \frac{E_{\varepsilon}(\mu^{\varepsilon}) - E_{\varepsilon}(-\mu^{\varepsilon})}{2} = \sum_{k=0}^{\infty}\frac{\mu^{(2k+1)\varepsilon}}{\Gamma(1+(2k+1)\varepsilon)}, \tag{1.94}$$

$$\cosh_{\varepsilon}\left(\mu^{\varepsilon}\right)=\frac{E_{\varepsilon}\left(\mu^{\varepsilon}\right)+E_{\varepsilon}\left(-\mu^{\varepsilon}\right)}{2}=\sum_{k=0}^{\infty}\frac{\mu^{2k\varepsilon}}{\Gamma\left(1+2k\varepsilon\right)}, \tag{1.95}$$

$$\tanh_{\varepsilon}\left(\mu^{\varepsilon}\right)=\frac{E_{\varepsilon}\left(\mu^{\varepsilon}\right)-E_{\varepsilon}\left(-\mu^{\varepsilon}\right)}{E_{\varepsilon}\left(\mu^{\varepsilon}\right)+E_{\varepsilon}\left(-\mu^{\varepsilon}\right)}, \tag{1.96}$$

$$\coth_{\varepsilon}\left(\mu^{\varepsilon}\right)=\frac{E_{\varepsilon}\left(\mu^{\varepsilon}\right)+E_{\varepsilon}\left(-\mu^{\varepsilon}\right)}{E_{\varepsilon}\left(\mu^{\varepsilon}\right)-E_{\varepsilon}\left(-\mu^{\varepsilon}\right)}, \tag{1.97}$$

$$\sec h_{\varepsilon}\left(\mu^{\varepsilon}\right)=\frac{2}{E_{\varepsilon}\left(\mu^{\varepsilon}\right)+E_{\varepsilon}\left(-\mu^{\varepsilon}\right)}, \tag{1.98}$$

$$\csc h_{\varepsilon}\left(\mu^{\varepsilon}\right)=\frac{2}{E_{\varepsilon}\left(\mu^{\varepsilon}\right)-E_{\varepsilon}\left(-\mu^{\varepsilon}\right)}. \tag{1.99}$$

The comparison plot of the nondifferentiable functions (1.94) and (1.95) when $\varepsilon=\ln 2/\ln 3$ is shown in Figure 1.6.

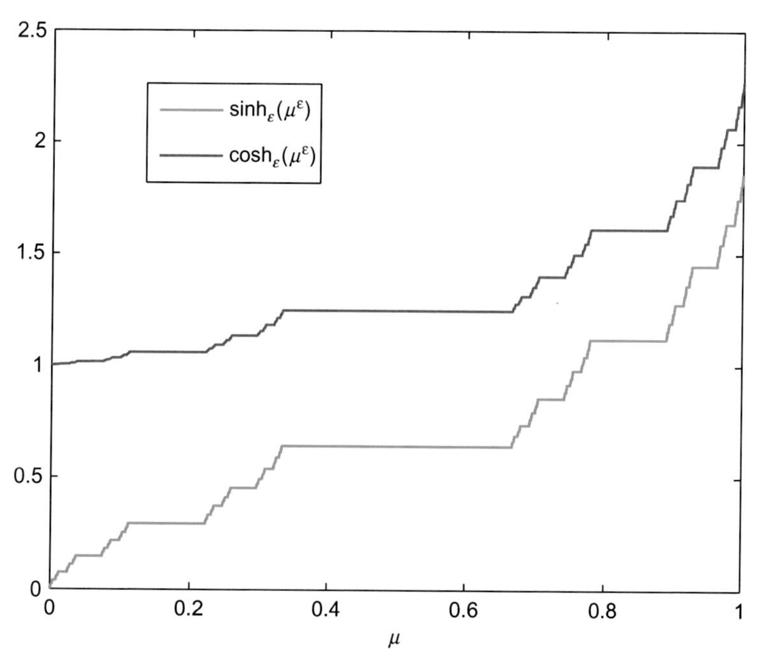

Figure 1.6 The comparisons of the nondifferentiable functions (1.94) and (1.95) when $\varepsilon=\ln 2/\ln 3$.

1.3 Definitions and properties of local fractional derivative

We discuss the definitions and prosperities of local fractional derivative for nondifferentiable functions defined on a fractal set.

1.3.1 Definitions of local fractional derivative

Definition 1.3. Suppose that $\Phi(\mu) \in C_\varepsilon(a,b)$ and $0 < \varepsilon \leq 1$. For $\sigma > 0$ and $0 < |\mu - \mu_0| < \delta$, the limit

$$D^{(\varepsilon)}\Phi(\mu_0) = \left.\frac{d^\varepsilon \Phi(\mu)}{d\mu^\varepsilon}\right|_{\mu=\mu_0} = \lim_{\mu\to\mu_0} \frac{\Delta^\varepsilon[\Phi(\mu) - \Phi(\mu_0)]}{(\mu-\mu_0)^\varepsilon} \tag{1.100}$$

exists and is finite, where $\Delta^\varepsilon[\Phi(\mu) - \Phi(\mu_0)] \cong \Gamma(1+\varepsilon)[\Phi(\mu) - \Phi(\mu_0)]$. In this case, $D^{(\varepsilon)}\Phi(\mu)$ is said to be the local fractional derivative of $\Phi(\mu)$ of order ε at $\mu = \mu_0$.

For our purposes, it is convenient to denote the local fractional derivative in the form $D^{(\varepsilon)}\Phi(\mu_0)$ or $\left.\frac{d^\varepsilon \Phi(\mu)}{d\mu^\varepsilon}\right|_{\mu=\mu_0}$.

If $\Phi(\mu)$ is defined on the interval $[\mu, b)$, the left-hand local fractional derivative of $\Phi(\mu)$ of order ε at $\mu = \mu_0$ is defined to be

$$\left.\frac{d^\varepsilon \Phi(\mu)}{d\mu^\varepsilon}\right|_{\mu=\mu_0^-} = \lim_{\mu\to\mu_0^-} \frac{\Delta^\varepsilon[\Phi(\mu) - \Phi(\mu_0)]}{(\mu-\mu_0)^\varepsilon}, \tag{1.101}$$

where

$$\Delta^\varepsilon[\Phi(\mu) - \Phi(\mu_0)] \cong \Gamma(1+\varepsilon)[\Phi(\mu) - \Phi(\mu_0)],$$

if the limit exists.

If $\Phi(\mu)$ is defined on $(a, \mu]$, the right-hand local fractional derivative of $\Phi(\mu)$ of order ε at $\mu = \mu_0$ is defined to be

$$\left.\frac{d^\varepsilon \Phi(\mu)}{d\mu^\varepsilon}\right|_{\mu=\mu_0^+} = \lim_{\mu\to\mu_0^+} \frac{\Delta^\varepsilon[\Phi(\mu) - \Phi(\mu_0)]}{(\mu-\mu_0)^\varepsilon}, \tag{1.102}$$

where

$$\Delta^\varepsilon[\Phi(\mu) - \Phi(\mu_0)] \cong \Gamma(1+\varepsilon)[\Phi(\mu) - \Phi(\mu_0)],$$

if the generalized limit exists.

Suppose that

$$\left.\frac{d^\varepsilon \Phi(\mu)}{d\mu^\varepsilon}\right|_{\mu=\mu_0^+}$$

and

$$\left.\frac{\mathrm{d}^{\varepsilon}\Phi(\mu)}{\mathrm{d}\mu^{\varepsilon}}\right|_{\mu=\mu_0^-}$$

exist and

$$\left.\frac{\mathrm{d}^{\varepsilon}\Phi(\mu)}{\mathrm{d}\mu^{\varepsilon}}\right|_{\mu=\mu_0^+} = \left.\frac{\mathrm{d}^{\varepsilon}\Phi(\mu)}{\mathrm{d}\mu^{\varepsilon}}\right|_{\mu=\mu_0^-}.$$

Then, we have

$$\left.\frac{\mathrm{d}^{\varepsilon}\Phi(\mu)}{\mathrm{d}\mu^{\varepsilon}}\right|_{\mu=\mu_0} = \left.\frac{\mathrm{d}^{\varepsilon}\Phi(\mu)}{\mathrm{d}\mu^{\varepsilon}}\right|_{\mu=\mu_0^+} = \left.\frac{\mathrm{d}^{\varepsilon}\Phi(\mu)}{\mathrm{d}\mu^{\varepsilon}}\right|_{\mu=\mu_0^-}. \tag{1.103}$$

For $0 < \varepsilon \leq 1$, the fractal increment of $\Phi(\mu)$ of order ε at $\mu = \mu_0$ is defined by

$$\Gamma(1+\varepsilon)\,\Delta^{\varepsilon}\Phi(\mu_0) = \Delta^{\varepsilon}[\Phi(\mu) - \Phi(\mu_0)] = D^{(\varepsilon)}\Phi(\mu_0)(\Delta\mu)^{\varepsilon} + \varpi(\Delta\mu)^{\varepsilon}, \tag{1.104}$$

where $\Delta\mu$ is increment of μ and $\varpi \to 0$ as $\Delta\mu \to 0$.

For $0 < \varepsilon \leq 1$, the local fractional differential of $\Phi(\mu)$ of order ε at $\mu = \mu_0$ is defined by

$$\mathrm{d}^{\varepsilon}\Phi(\mu_0) = D^{(\varepsilon)}\Phi(\mu_0)(\mathrm{d}\mu)^{\varepsilon} + \varpi(\mathrm{d}\mu)^{\varepsilon}. \tag{1.105}$$

Suppose that there exists any point $\mu \in (a, b)$ such that

$$\Phi^{(\varepsilon)}(\mu) = \frac{\mathrm{d}^{\varepsilon}\Phi(\mu)}{\mathrm{d}\mu^{\varepsilon}} = D^{(\varepsilon)}\Phi(\mu). \tag{1.106}$$

In this case, $D_{\varepsilon}(a, b)$ is called a ε-local fractional derivative set.

Property 1. *Suppose that $\Phi(\mu) \in D_{\varepsilon}(a, b)$. Then, $\Phi(\mu) \in C_{\varepsilon}(a, b)$.*

Proof. Using the formula (1.104), we arrive at

$$|\Phi(\mu)| = \left|D^{(\varepsilon)}\Phi(\mu_0)\frac{(\Delta\mu)^{\varepsilon}}{\Gamma(1+\varepsilon)} + \varpi\frac{(\Delta\mu)^{\varepsilon}}{\Gamma(1+\varepsilon)} + \Phi(\mu_0)\right|. \tag{1.107}$$

Taking the generalized limit of formula (1.107), we conclude

$$\lim_{\mu\to\mu_0}\Phi(\mu) = \Phi(\mu_0). \tag{1.108}$$

For any μ_0, we get the result. □

Property 2. *If $\Phi(\mu) \in D_{\varepsilon}(a, b)$, then $\Phi(\mu)$ is local fractional differentiable on the domain $I = (a, b)$.*

Proof. From (1.104), we have the relation

$$\Delta^{\varepsilon}\Phi(\mu_0) - D^{(\varepsilon)}\Phi(\mu_0)(\Delta\mu)^{\varepsilon} + \varpi(\Delta\mu)^{\varepsilon}, \tag{1.109}$$

where $\lim_{\mu\to\mu_0}\varpi = 0$.

If we replace $\Delta^{\varepsilon}\Phi(\mu_0)$ and $(\Delta\mu)^{\varepsilon}$ by $\mathrm{d}^{\varepsilon}\Phi(\mu_0)$ and $(\mathrm{d}\mu)^{\varepsilon}$ in (1.109), respectively, this identity yields

$$\mathrm{d}^{\varepsilon}\Phi(\mu_0) = D^{(\varepsilon)}\Phi(\mu_0)(\mathrm{d}\mu)^{\varepsilon} + \varpi(\mathrm{d}\mu)^{\varepsilon}. \tag{1.110}$$

Successively, making use of $\lim_{\mu\to\mu_0}\varpi = 0$ in (1.100), we deduce the result. □

Suppose that $\Phi(\mu), \Theta(\mu) \in D_{\varepsilon}(a,b)$. The local fractional differentiation rules of nondifferentiable functions defined on fractal sets are listed as follows:

(a) $D^{(\varepsilon)}[\Phi(\mu) \pm \Theta(\mu)] = D^{(\varepsilon)}\Phi(\mu) \pm D^{(\varepsilon)}\Theta(\mu)$;
(b) $D^{(\varepsilon)}[\Phi(\mu)\Theta(\mu)] = \left[D^{(\varepsilon)}\Phi(\mu)\right]\Theta(\mu) + \Phi(\mu)\left[D^{(\varepsilon)}\Theta(\mu)\right]$; and
(c) $D^{(\varepsilon)}[\Phi(\mu)/\Theta(\mu)] = \left\{\left[D^{(\varepsilon)}\Phi(\mu)\right]\Theta(\mu) - \Phi(\mu)\left[D^{(\varepsilon)}\Theta(\mu)\right]\right\}/\Theta^2(\mu)$, provided $\Theta(\mu) \neq 0$.

One observes that the formulas (a), (b), and (c) are presented to generalize the differentiation rules of the differentiable functions. These, in the Kolwankar-Gangal sense, are valid (e.g., [1–16]).

Setting $\Phi(\mu) \in D_{\varepsilon}^{n}(a,b)$ and $n = 2$, the interchanging operator of the order of the local fractional operators is defined as follows:

$$\left(\frac{\mathrm{d}^{\varepsilon}}{\mathrm{d}\mu^{\varepsilon}} \oplus \frac{\mathrm{d}^{\varepsilon}}{\mathrm{d}\mu^{\varepsilon}}\right)\Phi(\mu) = \frac{\mathrm{d}^{2\varepsilon}\Phi(\mu)}{\mathrm{d}\mu^{2\varepsilon}}. \tag{1.111}$$

There is one mechanism which may indeed be applied to the local fractional chain rule leading to the generalized chain rule of local fractional-order differential operator.

We present the local fractional chain rule via the interchanging operator of nondifferential functions as follows [1, 16, 21]:

Suppose that $\Phi(\mu) = (\phi \otimes \varphi)(\mu)$. Then, we have

$$\frac{\mathrm{d}^{\varepsilon}\Phi(\mu)}{\mathrm{d}\mu^{\varepsilon}} = \phi^{(\varepsilon)}(\varphi)\left[\varphi^{(1)}(\mu)\right]^{\varepsilon}, \tag{1.112}$$

if $\phi^{(\varepsilon)}(\varphi)$ and $\varphi^{(1)}(\mu)$ exist.

Let C be a constant. The local fractional derivative of some of nondifferentiable functions defined on fractal sets are listed in Table 1.1.

The above results devoted to local fractional derivative were listed in [1], and the proofs of them are also found in Appendix B.

In order to derive them, one start with the new series expansion in the form

$$(\phi + \varphi)^{n\varepsilon} = \sum_{i=0}^{\infty}\begin{pmatrix} n\varepsilon \\ i\varepsilon \end{pmatrix}\phi^{(n-i)\varepsilon}\varphi^{i\varepsilon} = \sum_{i=0}^{\infty}\begin{pmatrix} n\varepsilon \\ i\varepsilon \end{pmatrix}\phi^{i\varepsilon}\varphi^{(n-i)\varepsilon}, \tag{1.113}$$

where

$$\begin{pmatrix} n\varepsilon \\ i\varepsilon \end{pmatrix} = \frac{\Gamma(1+n\varepsilon)}{\Gamma(1+i\varepsilon)\Gamma(1+(n-i)\varepsilon)}. \tag{1.114}$$

Table 1.1 Basic operations of local fractional derivative of some of nondifferentiable functions defined on fractal sets

Original function	Transformed function
C	0
$\mu^{k\varepsilon}/\Gamma(1+k\varepsilon)$	$\mu^{(k-1)\varepsilon}/\Gamma(1+(k-1)\varepsilon)$
$E_\varepsilon(\mu^\varepsilon)$	$E_\varepsilon(\mu^\varepsilon)$
$E_\varepsilon(C\mu^\varepsilon)$	$CE_\varepsilon(C\mu^\varepsilon)$
$E_\varepsilon(-\mu^\varepsilon)$	$-E_\varepsilon(-\mu^\varepsilon)$
$E_\varepsilon\left(\mu^{2\varepsilon}\right)$	$(2\mu)^\varepsilon E_\varepsilon\left(\mu^{2\varepsilon}\right)$
$E_\varepsilon\left(C\mu^{2\varepsilon}\right)$	$(2\mu)^\varepsilon CE_\varepsilon\left(C\mu^{2\varepsilon}\right)$
$E_\varepsilon\left(-\mu^{2\varepsilon}\right)$	$-(2\mu)^\varepsilon E_\varepsilon\left(-\mu^{2\varepsilon}\right)$
$\sin_\varepsilon(\mu^\varepsilon)$	$\cos_\varepsilon(\mu^\varepsilon)$
$\sin_\varepsilon(C\mu^\varepsilon)$	$C\cos_\varepsilon(C\mu^\varepsilon)$
$\cos_\varepsilon(\mu^\varepsilon)$	$-\sin_\varepsilon(\mu^\varepsilon)$
$\cos_\varepsilon(C\mu^\varepsilon)$	$-C\sin_\varepsilon(C\mu^\varepsilon)$
$\sinh_\varepsilon(\mu^\varepsilon)$	$\cosh_\varepsilon(\mu^\varepsilon)$
$\sinh_\varepsilon(C\mu^\varepsilon)$	$C\cosh_\varepsilon(C\mu^\varepsilon)$
$\cosh_\varepsilon(\mu^\varepsilon)$	$-\sinh_\varepsilon(\mu^\varepsilon)$
$\cosh_\varepsilon(C\mu^\varepsilon)$	$-C\sinh_\varepsilon(C\mu^\varepsilon)$

In this case, we present three characters of the series expansion below:

(a) $(\phi+\varphi)^{n\varepsilon}=1$, when $n=0$;
(b) $(\phi+\varphi)^{n\varepsilon}=\phi^\varepsilon+\varphi^\varepsilon$, when $n=1$; and
(c) $(\phi+\varphi)^\varepsilon=(2\varphi)^\varepsilon=(2\phi)^\varepsilon$, when $\phi=\varphi$.

We notice (b) is true when it is defined on fractal sets [1].

When $n\varepsilon=\sigma$ is a real number, a fractional series expansion via arbitrary powers σ is presented as [69]

$$(\phi+\varphi)^\sigma=\sum_{i=0}^{\infty}\left(\begin{array}{c}\sigma\\ i\end{array}\right)\phi^{\sigma-k}\varphi^k=\sum_{i=0}^{\infty}\left(\begin{array}{c}\sigma\\ i\end{array}\right)\phi^k\varphi^{\sigma-k}, \tag{1.115}$$

where

$$\left(\begin{array}{c}\sigma\\ i\end{array}\right)=\frac{\Gamma(1+\sigma)}{\Gamma(1+i)\,\Gamma(1+\sigma-i)}. \tag{1.116}$$

With the help of (1.113), the nondifferential difference takes the form

$$\Delta^\varepsilon\left[\Phi(\mu)-\Phi(\mu_0)\right]=\Gamma(1+\varepsilon)\,\Delta^\varepsilon\Phi(\mu_0)\cong\Gamma(1+\varepsilon)\left[\Phi(\mu)-\Phi(\mu_0)\right], \tag{1.117}$$

where

$$\Delta^\varepsilon\Phi(\mu_0)=\sum_{i=0}^{\infty}(-1)^i\left(\begin{array}{c}\varepsilon\\ i\varepsilon\end{array}\right)\Phi(\mu-i\rho) \tag{1.118}$$

with $\rho=\mu-\mu_0$.

Adopting (1.117), we present two examples, namely,

$$\frac{d^{\varepsilon}}{d\mu^{\varepsilon}}\frac{\mu^{\varepsilon}}{\Gamma(1+\varepsilon)}=\lim_{\Delta\mu\to 0}\frac{1}{\Gamma(1+\varepsilon)}\frac{\Gamma(1+\varepsilon)\left[(\mu+\Delta\mu)^{\varepsilon}-\mu^{\varepsilon}\right]}{(\Delta\mu)^{\varepsilon}}=1. \tag{1.119}$$

$$\begin{aligned}\frac{d^{\varepsilon}}{d\mu^{\varepsilon}}\frac{\mu^{k\varepsilon}}{\Gamma(1+k\varepsilon)}&=\lim_{\Delta\mu\to 0}\left\{\frac{\Gamma(1+\varepsilon)}{\Gamma(1+k\varepsilon)}\frac{\left[(\mu+\Delta\mu)^{k\varepsilon}-\mu^{k\varepsilon}\right]}{(\Delta\mu)^{\varepsilon}}\right\}\\&=\lim_{\Delta\mu\to 0}\left\{\frac{\Gamma(1+\varepsilon)}{\Gamma(1+k\varepsilon)}\frac{\left[\mu^{k\varepsilon}+\frac{\Gamma(1+k\varepsilon)}{\Gamma(1+\varepsilon)\Gamma(1+(k-1)\varepsilon)}\mu^{(k-1)\varepsilon}(\Delta\mu)^{\varepsilon}+\cdots+-\mu^{k\varepsilon}\right]}{(\Delta\mu)^{\varepsilon}}\right\}\\&=\lim_{\Delta\mu\to 0}\left\{\frac{\Gamma(1+\varepsilon)}{\Gamma(1+k\varepsilon)}\frac{\left[\frac{\Gamma(1+k\varepsilon)}{\Gamma(1+\varepsilon)\Gamma(1+(k-1)\varepsilon)}\mu^{(k-1)\varepsilon}(\Delta\mu)^{\varepsilon}\right]}{(\Delta\mu)^{\varepsilon}}\right\}\\&=\frac{\mu^{(k-1)\varepsilon}}{\Gamma(1+(k-1)\varepsilon)}.\end{aligned} \tag{1.120}$$

In this case, from (1.120) we have

$$\frac{d^{\varepsilon}}{d\mu^{\varepsilon}}E_{\varepsilon}\left(\mu^{\varepsilon}\right)=\frac{d^{\varepsilon}}{d\mu^{\varepsilon}}\left(\sum_{k=0}^{\infty}\frac{\mu^{k\varepsilon}}{\Gamma(1+k\varepsilon)}\right)=1+\sum_{k=1}^{\infty}\frac{\mu^{k\varepsilon}}{\Gamma(1+k\varepsilon)}, \tag{1.121}$$

which leads to

$$1+\sum_{k=1}^{\infty}\frac{\mu^{k\varepsilon}}{\Gamma(1+k\varepsilon)}=\sum_{k=0}^{\infty}\frac{\mu^{k\varepsilon}}{\Gamma(1+k\varepsilon)}. \tag{1.122}$$

Therefore, we conclude that

$$\frac{d^{\varepsilon}}{d\mu^{\varepsilon}}E_{\varepsilon}\left(\mu^{\varepsilon}\right)=E_{\varepsilon}\left(\mu^{\varepsilon}\right). \tag{1.123}$$

1.3.2 Properties and theorems of local fractional derivatives

Theorem 1.2 (Local fractional Rolle's theorem)**.** *Suppose that* $\Phi(\mu)\in C_{\varepsilon}[a,b]$, $\Phi(\mu)\in D_{\varepsilon}(a,b)$, *and* $\Phi(a)=\Phi(b)$. *Then, there exists a point* $\mu_0\in(a,b)$ *and* $\varepsilon\in(0,1]$ *such that*

$$\Phi^{(\varepsilon)}(\mu_0)=0. \tag{1.124}$$

Proof.

(a) Let $\Phi(\mu)=0$ in $[a,b]$. Then, for all μ_0 in (a,b), there is $\Phi^{(\varepsilon)}(\mu_0)=0$.
(b) Let $\Phi(\mu)\neq 0$ in $[a,b]$.

Since $\Phi(\mu)$ is a local fractional continuous function in the domain $C_{\varepsilon}[a,b]$, there are points at which $\Phi(\mu)$ attains its maximum and minimum values, denoted by Π and T, respectively.

Because $\Phi(\mu)\neq 0$, at least one of the values Π, T is not zero.

Suppose, for instance, $\Pi \neq 0$ and that $\Phi(\mu_0) = \Pi$. In this case, we consider

$$\Phi(\mu_0 + \Delta\mu) \leq \Phi(\mu_0). \tag{1.125}$$

Assuming that $\Delta\mu > 0$, there is

$$\frac{\Delta^{\varepsilon}[\Phi(\mu_0 + \Delta\mu) - \Phi(\mu_0)]}{(\Delta\mu)^{\varepsilon}} \leq 0 \tag{1.126}$$

such that

$$\lim_{\Delta\mu \to 0} \frac{\Delta^{\varepsilon}[\Phi(\mu_0 + \Delta\mu) - \Phi(\mu_0)]}{(\Delta\mu)^{\varepsilon}} < 0. \tag{1.127}$$

In similar manner, we consider $\Delta\mu < 0$.

Considering $\Phi(\mu) \in D_{\varepsilon}(a,b)$ and applying (1.113), there is $\Phi^{(\varepsilon)}(\mu_0) = 0$. As similar argument can be applied in case of $\Pi = 0$ and $T \neq 0$.

Therefore, there is the formula $\Phi^{(\varepsilon)}(\mu_0) = 0$. □

There is a generalized local fractional Rolle's theorem devoted to the local fractional derivative in Kolwankar and Gangal sense.

Theorem 1.3. *Suppose $\Phi(\mu) \in C_{\varepsilon}[a,b]$ and $\Phi(\mu) \in D_{\varepsilon}(a,b)$. Then, there exists a point $\mu_0 \in (a,b)$ and $\varepsilon \in (0,1]$ such that*

$$\Phi(b) - \Phi(a) = \Phi^{(\varepsilon)}(\mu_0) \frac{(b-a)^{\varepsilon}}{\Gamma(1+\varepsilon)}. \tag{1.128}$$

Proof. Let us define the nondifferentiable function, which is given by

$$\Lambda(\mu) = \Gamma(1+\varepsilon)\left\{[\Phi(\mu) - \Phi(a)] - [\Phi(b) - \Phi(a)]\frac{(\mu-a)^{\varepsilon}}{(b-a)^{\varepsilon}}\right\} \tag{1.129}$$

with $\varepsilon \in (0,1]$.

Then, there are $\Lambda(a) = 0$ and $\Lambda(b) = 0$.

In this case, for $\mu_0 \in (a,b)$ there is the following identity in the form

$$\Lambda(\mu) = \Gamma(1+\varepsilon)\left\{[\Phi(\mu) - \Phi(a)] - [\Phi(b) - \Phi(a)]\frac{(\mu-a)^{\varepsilon}}{(b-a)^{\varepsilon}}\right\}. \tag{1.130}$$

Therefore, we have the result. □

Theorem 1.4. *Suppose that $\Phi(\mu) \in C_{\varepsilon}[a,b]$ and $\Phi(\mu) \in D_{\varepsilon}(a,b)$. Then, there exist $\lim_{\mu\to\mu_0}\Phi(\mu) = 0$ and $\lim_{\mu\to\mu_0}\Omega(\mu) = 0$, where K denotes either a real number or one of the symbols $-\infty, \infty$. Suppose that $\lim_{\mu\to\mu_0}\left[\Phi^{(\varepsilon)}(\mu)/\Omega^{(\varepsilon)}(\mu)\right] = K$. Then,*

$$\lim_{\mu\to\mu_0}[\Phi(\mu)/\Omega(\mu)] = K. \tag{1.131}$$

Proof. Let $\Phi(\mu) \in C_\varepsilon[a,b]$ and $\Phi(\mu) \in D_\varepsilon(a,b)$. There is $\mu_0 \in (a,b)$ such that $\Phi(\mu_0) = 0$ and $\Omega(\mu_0) = 0$.

There is $\eta \in (\mu_0, \mu)$ such that

$$\frac{\Phi(\mu)}{\Omega(\mu)} = \frac{\Phi(\mu) - \Phi(\mu_0)}{\Omega(\mu) - \Omega(\mu_0)} = \frac{\Phi^{(\varepsilon)}(\eta)}{\Omega^{(\varepsilon)}(\eta)}. \tag{1.132}$$

As $\mu \to \mu_0^+$, the identity

$$\lim_{\mu\to\mu_0^+} \frac{\Phi(\mu)}{\Omega(\mu)} = \lim_{\mu\to\mu_0^+} \frac{\Phi(\mu) - \Phi(\mu_0)}{\Omega(\mu) - \Omega(\mu_0)} = \lim_{\mu\to\mu_0^+} \frac{\Phi^{(\varepsilon)}(\mu_0)}{\Omega^{(\varepsilon)}(\mu_0)} = K \tag{1.133}$$

holds.

In similar manner, when $\mu \to \mu_0^-$, there is

$$\lim_{\mu\to\mu_0^-} \frac{\Phi(\mu)}{\Omega(\mu)} = \lim_{\mu\to\mu_0^-} \frac{\Phi^{(\varepsilon)}(\mu_0)}{\Omega^{(\varepsilon)}(\mu_0)} = K. \tag{1.134}$$

Therefore, we get the result. □

For more details regarding the proof of (1.121), we recommend to readers refs [1, 21, 70].

In order to demonstrate the above mechanism, we present elementary examples: Using (1.132), for $\mu \to 0$, we have

$$E_\varepsilon\left(\mu^\varepsilon\right) - 1 \approx \frac{\mu^\varepsilon}{\Gamma(1+\varepsilon)} \tag{1.135}$$

such that

$$\lim_{\mu\to 0} \frac{E_\varepsilon(\mu^\varepsilon) - 1}{\dfrac{\mu^\varepsilon}{\Gamma(1+\varepsilon)}} = \lim_{\mu\to 0} \frac{\dfrac{d^\varepsilon}{d\mu^\varepsilon}\left[\dfrac{\mu^\varepsilon}{\Gamma(1+\varepsilon)}\right]}{\dfrac{d^\varepsilon}{d\mu^\varepsilon}\left[\dfrac{\mu^\varepsilon}{\Gamma(1+\varepsilon)}\right]} = 1, \tag{1.136}$$

$$\lim_{\mu\to 0} \frac{\sin_\varepsilon(\mu^\varepsilon)}{\dfrac{\mu^\varepsilon}{\Gamma(1+\varepsilon)}} = \lim_{\mu\to 0} \frac{\dfrac{d^\varepsilon}{d\mu^\varepsilon}\left[\sin_\varepsilon(\mu^\varepsilon)\right]}{\dfrac{d^\varepsilon}{d\mu^\varepsilon}\left[\dfrac{\mu^\varepsilon}{\Gamma(1+\varepsilon)}\right]} = \lim_{\mu\to 0} \cos_\varepsilon\left(\mu^\varepsilon\right) = 1. \tag{1.137}$$

Similarly, for $\mu \to 0$ we conclude

$$1 - \cos_\varepsilon\left(\mu^\varepsilon\right) \approx \frac{\mu^{2\varepsilon}}{\Gamma(1+2\varepsilon)} \tag{1.138}$$

such that

$$\lim_{\mu\to 0}\frac{1-\cos_{\varepsilon}(\mu^{\varepsilon})}{\dfrac{\mu^{2\varepsilon}}{\Gamma(1+2\varepsilon)}}=\lim_{\mu\to 0}\frac{\dfrac{d^{\varepsilon}}{d\mu^{\varepsilon}}\left[\dfrac{\mu^{2\varepsilon}}{\Gamma(1+2\varepsilon)}\right]}{\dfrac{d^{\varepsilon}}{d\mu^{\varepsilon}}\left[\dfrac{\mu^{2\varepsilon}}{\Gamma(1+2\varepsilon)}\right]}=1. \tag{1.139}$$

1.4 Definitions and properties of local fractional integral

1.4.1 Definitions of local fractional integrals

Definition 1.4. Suppose $\varphi(\mu)\in C_{\varepsilon}[a,b]$. Then, we define the local fractional integral of $\varphi(\mu)$ of order $\varepsilon(0<\varepsilon\leq 1)$ by

$${}_{a}I_{b}^{(\varepsilon)}\varphi(\mu)=\frac{1}{\Gamma(1+\varepsilon)}\int_{a}^{b}\varphi(\mu)(d\mu)^{\varepsilon}=\frac{1}{\Gamma(1+\varepsilon)}\lim_{\Delta\mu_k\to 0}\sum_{k=0}^{N-1}\varphi(\mu_k)(\Delta\mu_k)^{\varepsilon}, \tag{1.140}$$

where $\Delta\mu_k=\mu_{k+1}-\mu_k$ with $\mu_0=a<\mu_1<\cdots<\mu_{N-1}<\mu_N=b$.

Suppose the local fractional integral of $\varphi(\mu)$ on the closed interval $[a,b]$ be equal to Θ.

For each $\rho>0$, there exists $0<|\Delta\mu_k|<\delta$ such that

$$\left|\Theta-\frac{1}{\Gamma(1+\varepsilon)}\lim_{\Delta\mu_k\to 0}\sum_{k=0}^{N-1}\varphi(\mu_k)(\Delta\mu_k)^{\varepsilon}\right|<\rho^{\varepsilon}. \tag{1.141}$$

In fact, we recall the condition of the Riemann integral that suppose $\varphi(\mu)$ is bounded on $[a,b]$, then, a necessary and sufficient condition for the existence of

$$\int_{a}^{b}\varphi(\mu)\,d\mu \tag{1.142}$$

is that $\varphi(\mu)$ has a Lebesgue measure zero.

Will the proposed procedure lead to Riemann integral on fractal sets? The answer is yes. The suggested mechanism may indeed be adapted to condition of the Riemann integral leading to generalized condition of the Riemann integral on fractal sets.

The Riemann integral on fractal sets is stated as follows [1, 16, 21]:

Let φ: $\wp\to\aleph$ be a function defined on a fractal set $\wp$ of fractal dimension $\varepsilon(0<\varepsilon<1)$. Suppose $\varphi(\mu)$ is bounded on $[a,b]$ (or $\varphi(\mu)\in C_{\varepsilon}[a,b]$). Then, a necessary and sufficient condition for the existence of

$$\frac{1}{\Gamma(1+\varepsilon)}\int_{a}^{b}\varphi(\mu)(d\mu)^{\varepsilon} \tag{1.143}$$

is that a fractal set of local fractional continuity of $\varphi(\mu)$ has a generalized Lebesgue measure zero.

We easily get the following result:

Suppose $\varphi(\mu) \in C_\varepsilon[a,b]$, then, $\varphi(\mu)$ is local fractional integral on $[a,b]$.

For convenience, we can write the following rules:

(a) ${}_aI_b^{(\varepsilon)}\varphi(\mu) = 0$ if $a = b$.
(b) ${}_aI_b^{(\varepsilon)}\varphi(\mu) = -{}_bI_a^{(\varepsilon)}\varphi(\mu)$ if $a < b$.
(c) ${}_aI_b^{(\varepsilon)}\varphi(\mu) = \varphi(\mu)$ if $\varepsilon = 0$.

1.4.2 Properties and theorems of local fractional integrals

Suppose $\varphi(\mu), \varphi_1(\mu)$, and $\varphi_2(\mu) \in C_\varepsilon[a,b]$, the local fractional integral rules of nondifferentiable functions defined on fractal sets are listed as follows [1, 16, 21]:

(a) ${}_aI_b^{(\varepsilon)}[\varphi_1(\mu) + \varphi_2(\mu)] = {}_aI_b^{(\varepsilon)}\varphi_1(\mu) + {}_aI_b^{(\varepsilon)}\varphi_2(\mu)$;
(b) ${}_aI_b^{(\varepsilon)}[C\varphi(\mu)] = C{}_aI_b^{(\varepsilon)}\varphi(\mu)$, provided a constant C;
(c) ${}_aI_b^{(\varepsilon)}1 = (b-a)^\varepsilon / \Gamma(1+\varepsilon)$;
(d) ${}_aI_b^{(\varepsilon)}\varphi(\mu) \geq 0$, provided $\varphi(\mu) \geq 0$;
(e) $\left|{}_aI_b^{(\varepsilon)}\varphi(\mu)\right| \leq {}_aI_b^{(\varepsilon)}|\varphi(\mu)|$;
(f) ${}_aI_b^{(\varepsilon)}\varphi(\mu) = {}_aI_c^{(\varepsilon)}\varphi(\mu) + {}_cI_b^{(\varepsilon)}\varphi(\mu)$, provided $a < c < b$; and
(g) ${}_aI_b^{(\varepsilon)}\varphi(\mu) \in \left[T(b-a)^\varepsilon / \Gamma(1+\varepsilon), \Pi(b-a)^\varepsilon / \Gamma(1+\varepsilon)\right]$, provided that the maximum and minimum values of $\varphi(\mu)$ are Π and T, respectively.

Theorem 1.5 (Mean value theorem for local fractional integrals). *Suppose that $\varphi(\mu) \in C_\varepsilon[a,b]$. Then, there exists a point ξ in (a,b) such that*

$$ {}_aI_b^{(\varepsilon)}\varphi(\mu) = \varphi(\xi)\frac{(b-a)^\varepsilon}{\Gamma(1+\varepsilon)}. \tag{1.144}$$

Proof. In view of $\varphi(\mu) \in C_\varepsilon[a,b]$, we have

$$ {}_aI_b^{(\varepsilon)}\varphi(\mu) \in \left[\mathrm{T}(b-a)^\varepsilon / \Gamma(1+\varepsilon), \Pi(b-a)^\varepsilon / \Gamma(1+\varepsilon)\right], \tag{1.145}$$

which leads us to

$$ \frac{{}_aI_b^{(\varepsilon)}\varphi(\mu)}{\frac{(b-a)^\varepsilon}{\Gamma(1+\varepsilon)}} \in [\mathrm{T}, \Pi]. \tag{1.146}$$

Therefore, for $\xi \in (a,b)$, we have

$$ \frac{{}_aI_b^{(\varepsilon)}\varphi(\mu)}{\frac{(b-a)^\varepsilon}{\Gamma(1+\varepsilon)}} = \varphi(\xi), \tag{1.147}$$

which yields the result. □

Theorem 1.6. *Suppose that $\varphi(\mu) \in C_\varepsilon[a,b]$. Then, for $\mu \in (a,b)$, there exists a function $\Phi(\mu)$ given by*

$$\Phi(\mu) = {}_aI_\mu^{(\varepsilon)}\varphi(\mu), \tag{1.148}$$

with the following local fractional derivative:

$$\frac{\partial^\varepsilon \Phi(\mu)}{\partial \mu^\varepsilon} = \varphi(\mu). \tag{1.149}$$

Proof. Let $\mu \in [a, b]$; then, there exists $\mu + \Delta\mu \in [a, b]$ such that

$$\Phi(\mu) = {}_aI_{\mu+\Delta\mu}^{(\varepsilon)}\varphi(\mu). \tag{1.150}$$

In this case, we present

$$\Delta^\varepsilon[\Phi(\mu + \Delta\mu) - \Phi(\mu)] = \int_a^{\mu+\Delta\mu} \varphi(\mu)(\mathrm{d}\mu)^\varepsilon - \int_a^\mu \varphi(\mu)(\mathrm{d}\mu)^\varepsilon, \tag{1.151}$$

which leads to

$$\Delta^\varepsilon[\Phi(\mu + \Delta\mu) - \Phi(\mu)] = \int_\mu^{\mu+\Delta\mu} \varphi(\mu)(\mathrm{d}\mu)^\varepsilon. \tag{1.152}$$

From (1.144), for $\xi \in (a, b)$, we present the formula

$${}_\mu I_{\mu+\Delta\mu}^{(\varepsilon)}\varphi(\mu) = \varphi(\xi)\frac{(\Delta\mu)^\varepsilon}{\Gamma(1+\varepsilon)}, \tag{1.153}$$

which yields that

$$\frac{{}_\mu I_{\mu+\Delta\mu}^{(\varepsilon)}\varphi(\mu)}{\frac{(\Delta\mu)^\varepsilon}{\Gamma(1+\varepsilon)}} = \varphi(\xi) \tag{1.154}$$

or

$$\frac{\Delta^\varepsilon[\Phi(\mu + \Delta\mu) - \Phi(\mu)]}{(\Delta\mu)^\varepsilon} = \varphi(\xi). \tag{1.155}$$

As $\Delta\mu \to 0$, we present

$$\lim_{\Delta\mu \to 0} \frac{\Delta^\varepsilon[\Phi(\mu + \Delta\mu) - \Phi(\mu)]}{(\Delta\mu)^\varepsilon} = \Phi^{(\varepsilon)}(\mu) = \varphi(\xi). \tag{1.156}$$

For $\Delta\mu > 0$, there exists a point $\mu = a$ such that

$$\Phi^{(\varepsilon)}(\mu)\Big|_{\mu=a^+} = \varphi\left(a^+\right). \tag{1.157}$$

In a similar manner, for $\Delta\mu < 0$, there exists a point $\mu = b$ such that

$$\Phi^{(\varepsilon)}(\mu)\Big|_{\mu=b^-} = \varphi\left(b^-\right). \tag{1.158}$$

Hence, we get the result. □

Theorem 1.7 (Newton–Leibniz formula of local fractional integrals)**.** *Suppose that*

$$\Phi^{(\varepsilon)}(\mu) = \varphi(\mu) \in C_\varepsilon[a,b].$$

Then

$$ {}_aI_b^{(\varepsilon)}\varphi(\mu) = \Phi(b) - \Phi(a). \tag{1.159}$$

Proof. Let us define the function $\Phi_0(\mu) = {}_aI_\mu^{(\varepsilon)}\varphi(\mu)$. Thus, we have

$$\frac{\partial^\varepsilon}{\partial\mu^\varepsilon}\left(\Phi_0(\mu) - \Phi(\mu)\right) = \frac{\partial^\varepsilon}{\partial\mu^\varepsilon}\Phi_0(\mu) - \frac{\partial^\varepsilon}{\partial\mu^\varepsilon}\Phi(\mu) = \varphi(\mu) - \varphi(\mu) = 0, \tag{1.160}$$

which leads to

$$\Phi_0(\mu) - \Phi(\mu) = C, \tag{1.161}$$

with C be a constant.

Therefore, from (1.160), we have the following identity

$$ {}_aI_b^{(\varepsilon)}\varphi(\mu) = \Phi_0(b) - \Phi_0(a) = \Phi(b) - \Phi(a). \tag{1.162}$$

Hence, we obtain the desired result. □

Theorem 1.8 (Local fractional integration by parts)**.** *Suppose that* $\varphi_1(\mu), \varphi_2(\mu) \in C_\varepsilon[a,b]$*, and* $\varphi_1(\mu), \varphi_2(\mu) \in D_\varepsilon(a,b)$*. Then,*

$$ {}_aI_b^{(\varepsilon)}\left\{\left[\frac{\partial^\varepsilon}{\partial\mu^\varepsilon}\varphi_1(\mu)\right]\varphi_2(\mu)\right\} = \left[\varphi_1(\mu)\varphi_2(\mu)\right]_a^b - {}_aI_b^{(\varepsilon)}\left\{\varphi_1(\mu)\left[\frac{\partial^\varepsilon}{\partial\mu^\varepsilon}\varphi_2(\mu)\right]\right\}. \tag{1.163}$$

Proof. We have

$$\left[\varphi_1(\mu)\varphi_2(\mu)\right]_a^b = {}_aI_b^{(\varepsilon)}\left\{\frac{\partial^\varepsilon}{\partial\mu^\varepsilon}\left[\varphi_1(\mu)\varphi_2(\mu)\right]\right\}. \tag{1.164}$$

Thus, there is

$$ {}_aI_b^{(\varepsilon)}\left\{\left[\frac{\partial^\varepsilon}{\partial\mu^\varepsilon}\varphi_1(\mu)\right]\varphi_2(\mu)\right\} = \left[\varphi_1(\mu)\varphi_2(\mu)\right]_a^b - {}_aI_b^{(\varepsilon)}\left\{\varphi_1(\mu)\left[\frac{\partial^\varepsilon}{\partial\mu^\varepsilon}\varphi_2(\mu)\right]\right\}. \tag{1.165}$$

Therefore, we obtain the desired result.
Suppose $D^{(k\varepsilon)}\varphi(\mu) \in C_\varepsilon(a,b)$, then, there is

$$D^{(k\varepsilon)}\left\{{}_{\mu_0}I_\mu^{(k\varepsilon)}\varphi(\mu)\right\} = \varphi(\mu), \tag{1.166}$$

where ${}_{\mu_0}I_{\mu}^{(k\varepsilon)}\varphi(\mu) = \overbrace{{}_{\mu_0}I_{\mu}^{(\varepsilon)} \cdots {}_{\mu_0}I_{\mu}^{(\varepsilon)}}^{k\text{-times}} \varphi(\mu)$ and $D^{(k\varepsilon)}\varphi(\mu) = \overbrace{D^{(\varepsilon)} \cdots D^{(\varepsilon)}}^{k\text{-times}} \varphi(\mu)$. □

Theorem 1.9. *Suppose that $D^{(k\varepsilon)}\varphi(\mu)$, $D^{((k+1)\varepsilon)}\varphi(\mu) \in C_{\varepsilon}(a,b)$. Then, for $0 < \varepsilon < 1$, there is a point $\mu_0 \in (a,b)$ such that*

$$ {}_{\mu_0}I_{\mu}^{(k\varepsilon)}\left[D^{(k\varepsilon)}\varphi(\mu)\right] - {}_{\mu_0}I_{\mu}^{((k+1)\varepsilon)}\left[D^{((k+1)\varepsilon)}\varphi(\mu)\right] = D^{(k\varepsilon)}\varphi(\mu_0)\frac{(\mu-\mu_0)^{k\varepsilon}}{\Gamma(1+k\varepsilon)}, \tag{1.167} $$

where ${}_{\mu_0}I_{\mu}^{(k\varepsilon)}\varphi(\mu) = \overbrace{{}_{\mu_0}I_{\mu}^{(\varepsilon)} \cdots {}_{\mu_0}I_{\mu}^{(\varepsilon)}}^{k\text{-times}} \varphi(\mu)$ and $D^{(k\varepsilon)}\varphi(\mu) = \overbrace{D^{(\varepsilon)} \cdots D^{(\varepsilon)}}^{k\text{-times}} \varphi(\mu)$.

Proof. We present the formula

$$ \begin{aligned} {}_{\mu_0}I_{\mu}^{((k+1)\varepsilon)}\left[D^{((k+1)\varepsilon)}\varphi(\mu)\right] &= {}_{\mu_0}I_{\mu}^{(k\varepsilon)}\left\{{}_{\mu_0}I_{\mu}^{(\varepsilon)}\left[D^{((k+1)\varepsilon)}\varphi(\mu)\right]\right\} \\ &= {}_{\mu_0}I_{\mu}^{(k\varepsilon)}\left\{D^{(k\varepsilon)}\varphi(\mu) - D^{(k\varepsilon)}\varphi(\mu_0)\right\} \\ &= {}_{\mu_0}I_{\mu}^{(k\varepsilon)}\left[D^{(k\varepsilon)}\varphi(\mu)\right] - {}_{\mu_0}I_{\mu}^{(k\varepsilon)}\left[D^{(k\varepsilon)}\varphi(\mu_0)\right]. \end{aligned} \tag{1.168} $$

Adopting the formula

$$ \begin{aligned} {}_{\mu_0}I_{\mu}^{(k\varepsilon)}\left[D^{(k\varepsilon)}\varphi(\mu_0)\right] &= D^{(k\varepsilon)}\varphi(\mu_0)\, {}_{\mu_0}I_{\mu}^{(k\varepsilon)}1 \\ &= D^{(k\varepsilon)}\varphi(\mu_0)\, {}_{\mu_0}I_{\mu}^{((k-1)\varepsilon)}\frac{(\mu-\mu_0)^{\varepsilon}}{\Gamma(1+\varepsilon)} \\ &= D^{(k\varepsilon)}\varphi(\mu_0)\frac{(\mu-\mu_0)^{k\varepsilon}}{\Gamma(1+k\varepsilon)}, \end{aligned} \tag{1.169} $$

there is

$$ {}_{\mu_0}I_{\mu}^{(k\varepsilon)}\left[D^{(k\varepsilon)}\varphi(\mu)\right] - {}_{\mu_0}I_{\mu}^{((k+1)\varepsilon)}\left[D^{((k+1)\varepsilon)}\varphi(\mu)\right] = D^{(k\varepsilon)}\varphi(\mu_0)\frac{(\mu-\mu_0)^{k\varepsilon}}{\Gamma(1+k\varepsilon)}. \tag{1.170} $$

Therefore, we proved the result. □

1.4.3 Local fractional Taylor's theorem for nondifferentiable functions

Theorem 1.10 (Local fractional Taylor's theorem). *Suppose that*

$$ D^{((k+1)\varepsilon)}\varphi(\mu) \in C_{\varepsilon}(a,b). $$

Then, for $k = 0, 1, \ldots, n$,

$$\varphi(\mu) = \sum_{k=0}^{n} \frac{D^{(k\varepsilon)}\varphi(\mu_0)}{\Gamma(1+k\varepsilon)}(\mu-\mu_0)^{k\varepsilon} + \frac{D^{((n+1)\varepsilon)}\varphi(\xi)}{\Gamma(1+(n+1)\varepsilon)}(\mu-\mu_0)^{(n+1)\varepsilon} \tag{1.171}$$

with $a < \mu_0 < \xi < \mu < b$, $\forall \mu \in (a,b)$, *where* $D^{(k\varepsilon)}\varphi(\mu) = \overbrace{D^{(\varepsilon)}\cdots D^{(\varepsilon)}}^{k\text{-times}}\varphi(\mu)$.

Proof. By making use of

$${}_{\mu_0}I_{\mu}^{(k\varepsilon)}\left[D^{(k\varepsilon)}\varphi(\mu)\right] - {}_{\mu_0}I_{\mu}^{((k+1)\varepsilon)}\left[D^{((k+1)\varepsilon)}\varphi(\mu)\right] = D^{(k\varepsilon)}\varphi(\mu_0)\frac{(\mu-\mu_0)^{k\varepsilon}}{\Gamma(1+k\varepsilon)}, \tag{1.172}$$

we conclude that

$$\begin{aligned}
&\sum_{k=0}^{n}\left\{{}_{\mu_0}I_{\mu}^{(k\varepsilon)}\left[D^{(k\varepsilon)}\varphi(\mu)\right] - {}_{\mu_0}I_{\mu}^{((k+1)\varepsilon)}\left[D^{((k+1)\varepsilon)}\varphi(\mu)\right]\right\} \\
&= \varphi(\mu) - {}_{\mu_0}I_{\mu}^{((k+1)\varepsilon)}\left[D^{((k+1)\varepsilon)}\varphi(\mu)\right] \\
&= \sum_{k=0}^{n}\left\{D^{(k\varepsilon)}\varphi(\mu_0)\frac{(\mu-\mu_0)^{k\varepsilon}}{\Gamma(1+k\varepsilon)}\right\}.
\end{aligned} \tag{1.173}$$

Thus, we show that

$$\begin{aligned}
{}_{\mu_0}I_{\mu}^{((k+1)\varepsilon)}\left[D^{((k+1)\varepsilon)}\varphi(\mu)\right] &= {}_{\mu_0}I_{\mu}^{(\varepsilon)}\left\{{}_{\mu_0}I_{\mu}^{(k\varepsilon)}\left[D^{((k+1)\varepsilon)}\varphi(\mu)\right]\right\} \\
&= D^{((k+1)\varepsilon)}\varphi(\xi)\,{}_{\mu_0}I_{\mu}^{((k+1)\varepsilon)}1 \\
&= D^{((k+1)\varepsilon)}\varphi(\xi)\frac{(\mu-\mu_0)^{(k+1)\varepsilon}}{\Gamma(1+(k+1)\varepsilon)},
\end{aligned} \tag{1.174}$$

where $\mu_0 < \xi < \mu$, $\forall \mu \in (a,b)$.

Therefore, we have proved the result. □

Theorem 1.11. *Suppose that*

$$D^{((k+1)\varepsilon)}\varphi(\mu) \in C_{\varepsilon}(a,b).$$

Then, for $k = 0, 1, \ldots, n$, *there is*

$$\varphi(\mu) = \sum_{k=0}^{n} \frac{D^{(k\varepsilon)}\varphi(\mu_0)}{\Gamma(1+k\varepsilon)}(\mu-\mu_0)^{k\varepsilon} + R_{n\varepsilon}(\mu-\mu_0) \tag{1.175}$$

with $a < \mu_0 < \xi < \mu < b$, $\forall \mu \in (a,b)$, *where* $D^{(k\varepsilon)}\varphi(\mu) = \overbrace{D^{(\varepsilon)} \cdots D^{(\varepsilon)}}^{k\text{-times}} \varphi(\mu)$ *and* $R_{n\varepsilon}(\mu - \mu_0) = O\left((\mu - \mu_0)^{n\varepsilon}\right)$.

Proof. Using (1.171), we can write

$$\left|\frac{R_{n\varepsilon}(\mu-\mu_0)}{(\mu-\mu_0)^{n\varepsilon}}\right| = \left|\frac{D^{((n+1)\varepsilon)}\varphi(\xi)}{\Gamma(1+(k+1)\varepsilon)}\frac{(\mu-\mu_0)^{(n+1)\varepsilon}}{(\mu-\mu_0)^{n\varepsilon}}\right| = \left|\frac{D^{((n+1)\varepsilon)}\varphi(\xi)}{\Gamma(1+(k+1)\varepsilon)}(\mu-\mu_0)^{\varepsilon}\right|. \tag{1.176}$$

Therefore, we conclude that

$$\left|\frac{R_{n\varepsilon}(\mu-\mu_0)}{(\mu-\mu_0)^{n\varepsilon}}\right| = \left|\frac{D^{((n+1)\varepsilon)}\varphi(\xi)}{\Gamma(1+(k+1)\varepsilon)}(\mu-\mu_0)^{\varepsilon}\right| = 0. \tag{1.177}$$

□

Theorem 1.12. *Suppose that*

$$D^{((k+1)\varepsilon)}\varphi(\mu) \in C_\varepsilon(a,b).$$

Then, for $k = 0, 1, \ldots, n$, *there is*

$$\varphi(\mu) = \sum_{k=0}^{n} \frac{D^{(k\varepsilon)}\varphi(0)}{\Gamma(1+k\varepsilon)}\mu^{k\varepsilon} + \frac{D^{((n+1)\varepsilon)}\varphi(\theta\mu)}{\Gamma(1+(n+1)\varepsilon)}\mu^{(n+1)\varepsilon} \tag{1.178}$$

with $0 < \theta < 1$, $\forall \mu \in (a,b)$, *where* $D^{(k\varepsilon)}\varphi(\mu) = \overbrace{D^{(\varepsilon)} \cdots D^{(\varepsilon)}}^{k\text{-times}} \varphi(\mu)$.

Proof. For $\mu_0 = 0$ and $\mu \in (a,b)$, from (1.175), we present

$$\varphi(\mu) = \sum_{k=0}^{n} \frac{D^{(k\varepsilon)}\varphi(0)}{\Gamma(1+k\varepsilon)}(\mu-\mu_0)^{k\varepsilon} + \frac{D^{((n+1)\varepsilon)}\varphi(\xi)}{\Gamma(1+(n+1)\varepsilon)}\mu^{(n+1)\varepsilon}, \tag{1.179}$$

where $a < \mu_0 < \xi < \mu < b$.

If $\xi = \theta\mu$ in (1.179), then, there is

$$\frac{D^{((n+1)\varepsilon)}\varphi(\xi)}{\Gamma(1+(n+1)\varepsilon)}\mu^{(n+1)\varepsilon} = \frac{D^{((n+1)\varepsilon)}\varphi(\theta\mu)}{\Gamma(1+(n+1)\varepsilon)}\mu^{(n+1)\varepsilon} \tag{1.180}$$

with $0 < \theta < 1$.

□

1.4.4 Local fractional Taylor's series for elementary functions

Theorem 1.13. *Suppose that*

$$D^{((k+1)\varepsilon)}\varphi(\mu) \in C_\varepsilon(a,b).$$

Then, for $k = 0, 1, \ldots, n$,

$$\varphi(\mu) = \sum_{k=0}^{\infty} \frac{D^{(k\varepsilon)}\varphi(\mu_0)}{\Gamma(1+k\varepsilon)} (\mu - \mu_0)^{k\varepsilon} \tag{1.181}$$

with $a < \mu_0 < \mu < b, \forall \mu \in (a,b)$, *where* $D^{(k\varepsilon)}\varphi(\mu) = \overbrace{D^{(\varepsilon)} \cdots D^{(\varepsilon)}}^{k\text{-times}} \varphi(\mu)$.

Proof. According to local fractional Taylor's theorem, from (1.171), there is

$$\begin{aligned}
\varphi(\mu) &= \lim_{\mu \to \mu_0} \left\{ \sum_{k=0}^{n} \frac{D^{(k\varepsilon)}\varphi(\mu_0)}{\Gamma(1+k\varepsilon)} (\mu - \mu_0)^{k\varepsilon} + \frac{D^{((n+1)\varepsilon)}\varphi(\xi)}{\Gamma(1+(n+1)\varepsilon)} (\mu - \mu_0)^{(n+1)\varepsilon} \right\} \\
&= \sum_{k=0}^{\infty} \frac{D^{(k\varepsilon)}\varphi(\mu_0)}{\Gamma(1+k\varepsilon)} (\mu - \mu_0)^{k\varepsilon}
\end{aligned} \tag{1.182}$$

with $a < \mu_0 < \xi < \mu < b, \forall \mu \in (a,b)$, where $D^{(k\varepsilon)}\varphi(\mu) = \overbrace{D^{(\varepsilon)} \cdots D^{(\varepsilon)}}^{k\text{-times}} \varphi(\mu)$. □

In this case, we present the following result.

Suppose $D^{((k+1)\varepsilon)}\varphi(\mu) \in C_\varepsilon(a,b)$. Then, for $k = 0, 1, \ldots, n$, there is

$$\varphi(\mu) = \sum_{k=0}^{\infty} \frac{D^{(k\varepsilon)}\varphi(0)}{\Gamma(1+k\varepsilon)} \mu^{k\varepsilon} \tag{1.183}$$

with $a < 0 < \mu < b, \forall \mu \in (a,b)$, where $D^{(k\varepsilon)}\varphi(\mu) = \overbrace{D^{(\varepsilon)} \cdots D^{(\varepsilon)}}^{k\text{-times}} \varphi(\mu)$.

This series is said to be local fractional MacLaurin's series of the function $\varphi(\mu)$.

In this case, we present the following local fractional MacLaurin's series of elementary functions:

(a) $E_\varepsilon(\mu^\varepsilon) = \sum_{k=0}^{\infty} \frac{\mu^{k\varepsilon}}{\Gamma(1+k\varepsilon)}$;

(b) $E_\varepsilon(-\mu^\varepsilon) = \sum_{k=0}^{\infty} \frac{(-1)^k \mu^{k\varepsilon}}{\Gamma(1+k\varepsilon)}$;

(c) $\sin_\varepsilon(\mu^\varepsilon) = \sum_{k=0}^{\infty} \frac{(-1)^k \mu^{(2k+1)\varepsilon}}{\Gamma(1+(2k+1)\varepsilon)}$;

(d) $\cos_\varepsilon(\mu^\varepsilon) = \sum_{k=0}^{\infty} \frac{(-1)^k \mu^{2k\varepsilon}}{\Gamma(1+2k\varepsilon)}$;

(e) $\sinh_\varepsilon(\mu^\varepsilon) = \sum_{k=0}^{\infty} \frac{\mu^{(2k+1)\varepsilon}}{\Gamma(1+(2k+1)\varepsilon)}$; and

(f) $\cosh_\varepsilon(\mu^\varepsilon) = \sum_{k=0}^{\infty} \frac{\mu^{2k\varepsilon}}{\Gamma(1+2k\varepsilon)}$.

The proofs of them are listed in Appendix C. Let C be a constant. The local fractional integrals of some of nondifferentiable functions defined on fractal sets are listed in Table 1.2.

Let $m, n\ (m \neq n)$ be integrals. The local fractional integrals of some of nondifferentiable functions via Mittag–Leffler function defined on fractal sets are listed in Table 1.3.

Table 1.2 Basic operations of local fractional integral of some of nondifferentiable functions defined on fractal sets

Original function	Transformed function
C	$C\mu^{\varepsilon}/\Gamma(1+\varepsilon)$
$\mu^{k\varepsilon}/\Gamma(1+k\varepsilon)$	$\mu^{(k+1)\varepsilon}/\Gamma(1+(k+1)\varepsilon)$
$E_{\varepsilon}(\mu^{\varepsilon})$	$E_{\varepsilon}(\mu^{\varepsilon})-1$
$E_{\varepsilon}(C\mu^{\varepsilon})$	$\frac{E_{\varepsilon}(C\mu^{\varepsilon})-1}{C}$
$\sin_{\varepsilon}(\mu^{\varepsilon})$	$-[\cos_{\varepsilon}(\mu^{\varepsilon})-1]$
$\sin_{\varepsilon}(C\mu^{\varepsilon})$	$\frac{-[\cos_{\varepsilon}(\mu^{\varepsilon})-1]}{C}$
$\cos_{\varepsilon}(\mu^{\varepsilon})$	$\sin_{\varepsilon}(\mu^{\varepsilon})$
$\cos_{\varepsilon}(C\mu^{\varepsilon})$	$\frac{\sin_{\varepsilon}(C\mu^{\varepsilon})}{C}$
$\frac{\mu^{\varepsilon}}{\Gamma(1+\varepsilon)}\sin_{\varepsilon}(C\mu^{\varepsilon})$	$-\frac{1}{C}\left[\frac{\mu^{\varepsilon}}{\Gamma(1+\varepsilon)}\cos_{\varepsilon}(C\mu^{\varepsilon})-\frac{1}{C}\sin_{\varepsilon}(C\mu^{\varepsilon})\right]$
$\frac{\mu^{\varepsilon}}{\Gamma(1+\varepsilon)}\cos_{\varepsilon}(C\mu^{\varepsilon})$	$\frac{1}{C}\left\{\frac{\mu^{\varepsilon}}{\Gamma(1+\varepsilon)}\sin_{\varepsilon}(C\mu^{\varepsilon})-\frac{1}{C}[\cos_{\varepsilon}(C\mu^{\varepsilon})-1]\right\}$
$E_{\varepsilon}(\mu^{\varepsilon})\sin_{\varepsilon}(C\mu^{\varepsilon})$	$\frac{E_{\varepsilon}(\mu^{\varepsilon})[\sin_{\varepsilon}(C\mu^{\varepsilon})-C\cos_{\varepsilon}(C\mu^{\varepsilon})]+C}{1+C^2}$
$E_{\varepsilon}(\mu^{\varepsilon})\cos_{\varepsilon}(C\mu^{\varepsilon})$	$\frac{E_{\varepsilon}(\mu^{\varepsilon})[\cos_{\varepsilon}(C\mu^{\varepsilon})+C\sin_{\varepsilon}(C\mu^{\varepsilon})]-1}{1+C^2}$

Table 1.3 Basic operations of local fractional integral of some of nondifferentiable functions via Mittag–Leffler function defined on fractal sets

Original function	Transformed function
$\sin_{\varepsilon}(\mu^{\varepsilon})$	0
$\cos_{\varepsilon}(\mu^{\varepsilon})$	0
$\sin_{\varepsilon}(m^{\varepsilon}\mu^{\varepsilon})$	0
$\cos_{\varepsilon}(m^{\varepsilon}\mu^{\varepsilon})$	0
$\sin_{\varepsilon}(m^{\varepsilon}\mu^{\varepsilon})\cos_{\varepsilon}(n^{\varepsilon}\mu^{\varepsilon})$	0
$\sin_{\varepsilon}(m^{\varepsilon}\mu^{\varepsilon})\cos_{\varepsilon}(m^{\varepsilon}\mu^{\varepsilon})$	0
$\sin_{\varepsilon}(m^{\varepsilon}\mu^{\varepsilon})\sin_{\varepsilon}(m^{\varepsilon}\mu^{\varepsilon})$	$\pi^{\varepsilon}/\Gamma(1+\varepsilon)$
$\cos_{\varepsilon}(n^{\varepsilon}\mu^{\varepsilon})\cos_{\varepsilon}(n^{\varepsilon}\mu^{\varepsilon})$	$\pi^{\varepsilon}/\Gamma(1+\varepsilon)$
$\frac{\sin_{\varepsilon}[(2n+1)\mu/2]^{\varepsilon}}{2^{\varepsilon}\sin_{\varepsilon}(\mu/2)^{\varepsilon}}$	$\pi^{\varepsilon}/\Gamma(1+\varepsilon)$

1.5 Local fractional partial differential equations in mathematical physics

1.5.1 Local fractional partial derivatives

The general equation of the circle of Cantor type with fractal dimension ε $(0 < \varepsilon \leq 1)$ is given by

$$\mu^{2\varepsilon} + \eta^{2\varepsilon} = a^{2\varepsilon}, \tag{1.184}$$

where a is the radius of the circle.

Let Φ: $\wp \to \aleph$ be a function defined on a fractal set $\wp$ of fractal dimension $\varepsilon(0 < \varepsilon < 1)$. A function $\Phi(\mu, \eta)$ is local fractional continuous at the point (μ_0, η_0) if there is a number $\tau > 0$ such that

$$|\Phi(\mu, \eta) - \Phi(\mu_0, \eta_0)| < \tau^{\varepsilon}, \tag{1.185}$$

where its circular δ neighborhood of (μ_0, η_0) is

$$(\mu - \mu_0)^{2\varepsilon} + (\eta - \eta_0)^{2\varepsilon} < \delta^{2\varepsilon}. \tag{1.186}$$

It is said to be the local fractional continuous if there is

$$\lim_{(\mu,\eta)\to(\mu_0,\eta_0)} \Phi(\mu, \eta) = \Phi(\mu_0, \eta_0). \tag{1.187}$$

Let $\Phi(\mu, \eta)$ be defined in the domain $\wp$ of the $\mu\eta$-plane. The local fractional partial derivative operator of $\Phi(\mu, \eta)$ of order $\varepsilon(0 < \varepsilon < 1)$ with respect to μ in the domain $\wp$ is defined as follows:

$$\Phi^{(\varepsilon)}(\mu_0, \eta) = \left.\frac{\partial^{\varepsilon}\Phi(\mu, \eta)}{\partial\mu^{\varepsilon}}\right|_{\mu=\mu_0} = \lim_{\mu\to\mu_0} \frac{\Delta^{\varepsilon}[\Phi(\mu, \eta) - \Phi(\mu_0, \eta)]}{(\mu_0 - \mu_0)^{\varepsilon}}, \tag{1.188}$$

where $\Delta^{\varepsilon}[\Phi(\mu, \eta) - \Phi(\mu_0, \eta)] \cong \Gamma(1+\varepsilon)[\Phi(\mu, \eta) - \Phi(\mu_0, \eta)]$.

The local fractional partial derivative operator of $\Phi(\mu, \eta)$ of order $\varepsilon(0 < \varepsilon < 1)$ with respect to η in the domain $\wp$ is defined as follows:

$$\Phi^{(\varepsilon)}(\mu, \eta_0) = \left.\frac{\partial^{\varepsilon}\Phi(\mu, \eta)}{\partial\eta^{\varepsilon}}\right|_{\eta=\eta_0} = \lim_{\eta\to\eta_0} \frac{\Delta^{\varepsilon}[\Phi(\mu, \eta) - \Phi(\mu, \eta_0)]}{(\eta_0 - \eta_0)^{\varepsilon}}, \tag{1.189}$$

where $\Delta^{\varepsilon}[\Phi(\mu, \eta) - \Phi(\mu, \eta_0)] \cong \Gamma(1+\varepsilon)[\Phi(\mu, \eta) - \Phi(\mu, \eta_0)]$.

The local fractional partial derivative operator of $\Phi(\mu, \eta)$ of higher order $(m+n)\varepsilon(0 < \varepsilon < 1)$ with respect to η and μ in the domain $\wp$ is defined as follows:

$$\underbrace{\frac{\partial^{\varepsilon}}{\partial\mu^{\varepsilon}}\cdots\frac{\partial^{\varepsilon}}{\partial\mu^{\varepsilon}}}_{n\text{-times}}\underbrace{\frac{\partial^{\varepsilon}}{\partial\eta^{\varepsilon}}\cdots\frac{\partial^{\varepsilon}}{\partial\eta^{\varepsilon}}}_{m\text{-times}}\Phi(\mu, \eta) = \frac{\partial^{(m+m)\varepsilon}\Phi(\mu, \eta)}{\underbrace{\partial\mu^{\varepsilon}\ldots\partial\mu^{\varepsilon}}_{n\text{-times}}\underbrace{\partial\eta^{\varepsilon}\ldots\partial\eta^{\varepsilon}}_{m\text{-times}}} = \Phi^{(m+n)\alpha}_{\eta^m\mu^n}(\mu, \eta), \tag{1.190}$$

where m and n are positive integers.

We have

$$\Phi(\mu,\eta) \in C_{\varepsilon}^{m+n}, \tag{1.191}$$

if (1.190) holds.

The local fractional gradient and Laplace operators of a local fractional scalar field $\varphi(\mu,\eta,\sigma)$ in 3 fractal dimensional space are presented as

$$\nabla^{\varepsilon}\varphi(\mu,\eta,\sigma) = \frac{\partial^{\varepsilon}\varphi(\mu,\eta,\sigma)}{\partial\mu^{\varepsilon}}\mathbf{e}_1^{\varepsilon} + \frac{\partial^{\varepsilon}\varphi(\mu,\eta,\sigma)}{\partial\eta^{\varepsilon}}\mathbf{e}_2^{\varepsilon} + \frac{\partial^{\varepsilon}\varphi(\mu,\eta,\sigma)}{\partial\sigma^{\varepsilon}}\mathbf{e}_3^{\varepsilon}$$

and

$$\nabla^{2\varepsilon}\varphi(\mu,\eta,\sigma) = \frac{\partial^{2\varepsilon}\varphi(\mu,\eta,\sigma)}{\partial\mu^{2\varepsilon}} + \frac{\partial^{2\varepsilon}\varphi(\mu,\eta,\sigma)}{\partial\eta^{2\varepsilon}} + \frac{\partial^{2\varepsilon}\varphi(\mu,\eta,\sigma)}{\partial\sigma^{2\varepsilon}},$$

respectively.

The local fractional gradient and Laplace operators of a local fractional scalar field $\varphi(\mu,\sigma)$ in 2 fractal dimensional space are presented as

$$\nabla^{\varepsilon}\varphi(\mu,\sigma) = \frac{\partial^{\varepsilon}\varphi(\mu,\sigma)}{\partial\mu^{\varepsilon}}\mathbf{e}_1^{\varepsilon} + \frac{\partial^{\varepsilon}\varphi(\mu,\sigma)}{\partial\sigma^{\varepsilon}}\mathbf{e}_2^{\varepsilon}$$

and

$$\nabla^{2\varepsilon}\varphi(\mu,\sigma) = \frac{\partial^{2\varepsilon}\varphi(\mu,\sigma)}{\partial\mu^{2\varepsilon}} + \frac{\partial^{2\varepsilon}\varphi(\mu,\sigma)}{\partial\eta^{2\varepsilon}},$$

respectively.

The local fractional gradient and Laplace operators of a local fractional scalar field $\varphi(\mu,\sigma)$ in 1 fractal dimensional space are presented as

$$\nabla^{\varepsilon}\varphi(\mu) = \frac{\partial^{\varepsilon}\varphi(\mu)}{\partial\mu^{\varepsilon}}\mathbf{e}_1^{\varepsilon}$$

and

$$\nabla^{2\varepsilon}\varphi(\mu) = \frac{\partial^{2\varepsilon}\varphi(\mu,\sigma)}{\partial\mu^{2\varepsilon}},$$

respectively.

Here, we do not refer to Jacobian and inequality theory via local fractional partial derivative operator [1, 16, 21, 70–72].

1.5.2 Linear and nonlinear partial differential equations in mathematical physics

In mathematical physics, the partial differential equations describing the physical phenomena were always derived from the calculus involving the different kernel functions of differentiability and nondifferentiability. Theory of local fractional calculus was applied to solve the mathematical models from science and engineering, such as vibrating strings, traffic flow, and mass and heat transfer in fractal dimensional time-space. Here, we consider the local fractional partial differential equations in sense of the nondifferentiable characteristics [1, 73–88]. Here, we will put our work upon linear and nonlinear local fractional partial differential equations in 1 + 1 fractal dimensional space and in 1 + 3 fractal dimensional space, such as heat equation, wave equation, the Laplace equation, the Klein–Gordon equation, the Schrödinger equation, diffusion equation, transport equation, the Poisson equation, the linear Korteweg–de Vries equation, the Tricomi equation, the Fokker–Planck equation, the Lighthill–Whitham–Richards equation, the Helmholtz equation, damped wave equation, dissipative wave equation, the Boussinesq equation, nonlinear wave equation, the Burgers equation, the forced Burgers equation, the inviscid Burgers equation, the nonlinear Korteweg–de Vries equation, the modified Korteweg–de Vries equation, the generalized Korteweg–de Vries equation, the nonlinear Klein–Gordon equation, Maxwell's equation, the Navier–Stokes equation, and Euler's equation involving the local fractional partial derivative operator.

We now present some linear local fractional partial differential equations that are of important concern:

The local fractional heat equation in 1 + 1 fractal dimensional space takes the form

$$\frac{\partial^{\varepsilon}\Phi(\mu,\tau)}{\partial\tau^{\varepsilon}} - \kappa\frac{\partial^{2\varepsilon}\Phi(\mu,\tau)}{\partial\mu^{2\varepsilon}} = \Theta(\mu,\tau), \tag{1.192}$$

where κ is the thermal conductivity coefficient (a positive constant) and $\Theta(\mu,\tau)$ is a nondifferentiable heat source.

The local fractional wave equation in 1 + 1 fractal dimensional space takes the form

$$\frac{\partial^{2\varepsilon}\Phi(\mu,\tau)}{\partial\tau^{2\varepsilon}} - \varpi\frac{\partial^{2\varepsilon}\Phi(\mu,\tau)}{\partial\mu^{2\varepsilon}} = 0, \tag{1.193}$$

where ϖ is a constant.

The local fractional Laplace equation in 1 + 1 fractal dimensional space takes the form

$$\frac{\partial^{2\varepsilon}\Phi(\mu,\eta)}{\partial\mu^{2\varepsilon}} + \frac{\partial^{2\varepsilon}\Phi(\mu,\eta)}{\partial\eta^{2\varepsilon}} = 0. \tag{1.194}$$

The local fractional Klein–Gordon equation in 1 + 1 fractal dimensional space takes the form

$$\frac{\partial^{\varepsilon}\Phi(\mu,\tau)}{\partial\tau^{\varepsilon}} - \frac{\partial^{2\varepsilon}\Phi(\mu,\tau)}{\partial\mu^{2\varepsilon}} = \Phi(\mu,\tau). \tag{1.195}$$

The local fractional Schrödinger equation in 1 + 1 fractal dimensional space takes the form

$$i^{\varepsilon}h_{\varepsilon}\frac{\partial^{\varepsilon}\Phi(\mu,\tau)}{\partial\tau^{\varepsilon}} = -\frac{h_{\varepsilon}^{2}}{2m}\frac{\partial^{2\varepsilon}\Phi(\mu,\tau)}{\partial\mu^{2\varepsilon}}, \tag{1.196}$$

where m and h_{ε} are constants.

Local fractional diffusion equation in 1 + 1 fractal dimensional space takes the form

$$\frac{\partial^{\varepsilon}\Phi(\mu,\tau)}{\partial\tau^{\varepsilon}}-D\frac{\partial^{2\varepsilon}\Phi(\mu,\tau)}{\partial\mu^{2\varepsilon}}=0, \tag{1.197}$$

where D is a diffusive coefficient.

The linear local fractional transport equation in 1 + 1 fractal dimensional space takes the form

$$\frac{\partial^{\varepsilon}\Phi(\mu,\tau)}{\partial\tau^{\varepsilon}}+\frac{\partial^{\varepsilon}\Phi(\mu,\tau)}{\partial\mu^{\varepsilon}}=0. \tag{1.198}$$

The local fractional Poisson equation in 1 fractal dimensional space takes the form

$$\frac{\partial^{2\varepsilon}\Phi(\mu,\eta)}{\partial\mu^{2\varepsilon}}+\frac{\partial^{2\varepsilon}\Phi(\mu,\eta)}{\partial\eta^{2\varepsilon}}=\Theta(\mu,\eta), \tag{1.199}$$

where $\Theta(\mu,\eta)$ is a nondifferentiable function.

The linear local fractional Korteweg–de Vries equation in 1 + 1 fractal dimensional space takes the form

$$\frac{\partial^{\varepsilon}\Phi(\mu,\tau)}{\partial\tau^{\varepsilon}}+\frac{\partial^{\varepsilon}\Phi(\mu,\tau)}{\partial\mu^{\varepsilon}}+\frac{\partial^{3\varepsilon}\Phi(\mu,\tau)}{\partial\mu^{3\varepsilon}}=0. \tag{1.200}$$

The local fractional wave equation of fractal transverse vibration of a beam takes the form

$$\frac{\partial^{2\varepsilon}\Phi(\mu,\eta)}{\partial\mu^{2\varepsilon}}+\frac{\partial^{4\varepsilon}\Phi(\mu,\eta)}{\partial\eta^{4\varepsilon}}=0. \tag{1.201}$$

The local fractional Tricomi equation in 1 + 1 fractal dimensional space takes the form

$$\frac{\eta^{\varepsilon}}{\Gamma(1+\varepsilon)}\frac{\partial^{2\varepsilon}\Phi(\mu,\eta)}{\partial\mu^{2\varepsilon}}+\frac{\partial^{2\varepsilon}\Phi(\mu,\eta)}{\partial\eta^{2\varepsilon}}=0. \tag{1.202}$$

The local fractional Fokker–Planck equation in 1 + 1 fractal dimensional space takes the form

$$\frac{\partial^{\varepsilon}\Phi(\mu,\tau)}{\partial\tau^{\varepsilon}}=\frac{\partial^{2\varepsilon}\Phi(\mu,\tau)}{\partial\mu^{2\varepsilon}}-\frac{\partial^{\varepsilon}\Phi(\mu,\tau)}{\partial\mu^{\varepsilon}}. \tag{1.203}$$

The linear local fractional Lighthill–Whitham–Richards equation on a finite length highway is given by

$$\frac{\partial^{\varepsilon}\Phi(\mu,\tau)}{\partial\tau^{\varepsilon}}+\mu\frac{\partial^{\varepsilon}\Phi(\mu,\tau)}{\partial\mu^{\varepsilon}}=0, \tag{1.204}$$

where μ is a constant.

The linear local fractional homogeneous Helmholtz equation in 1 fractal dimensional space takes the form

$$\frac{\partial^{2\varepsilon}\Phi(\mu,\eta)}{\partial\mu^{2\varepsilon}}+\frac{\partial^{2\varepsilon}\Phi(\mu,\eta)}{\partial\eta^{2\varepsilon}}+\varpi\Phi(\mu,\eta)=0, \tag{1.205}$$

where ϖ is a constant.

The linear local fractional inhomogeneous Helmholtz equation in 1 fractal dimensional space with nondifferentiable inhomogeneous term takes the form

$$\frac{\partial^{2\varepsilon}\Phi(\mu,\eta)}{\partial\mu^{2\varepsilon}}+\frac{\partial^{2\varepsilon}\Phi(\mu,\eta)}{\partial\eta^{2\varepsilon}}+\varpi\Phi(\mu,\eta)=\Theta(\mu,\eta), \tag{1.206}$$

where ϖ is a constant and $\Theta(\mu,\eta)$ is a differentiable function.

The linear local damped wave equation in 1 + 1 fractal dimensional space with nondifferentiable inhomogeneous term takes the form

$$\frac{\partial^{2\varepsilon}\Phi(\mu,\tau)}{\partial\tau^{2\varepsilon}}-\frac{\partial^{\varepsilon}\Phi(\mu,\tau)}{\partial\tau^{\varepsilon}}-\frac{\partial^{2\varepsilon}\Phi(\mu,\tau)}{\partial\mu^{2\varepsilon}}=\Theta(\mu,\tau), \tag{1.207}$$

where $\Theta(\mu,\tau)$ is a nondifferentiable inhomogeneous term.

The linear local homogeneous damped wave equation of fractal strings in 1 + 1 fractal dimensional space takes the form

$$\frac{\partial^{2\varepsilon}\Phi(\mu,\tau)}{\partial\tau^{2\varepsilon}}-\frac{\partial^{\varepsilon}\Phi(\mu,\tau)}{\partial\tau^{\varepsilon}}-\frac{\partial^{2\varepsilon}\Phi(\mu,\tau)}{\partial\mu^{2\varepsilon}}=0. \tag{1.208}$$

The local fractional inhomogeneous dissipative wave equation of fractal strings in 1 + 1 fractal dimensional space takes the form

$$\frac{\partial^{2\varepsilon}\Phi(\mu,\tau)}{\partial\tau^{2\varepsilon}}-\frac{\partial^{\varepsilon}\Phi(\mu,\tau)}{\partial\tau^{\varepsilon}}-\frac{\partial^{\varepsilon}\Phi(\mu,\tau)}{\partial\mu^{\varepsilon}}-\frac{\partial^{2\varepsilon}\Phi(\mu,\tau)}{\partial\mu^{2\varepsilon}}=\Theta(\mu,\tau), \tag{1.209}$$

where $\Theta(\mu,\tau)$ is a nondifferentiable inhomogeneous term.

The local fractional inhomogeneous dissipative wave equation of fractal strings in 1 + 1 fractal dimensional space takes the form

$$\frac{\partial^{2\varepsilon}\Phi(\mu,\tau)}{\partial\tau^{2\varepsilon}}-\frac{\partial^{\varepsilon}\Phi(\mu,\tau)}{\partial\tau^{\varepsilon}}-\frac{\partial^{\varepsilon}\Phi(\mu,\tau)}{\partial\mu^{\varepsilon}}-\frac{\partial^{2\varepsilon}\Phi(\mu,\tau)}{\partial\mu^{2\varepsilon}}=0. \tag{1.210}$$

The linear local fractional Boussinesq equation of fractal long water waves in 1 + 1 fractal dimensional space takes the form

$$\frac{\partial^{2\varepsilon}\Phi(\mu,\tau)}{\partial\tau^{2\varepsilon}}-\frac{\partial^{2\varepsilon}\Phi(\mu,\tau)}{\partial\mu^{2\varepsilon}}-\frac{\partial^{4\varepsilon}\Phi(\mu,\tau)}{\partial\mu^{2\varepsilon}\partial\tau^{2\varepsilon}}=0. \tag{1.211}$$

Here, we present some nonlinear local fractional partial differential equations that are of important concern:

The local fractional nonlinear wave equation for the velocity potential of fluid flow in 1 + 1 fractal dimensional space takes the form

$$\frac{\partial^{2\varepsilon}\Phi(\mu,\tau)}{\partial\tau^{2\varepsilon}} = \omega\frac{\partial^{2\varepsilon}\Phi(\mu,\tau)}{\partial\mu^{2\varepsilon}} + \varpi\phi\frac{\partial^{2\varepsilon}\Phi(\mu,\tau)}{\partial\mu^{2\varepsilon}}, \tag{1.212}$$

where ω and ϖ are constants.

The nonlinear local fractional Burgers equation in 1 + 1 fractal dimensional space is given by

$$\frac{\partial^{\varepsilon}\Phi(\mu,\tau)}{\partial\tau^{\varepsilon}} + \Phi(\mu,\tau)\frac{\partial^{\varepsilon}\Phi(\mu,\tau)}{\partial\mu^{\varepsilon}} = \kappa\frac{\partial^{2\varepsilon}\Phi(\mu,\tau)}{\partial\mu^{2\varepsilon}}, \tag{1.213}$$

where κ is a constant.

The nonlinear local fractional forced Burgers equation in 1 + 1 fractal dimensional space is given by

$$\frac{\partial^{\varepsilon}\Phi(\mu,\tau)}{\partial\tau^{\varepsilon}} + \Phi(\mu,\tau)\frac{\partial^{\varepsilon}\Phi(\mu,\tau)}{\partial\mu^{\varepsilon}} = \kappa\frac{\partial^{2\varepsilon}\Phi(\mu,\tau)}{\partial\mu^{2\varepsilon}} + \Theta(\mu,\tau), \tag{1.214}$$

where $\Theta(\mu,\tau)$ is a forced source.

The nonlinear local fractional inviscid Burgers equation in 1 + 1 fractal dimensional space is given by

$$\frac{\partial^{\varepsilon}\Phi(\mu,\tau)}{\partial\tau^{\varepsilon}} + \Phi(\mu,\tau)\frac{\partial^{\varepsilon}\Phi(\mu,\tau)}{\partial\mu^{\varepsilon}} = 0. \tag{1.215}$$

The nonlinear local fractional transport equation in 1 + 1 fractal dimensional space is given by

$$\frac{\partial^{\varepsilon}\Phi(\mu,\tau)}{\partial\tau^{\varepsilon}} + \Phi(\mu,\tau)\frac{\partial^{\varepsilon}\Phi(\mu,\tau)}{\partial\mu^{\varepsilon}} = \Theta(\mu,\tau), \tag{1.216}$$

where $\Theta(\mu,\tau)$ is a forced source.

The nonlinear local fractional Korteweg–de Vries equation in 1 + 1 fractal dimensional space is given by

$$\frac{\partial^{\varepsilon}\Phi(\mu,\tau)}{\partial\tau^{\varepsilon}} - R\Phi(\mu,\tau)\frac{\partial^{\varepsilon}\Phi(\mu,\tau)}{\partial\mu^{\varepsilon}} + \frac{\partial^{3\varepsilon}\Phi(\mu,\tau)}{\partial\mu^{3\varepsilon}} + \frac{\partial^{\varepsilon}\Phi(\mu,\tau)}{\partial\mu^{\varepsilon}} = 0 \tag{1.217}$$

or

$$\frac{\partial^{\varepsilon}\Phi(\mu,\tau)}{\partial\tau^{\varepsilon}} + S\Phi(\mu,\tau)\frac{\partial^{\varepsilon}\Phi(\mu,\tau)}{\partial\mu^{\varepsilon}} - \frac{\partial^{3\varepsilon}\Phi(\mu,\tau)}{\partial\mu^{3\varepsilon}} = 0, \tag{1.218}$$

where R and S are constants.

The nonlinear local fractional modified Korteweg–de Vries equation in 1 + 1 fractal dimensional space is given by

$$\frac{\partial^{\varepsilon}\Phi(\mu,\tau)}{\partial\tau^{\varepsilon}} + \frac{\partial^{3\varepsilon}\Phi(\mu,\tau)}{\partial\mu^{3\varepsilon}} \pm S\Phi^{2}(\mu,\tau)\frac{\partial^{\varepsilon}\Phi(\mu,\tau)}{\partial\mu^{\varepsilon}} = 0, \tag{1.219}$$

where S is a constant.

The nonlinear local fractional generalized Korteweg–de Vries equation in 1 + 1 fractal dimensional space is given by

$$\frac{\partial^{\varepsilon}\Phi(\mu,\tau)}{\partial\tau^{\varepsilon}}+S\Phi(\mu,\tau)\frac{\partial^{\varepsilon}\Phi(\mu,\tau)}{\partial\mu^{\varepsilon}}-\frac{\partial^{5\varepsilon}\Phi(\mu,\tau)}{\partial\mu^{5\varepsilon}}=0, \tag{1.220}$$

where S is a constant.

The nonlinear local fractional Klein–Gordon equation in 1 + 1 fractal dimensional space is given by

$$\frac{\partial^{2\varepsilon}\Phi(\mu,\tau)}{\partial\tau^{2\varepsilon}}-\frac{\partial^{2\varepsilon}\Phi(\mu,\tau)}{\partial\mu^{2\varepsilon}}=\Theta\left(\Phi(\mu,\tau)\right), \tag{1.221}$$

where $\Theta\left(\Phi(\mu,\tau)\right)$ is a nonlinear term related to $\Phi(\mu,\tau)$.

The nonlinear local fractional Lighthill–Whitham–Richards equation on a finite length highway is given by

$$\frac{\partial^{2\varepsilon}\Phi(\mu,\tau)}{\partial\tau^{2\varepsilon}}+\xi\frac{\partial^{\varepsilon}\Phi(\mu,\tau)}{\partial\mu^{\varepsilon}}+\eta\Phi(\mu,\tau)\frac{\partial^{\varepsilon}\Phi(\mu,\tau)}{\partial\mu^{\varepsilon}}=0, \tag{1.222}$$

where ξ and η are constants.

Here, we present some nonlinear local fractional partial differential equations in 1 + 3 fractal dimensional space that are of important concern:

The linear local fractional wave equation for the velocity potential of fluid flow in 1 + 3 fractal dimensional space takes the form

$$\omega\nabla^{2\varepsilon}\Phi(\mu,\eta,\sigma,\tau)-\frac{\partial^{2\varepsilon}\Phi(\mu,\eta,\sigma,\tau)}{\partial\tau^{2\varepsilon}}=0, \tag{1.223}$$

where ω are a constant.

The local fractional Laplace equation arising in fractal electrostatics in 1 + 3 fractal dimensional space takes the form

$$\nabla^{2\varepsilon}\Phi(\mu,\eta,\sigma)=0. \tag{1.224}$$

The local fractional Poisson equation in 3 fractal dimensional space takes the form

$$\nabla^{2\varepsilon}\Phi(\mu,\eta,\sigma)=\Theta(\mu,\eta,\sigma), \tag{1.225}$$

where $\Theta(\mu,\eta,\sigma)$ is a nondifferentiable function.

The linear local fractional inhomogeneous Helmholtz equation in 3 fractal dimensional space with nondifferentiable inhomogeneous term takes the form

$$\nabla^{2\varepsilon}\Phi(\mu,\eta,\sigma)+\varpi\Phi(\mu,\eta,\sigma)=\Theta(\mu,\eta,\sigma), \tag{1.226}$$

where ϖ is a constant and $\Theta(\mu,\eta,\sigma)$ is a differentiable function.

The local fractional heat-conduction equation in 1 + 3 fractal dimensional space takes the form

$$\frac{\partial^{\varepsilon}\Phi(\mu,\eta,\sigma,\tau)}{\partial\tau^{\varepsilon}}-\varpi\nabla^{2\varepsilon}\Phi(\mu,\eta,\sigma,\tau)=H(\mu,\eta,\sigma,\tau), \tag{1.227}$$

where ϖ is the thermal conductivity coefficient and $H(\mu, \eta, \sigma, \tau)$ is the heat source.

The linear local homogeneous damped wave equation of fractal strings in 1 + 3 fractal dimensional space takes the form

$$\frac{\partial^{2\varepsilon}\Phi(\mu,\eta,\sigma,\tau)}{\partial\tau^{2\varepsilon}}-\frac{\partial^{\varepsilon}\Phi(\mu,\eta,\sigma,\tau)}{\partial\tau^{\varepsilon}}-\nabla^{2\varepsilon}\Phi(\mu,\eta,\sigma,\tau)=0. \tag{1.228}$$

The local fractional inhomogeneous dissipative wave equation of fractal strings in 1 + 3 fractal dimensional space takes the form

$$\frac{\partial^{2\varepsilon}\Phi(\mu,\eta,\sigma,\tau)}{\partial\tau^{2\varepsilon}}-\frac{\partial^{\varepsilon}\Phi(\mu,\eta,\sigma,\tau)}{\partial\tau^{\varepsilon}}-\nabla^{2\varepsilon}\Phi(\mu,\eta,\sigma,\tau)-\nabla^{\varepsilon}\Phi(\mu,\eta,\sigma,\tau)=0. \tag{1.229}$$

The local fractional diffusion equation in 1 + 3 fractal dimensional space takes the form

$$\frac{\mathrm{d}^{\varepsilon}\Phi(\mu,\eta,\sigma,\tau)}{\mathrm{d}\tau^{\varepsilon}}-\nabla^{\varepsilon}D(\phi)\nabla^{\varepsilon}\Phi(\mu,\eta,\sigma,\tau)-D(\phi)\nabla^{2\varepsilon}\Phi(\mu,\eta,\sigma,\tau)=0, \tag{1.230}$$

where $D(\phi)$ is diffusion coefficient related to $\Phi(\mu, \eta, \sigma, \tau)$.

The local fractional Schrödinger equation with the nondifferentiable potential function in 1 + 3 fractal dimensional space takes the form

$$i^{\alpha}h_{\alpha}\frac{\partial^{\alpha}\Phi(\mu,\eta,\sigma,\tau)}{\partial\tau^{\alpha}}=-\frac{h_{\alpha}^{2}}{2m}\nabla^{2\varepsilon}\Phi(\mu,\eta,\sigma,\tau)+\Theta(\mu,\eta,\sigma)\Phi(\mu,\eta,\sigma,\tau), \tag{1.231}$$

where $\Theta(\mu, \eta, \sigma)$ is the nondifferentiable potential function.

The nonlinear local fractional wave equation for the velocity potential of fluid flow in 1 + 3 fractal dimensional space takes the form

$$\frac{\partial^{2\varepsilon}\Phi(\mu,\eta,\sigma,\tau)}{\partial\tau^{2\varepsilon}}=\omega\nabla^{2\varepsilon}\Phi(\mu,\eta,\sigma,\tau)+\varpi\Phi(\mu,\eta,\sigma,\tau)\nabla^{2\varepsilon}\Phi(\mu,\eta,\sigma,\tau), \tag{1.232}$$

where ω and ϖ are constants.

Systems of local fractional Maxwell's equations in 1 + 3 fractal dimensional space take the form

$$\nabla^{\varepsilon}\cdot D(\mu,\eta,\sigma,\tau)=\rho(\mu,\eta,\sigma,\tau), \tag{1.233}$$

$$\nabla^{\varepsilon}\times H(\mu,\eta,\sigma,\tau)=\mathrm{J}_{\varepsilon}(\mu,\eta,\sigma,\tau)+\frac{\partial^{\varepsilon}D(\mu,\eta,\sigma,\tau)}{\partial\tau^{\varepsilon}}, \tag{1.234}$$

$$\nabla^{\varepsilon}\times E(\mu,\eta,\sigma,\tau)=-\frac{\partial^{\varepsilon}B(\mu,\eta,\sigma,\tau)}{\partial\tau^{\varepsilon}}, \tag{1.235}$$

$$\nabla^{\varepsilon}\cdot B(\mu,\eta,\sigma,\tau)=0, \tag{1.236}$$

where $\rho(\mu,\eta,\sigma,\tau)$ is the fractal electric charge density, $D(\mu,\eta,\sigma,\tau)$ is electric displacement in the fractal electric field, $H(\mu,\eta,\sigma,\tau)$ is the magnetic field strength in the fractal field, $E(\mu,\eta,\sigma,\tau)$ is the electric field strength in the fractal field, $J_\varepsilon(\mu,\eta,\sigma,\tau)$ is the conductive current, and $B(\mu,\eta,\sigma,\tau)$ is the magnetic induction in the fractal field, and the constitutive relationships in fractal electromagnetic can be written as

$$D(\mu,\eta,\sigma,\tau) = \varepsilon_f E(\mu,\eta,\sigma,\tau) \tag{1.237}$$

and

$$H(\mu,\eta,\sigma,\tau) = \mu_f B(\mu,\eta,\sigma,\tau), \tag{1.238}$$

with the fractal dielectric permittivity ε_f and the fractal magnetic permeability μ_f.

Systems of the local fractional compressible Navier–Stokes equations in 1 + 3 fractal dimensional space take the form

$$\frac{\partial^\varepsilon \rho}{\partial \tau^\varepsilon} + \nabla^\varepsilon \cdot (\rho \boldsymbol{v}) = 0, \tag{1.239}$$

$$\rho\left(\frac{\partial^\varepsilon v}{\partial \tau^\varepsilon} + v \cdot \nabla^\alpha v\right) = -\nabla^\varepsilon p + \frac{1}{3}\mu\nabla^\varepsilon\left(\nabla^\varepsilon \cdot \boldsymbol{v}\right) + \mu\nabla^{2\varepsilon}\boldsymbol{v} + \rho\mathbf{b}, \tag{1.240}$$

$$\rho\left[\frac{\partial^\varepsilon(\theta+\phi)}{\partial\tau^\varepsilon} + v \cdot \nabla^\varepsilon(\theta+\phi)\right] = -\nabla^\varepsilon\cdot(p\boldsymbol{v}) + \boldsymbol{v}\cdot\left(\nabla^\varepsilon\cdot\mathbf{J}\right) + \rho\mathbf{b}\cdot\boldsymbol{v} + K^{2\varepsilon}\nabla^\varepsilon\cdot\mathbf{q}, \tag{1.241}$$

where $\boldsymbol{v}(\mu,\eta,\sigma,\tau)$ is the fractal fluid velocity, μ is the fractal shear moduli of viscosity, $p(\mu,\eta,\sigma,\tau)$ is the thermodynamic pressure, $\rho(\mu,\eta,\sigma,\tau)$ is the fractal fluid density, $\phi(\mu,\eta,\sigma,\tau)$ is the kinetic energy per unit of mass, $\mathbf{b}(\mu,\eta,\sigma,\tau)$ is the external force per unit of mass, $\mathbf{J}(\mu,\eta,\sigma,\tau)$ is the fractal Cauchy stress tensor, and $\theta(\mu,\eta,\sigma,\tau)$ is the internal energy per unit of mass.

Systems of the local fractional incompressible Navier–Stokes equations in 1 + 3 fractal dimensional space take the form

$$\nabla^\varepsilon \cdot \boldsymbol{v} = 0, \tag{1.242}$$

$$\rho\left(\frac{\partial^\varepsilon v}{\partial \tau^\varepsilon} + v \cdot \nabla^\varepsilon v\right) = -\nabla^\varepsilon p + \mu\nabla^{2\varepsilon}\boldsymbol{v} + \rho\mathbf{b}, \tag{1.243}$$

$$\rho\left[\frac{\partial^\varepsilon(\theta+\phi)}{\partial\tau^\varepsilon} + v \cdot \nabla^\varepsilon(\theta+\phi)\right] = -\nabla^\varepsilon\cdot(p\boldsymbol{v}) + \boldsymbol{v}\cdot\left(\nabla^\varepsilon\cdot\mathbf{J}\right) + \rho\mathbf{b}\cdot\boldsymbol{v} + K^{2\varepsilon}\nabla^\varepsilon\cdot\mathbf{q}, \tag{1.244}$$

where $\boldsymbol{v}(\mu,\eta,\sigma,\tau)$ is the fractal fluid velocity, μ is the fractal shear moduli of viscosity, $p(\mu,\eta,\sigma,\tau)$ is the thermodynamic pressure, $\rho(\mu,\eta,\sigma,\tau)$ is the fractal fluid density, $\phi(\mu,\eta,\sigma,\tau)$ is the kinetic energy per unit of mass, $\mathbf{b}(\mu,\eta,\sigma,\tau)$ is the external force per unit of mass, $\mathbf{J}(\mu, \eta, \sigma, \tau)$ is the fractal Cauchy stress tensor, and $\theta(\mu, \eta, \sigma, \tau)$ is the internal energy per unit of mass.

Systems of local fractional compressible Euler's equation in 1 + 3 fractal dimensional space take the form

$$\frac{\partial^{\varepsilon}\rho}{\partial\tau^{\varepsilon}} + \nabla^{\varepsilon} \cdot (\rho \boldsymbol{v}) = 0, \tag{1.245}$$

$$\rho\left(\frac{\partial^{\varepsilon}\upsilon}{\partial\tau^{\varepsilon}} + \upsilon \cdot \nabla^{\varepsilon}\upsilon\right) = -\nabla^{\varepsilon}p + \rho\mathbf{b}, \tag{1.246}$$

$$\rho\left[\frac{\partial^{\varepsilon}(\theta+\phi)}{\partial\tau^{\varepsilon}} + \upsilon \cdot \nabla^{\varepsilon}(\theta+\phi)\right] = -\nabla \cdot (p\boldsymbol{v}), \tag{1.247}$$

where $\boldsymbol{v}(\mu, \eta, \sigma, \tau)$ is the fractal fluid velocity, $p(\mu, \eta, \sigma, \tau)$ is the thermodynamic pressure, $\rho(\mu, \eta, \sigma, \tau)$ is the fractal fluid density, $\phi(\mu, \eta, \sigma, \tau)$ is the kinetic energy per unit of mass, $\mathbf{b}(\mu, \eta, \sigma, \tau)$ is the external force per unit of mass, and $\theta(\mu, \eta, \sigma, \tau)$ is the internal energy per unit of mass.

Systems of local fractional incompressible Euler's equation in 1 + 3 fractal dimensional space take the form

$$\frac{\partial^{\varepsilon}\rho}{\partial\tau^{\varepsilon}} + \upsilon \cdot \nabla^{\varepsilon}\rho = 0, \tag{1.248}$$

$$\nabla^{\varepsilon} \cdot \boldsymbol{v} = 0, \tag{1.249}$$

$$\rho\left(\frac{\partial^{\varepsilon}\upsilon}{\partial\tau^{\varepsilon}} + \upsilon \cdot \nabla^{\varepsilon}\upsilon\right) = -\nabla^{\varepsilon}p + \rho\mathbf{b}, \tag{1.250}$$

where $\boldsymbol{v}(\mu, \eta, \sigma, \tau)$ is the fractal fluid velocity, $p(\mu, \eta, \sigma, \tau)$ is the thermodynamic pressure, $\rho(\mu, \eta, \sigma, \tau)$ is the fractal fluid density, and $\mathbf{b}(\mu, \eta, \sigma, \tau)$ is the external force per unit of mass.

1.5.3 Applications of local fractional partial derivative operator to coordinate systems

One of interesting things in coordinate systems is that many three-dimensional coordinate systems may be used to convert between them. The Cantorian coordinate system is an analogous version of the Cartesian coordinate system on fractal sets. Similarity, we may transfer Cantorian coordinate system into Cantor-type cylindrical coordinates and Cantor-type spherical coordinates via Mittag–Leffler function defined on the fractal sets [22, 88, 89]. Here, we present the basic theory of Cantor-type circular coordinates, Cantor-type cylindrical coordinates, and Cantor-type spherical coordinates as follows.

For $R \in (0, +\infty)$ and $\theta \in (0, 2\pi)$, the Cantor-type circular coordinate system is written as

$$\begin{cases} \mu^{\varepsilon} = R^{\varepsilon} \cos_{\varepsilon}(\theta^{\varepsilon}) \\ \sigma^{\varepsilon} = R^{\varepsilon} \sin_{\varepsilon}(\theta^{\varepsilon}) \end{cases}, \tag{1.251}$$

where $R > 0$ and $0 < \theta < 2\pi$.

A local fractional vector is written as

$$\begin{aligned} r =& R^{\varepsilon} \cos_{\varepsilon}\left(\theta^{\varepsilon}\right) e_1^{\varepsilon} + R^{\varepsilon} \sin_{\varepsilon}\left(\theta^{\varepsilon}\right) e_2^{\varepsilon} \\ =& r_R e_R^{\varepsilon} + r_{\theta} e_{\theta}^{\varepsilon}. \end{aligned} \tag{1.252}$$

Hence, we have a local fractional vector

$$\begin{cases} e_R^{\varepsilon} = \cos_{\varepsilon}(\theta^{\varepsilon}) e_1^{\varepsilon} + \sin_{\varepsilon}(\theta^{\varepsilon}) e_2^{\varepsilon}, \\ e_{\theta}^{\varepsilon} = -\sin_{\varepsilon}(\theta^{\varepsilon}) e_1^{\varepsilon} + \cos_{\varepsilon}(\theta^{\varepsilon}) e_2^{\varepsilon} \end{cases} \tag{1.253}$$

such that the local fractional gradient operator and local fractional Laplace operator in the Cantor-type circular coordinate system is presented as

$$\nabla^{\varepsilon} \varphi(R, \theta) = e_R^{\varepsilon} \frac{\partial^{\varepsilon} \varphi}{\partial R^{\varepsilon}} + e_{\theta}^{\varepsilon} \frac{1}{R^{\varepsilon}} \frac{\partial^{\varepsilon} \varphi}{\partial \theta^{\varepsilon}}, \tag{1.254}$$

$$\nabla^{2\varepsilon} \varphi(R, \theta) = \frac{\partial^{2\varepsilon} \varphi}{\partial R^{2\varepsilon}} + \frac{1}{R^{2\varepsilon}} \frac{\partial^{2\varepsilon} \varphi}{\partial \theta^{2\varepsilon}} + \frac{1}{R^{\varepsilon}} \frac{\partial^{\varepsilon} \varphi}{\partial R^{\varepsilon}}. \tag{1.255}$$

The Cantor-type cylindrical coordinates can be written as follows:

$$\begin{cases} \mu^{\varepsilon} = R^{\varepsilon} \cos_{\varepsilon}(\theta^{\varepsilon}) \\ \eta^{\varepsilon} = R^{\varepsilon} \sin_{\varepsilon}(\theta^{\varepsilon}) \\ \sigma^{\varepsilon} = \sigma^{\varepsilon} \end{cases} \tag{1.256}$$

with $R \in (0, +\infty)$, $z \in (-\infty, +\infty)$, $\theta \in (0, \pi]$, and $\mu^{2\varepsilon} + \eta^{2\varepsilon} = R^{2\varepsilon}$.

Adopting (1.256), we have

$$\nabla^{\varepsilon} \cdot r = \frac{\partial^{\varepsilon} r_R}{\partial R^{\varepsilon}} + \frac{1}{R^{\varepsilon}} \frac{\partial^{\varepsilon} r_{\theta}}{\partial \theta^{\varepsilon}} + \frac{r_R}{R^{\varepsilon}} + \frac{\partial^{\alpha} r_z}{\partial \sigma^{\varepsilon}} \tag{1.257}$$

and

$$\nabla^{\varepsilon} \times r = \left(\frac{1}{R^{\varepsilon}} \frac{\partial^{\varepsilon} r_{\theta}}{\partial \theta^{\alpha}} - \frac{\partial^{\varepsilon} r_{\theta}}{\partial \sigma^{\varepsilon}} \right) e_R^{\varepsilon} + \left(\frac{\partial^{\varepsilon} r_R}{\partial \sigma^{\varepsilon}} - \frac{\partial^{\varepsilon} r_z}{\partial R^{\varepsilon}} \right) e_{\theta}^{\varepsilon} + \left(\frac{\partial^{\varepsilon} r_{\theta}}{\partial R^{\varepsilon}} + \frac{r_R}{R^{\varepsilon}} - \frac{1}{R^{\varepsilon}} \frac{\partial^{\varepsilon} r_R}{\partial \theta^{\varepsilon}} \right) e_{\sigma}^{\varepsilon}, \tag{1.258}$$

where

$$\begin{aligned} r =& R^{\varepsilon} \cos_{\varepsilon}\left(\theta^{\varepsilon}\right) e_1^{\varepsilon} + R^{\varepsilon} \sin_{\varepsilon}\left(\theta^{\varepsilon}\right) e_2^{\varepsilon} + \sigma^{\varepsilon} e_3^{\varepsilon} \\ =& r_R e_R^{\varepsilon} + r_{\theta} e_{\theta}^{\varepsilon} + r_{\sigma} e_{\sigma}^{\varepsilon}. \end{aligned} \tag{1.259}$$

Hence, we get the local fractional gradient operator and local fractional Laplace operator in the Cantor-type cylindrical coordinate system

$$\nabla^{\varepsilon}\varphi(R,\theta,\sigma) = e_R^{\varepsilon}\frac{\partial^{\varepsilon}\varphi}{\partial R^{\varepsilon}} + e_{\theta}^{\varepsilon}\frac{1}{R^{\varepsilon}}\frac{\partial^{\varepsilon}\varphi}{\partial\theta^{\varepsilon}} + e_{\sigma}^{\varepsilon}\frac{\partial^{\varepsilon}\varphi}{\partial\sigma^{\varepsilon}}, \tag{1.260}$$

$$\nabla^{2\varepsilon}\varphi(R,\theta,\sigma) = \frac{\partial^{2\varepsilon}\varphi}{\partial R^{2\varepsilon}} + \frac{1}{R^{2\varepsilon}}\frac{\partial^{2\varepsilon}\varphi}{\partial\theta^{2\varepsilon}} + \frac{1}{R^{\varepsilon}}\frac{\partial^{\varepsilon}\varphi}{\partial R^{\varepsilon}} + \frac{\partial^{2\varepsilon}\varphi}{\partial\sigma^{2\varepsilon}}, \tag{1.261}$$

where a local fractional vector is given as

$$\begin{cases} e_R^{\varepsilon} = \cos_{\varepsilon}(\theta^{\varepsilon})\, e_1^{\varepsilon} + \sin_{\alpha}(\theta^{\varepsilon})\, e_2^{\varepsilon}, \\ e_{\theta}^{\varepsilon} = -\sin_{\varepsilon}(\theta^{\varepsilon})\, e_1^{\varepsilon} + \cos_{\alpha}(\theta^{\varepsilon})\, e_2^{\varepsilon}, \\ e_{\sigma}^{\varepsilon} = e_3^{\varepsilon}. \end{cases} \tag{1.262}$$

For $R \in (0,+\infty)$, $\eta \in (0,\pi)$, $\theta \in (0,2\pi)$, and $\mu^{2\varepsilon} + \eta^{2\varepsilon} + \sigma^{2\varepsilon} = R^{2\varepsilon}$, the Cantor-type spherical coordinate system is written as

$$\begin{cases} \mu^{\varepsilon} = R^{\varepsilon}\cos_{\varepsilon}(\vartheta^{\varepsilon})\cos_{\varepsilon}(\theta^{\varepsilon}), \\ \eta^{\varepsilon} = R^{\varepsilon}\cos_{\varepsilon}(\vartheta^{\varepsilon})\sin_{\varepsilon}(\theta^{\varepsilon}), \\ \sigma^{\varepsilon} = R^{\varepsilon}\sin_{\varepsilon}(\vartheta^{\varepsilon}). \end{cases} \tag{1.263}$$

A local fractional vector is written as

$$\begin{aligned} r &= R^{\varepsilon}\cos_{\varepsilon}\left(\vartheta^{\varepsilon}\right)\cos_{\varepsilon}\left(\theta^{\varepsilon}\right) e_1^{\varepsilon} + R^{\varepsilon}\cos_{\varepsilon}\left(\vartheta^{\varepsilon}\right)\sin_{\varepsilon}\left(\theta^{\varepsilon}\right) e_2^{\varepsilon} + R^{\varepsilon}\sin_{\varepsilon}\left(\vartheta^{\varepsilon}\right) e_3^{\varepsilon} \\ &= r_R e_R^{\varepsilon} + r_{\vartheta} e_{\vartheta}^{\varepsilon} + r_{\theta} e_{\theta}^{\varepsilon}. \end{aligned} \tag{1.264}$$

Hence, we get the local fractional gradient operator and local fractional Laplace operator in the Cantor-type spherical coordinate system

$$\nabla^{\varepsilon}\varphi(R,\vartheta,\theta) = e_R^{\varepsilon}\frac{\partial^{\varepsilon}\varphi}{\partial R^{\varepsilon}} + e_{\vartheta}^{\varepsilon}\frac{1}{R^{\varepsilon}}\frac{\partial^{\varepsilon}\varphi}{\partial\vartheta^{\varepsilon}} + e_{\theta}^{\varepsilon}\frac{1}{R^{\varepsilon}}\frac{1}{\sin_{\varepsilon}(\vartheta^{\varepsilon})}\frac{\partial^{\varepsilon}\varphi}{\partial\theta^{\varepsilon}}, \tag{1.265}$$

$$\begin{aligned} \nabla^{2\varepsilon}\varphi(R,\vartheta,\theta) =& \frac{\partial^{2\varepsilon}\varphi}{\partial R^{2\varepsilon}} + \frac{1}{R^{2\varepsilon}}\frac{1}{\sin_{\varepsilon}(\vartheta^{\varepsilon})}\frac{\partial^{\varepsilon}}{\partial\vartheta^{\varepsilon}}\left(\sin_{\varepsilon}\left(\vartheta^{\varepsilon}\right)\frac{\partial^{\varepsilon}\varphi}{\partial\vartheta^{\varepsilon}}\right) \\ &+ \frac{2}{R^{\varepsilon}}\frac{\partial^{\varepsilon}\varphi}{\partial R^{\varepsilon}} + \frac{1}{R^{2\varepsilon}}\frac{1}{\sin_{\varepsilon}^{2}(\vartheta^{\varepsilon})}\frac{\partial^{2\varepsilon}\varphi}{\partial\theta^{2\varepsilon}}, \end{aligned} \tag{1.266}$$

where a local fractional vector is determined by

$$\begin{cases} e_R^{\varepsilon} = \sin_{\varepsilon}(\vartheta^{\varepsilon})\cos_{\varepsilon}(\theta^{\varepsilon})\, e_1^{\varepsilon} + \sin_{\varepsilon}(\vartheta^{\varepsilon})\sin_{\varepsilon}(\theta^{\varepsilon})\, e_2^{\varepsilon} + \cos_{\varepsilon}(\vartheta^{\varepsilon})\, e_3^{\varepsilon}, \\ e_{\vartheta}^{\varepsilon} = \cos_{\varepsilon}(\vartheta^{\varepsilon})\cos_{\varepsilon}(\theta^{\varepsilon})\, e_1^{\varepsilon} + \cos_{\varepsilon}(\vartheta^{\varepsilon})\sin_{\varepsilon}(\theta^{\varepsilon})\, e_2^{\varepsilon} - \sin_{\varepsilon}(\vartheta^{\varepsilon})\, e_3^{\varepsilon}, \\ e_{\theta}^{\varepsilon} = -\sin_{\varepsilon}(\theta^{\varepsilon})\, e_1^{\varepsilon} + \cos_{\varepsilon}(\theta^{\varepsilon})\, e_2^{\varepsilon}. \end{cases} \tag{1.267}$$

The local fractional gradient operator of $\varphi(R)$ in the Cantor-type cylindrical symmetry form is written as

$$\nabla^{\varepsilon}\varphi(R) = e_R^{c}\frac{\partial^{\varepsilon}\varphi}{\partial R^{\varepsilon}}. \tag{1.268}$$

Similarly, the local fractional Laplace operator in the Cantor-type cylindrical symmetry form can be written as

$$\nabla^{2\varepsilon}\varphi(R)=\frac{\partial^{2\varepsilon}\varphi}{\partial R^{2\varepsilon}}+\frac{1}{R^{\varepsilon}}\frac{\partial^{\varepsilon}\varphi}{\partial R^{\varepsilon}}. \tag{1.269}$$

The local fractional gradient operator of $\varphi(R)$ in the Cantor-type spherical symmetry form is written as

$$\nabla^{\varepsilon}\varphi(R)=e_R^{\varepsilon}\frac{\partial^{\varepsilon}\varphi}{\partial R^{\varepsilon}}. \tag{1.270}$$

Similarly, the local fractional Laplace operator takes in the Cantor-type spherical symmetry form

$$\nabla^{2\varepsilon}\varphi(R)=\frac{\partial^{2\varepsilon}\varphi}{\partial R^{2\varepsilon}}+\frac{2}{R^{\varepsilon}}\frac{\partial^{\varepsilon}\varphi}{\partial R^{\varepsilon}}. \tag{1.271}$$

The theory of the Cantor-type circular coordinates, Cantor-type cylindrical coordinates, and Cantor-type spherical coordinates is presented in Appendix D.

1.5.4 Alternative observations of local fractional partial differential equations

Based upon the basic theory of Cantor-type circular coordinates, Cantor-type cylindrical coordinates, and Cantor-type spherical coordinates, we will discuss the local fractional partial differential equations via local fractional partial derivative operator.

We now present some applications of the Cantor-type circular coordinates for adopting wave equation, the Laplace equation, the Poisson equation, the Helmholtz equation, heat-conduction equation, damped wave equation, dissipative wave equation, diffusion equation, and Maxwell's equations in fractal dimensional space.

The linear local fractional wave equation for the velocity potential of fluid flow in 1 + 2 fractal dimensional space takes the form

$$\omega\nabla^{2\varepsilon}\Phi(\mu,\sigma,\tau)-\frac{\partial^{2\varepsilon}\Phi(\mu,\sigma,\tau)}{\partial\tau^{2\varepsilon}}=0, \tag{1.272}$$

where ω is a constant.

The linear local fractional wave equation for the velocity potential of fluid flow in Cantor-type circular coordinate system takes the form

$$\omega\left(\frac{\partial^{2\varepsilon}\Phi(R,\theta,\tau)}{\partial R^{2\varepsilon}}+\frac{1}{R^{2\varepsilon}}\frac{\partial^{2\varepsilon}\Phi(R,\theta,\tau)}{\partial\theta^{2\varepsilon}}+\frac{1}{R^{\varepsilon}}\frac{\partial^{\varepsilon}\Phi(R,\theta,\tau)}{\partial R^{\varepsilon}}\right)-\frac{\partial^{2\varepsilon}\Phi(R,\theta,\tau)}{\partial\tau^{2\varepsilon}}=0, \tag{1.273}$$

where ω is a constant.

The local fractional Laplace equation arising in fractal electrostatics in 1 + 2 fractal dimensional space takes the form

$$\nabla^{2\varepsilon}\Phi(\mu,\sigma)=0. \tag{1.274}$$

The linear local fractional Laplace equation in Cantor-type circular coordinate system takes the form

$$\frac{\partial^{2\varepsilon}\Phi(R,\theta)}{\partial R^{2\varepsilon}}+\frac{1}{R^{2\varepsilon}}\frac{\partial^{2\varepsilon}\Phi(R,\theta)}{\partial\theta^{2\varepsilon}}+\frac{1}{R^{\varepsilon}}\frac{\partial^{\varepsilon}\Phi(R,\theta)}{\partial R^{\varepsilon}}=0. \tag{1.275}$$

The local fractional Poisson equation in 2 fractal dimensional space takes the form

$$\nabla^{2\varepsilon}\Phi(\mu,\sigma)=\Theta(\mu,\sigma), \tag{1.276}$$

where $\Theta(\mu,\sigma)$ is a nondifferentiable function.

The linear local fractional Poisson equation in Cantor-type circular coordinate system takes the form

$$\frac{\partial^{2\varepsilon}\Phi(R,\theta)}{\partial R^{2\varepsilon}}+\frac{1}{R^{2\varepsilon}}\frac{\partial^{2\varepsilon}\Phi(R,\theta)}{\partial\theta^{2\varepsilon}}+\frac{1}{R^{\varepsilon}}\frac{\partial^{\varepsilon}\Phi(R,\theta)}{\partial R^{\varepsilon}}=\Phi(R,\theta). \tag{1.277}$$

The linear local fractional homogeneous Helmholtz equation in 2 fractal dimensional space with nondifferentiable inhomogeneous term takes the form

$$\nabla^{2\varepsilon}\Phi(\mu,\sigma)+\varpi\Phi(\mu,\sigma)=0, \tag{1.278}$$

where ϖ is a constant and $\Theta(\mu,\sigma)$ is a differentiable function.

The linear local fractional homogeneous Helmholtz equation in Cantor-type circular coordinate system takes the form

$$\frac{\partial^{2\varepsilon}\Phi(R,\theta)}{\partial R^{2\varepsilon}}+\frac{1}{R^{2\varepsilon}}\frac{\partial^{2\varepsilon}\Phi(R,\theta)}{\partial\theta^{2\varepsilon}}+\frac{1}{R^{\varepsilon}}\frac{\partial^{\varepsilon}\Phi(R,\theta)}{\partial R^{\varepsilon}}+\varpi\Phi(R,\theta)=0, \tag{1.279}$$

where ϖ is a constant.

The local fractional heat-conduction equation in 1 + 2 fractal dimensional space takes the form

$$\frac{\partial^{\varepsilon}\Phi(\mu,\sigma,\tau)}{\partial\tau^{\varepsilon}}-\varpi\nabla^{2\varepsilon}\Phi(\mu,\sigma,\tau)=H(\mu,\sigma,\tau), \tag{1.280}$$

where ϖ is the thermal conductivity coefficient and $H(\mu,\sigma,\tau)$ is the heat source.

The local fractional heat-conduction equation in Cantor-type circular coordinate system takes the form

$$\begin{aligned}&\frac{\partial^{\varepsilon}\Phi(R,\theta,\tau)}{\partial\tau^{\varepsilon}}-\varpi\left(\frac{\partial^{2\varepsilon}\Phi(R,\theta,\tau)}{\partial R^{2\varepsilon}}+\frac{1}{R^{2\varepsilon}}\frac{\partial^{2\varepsilon}\Phi(R,\theta,\tau)}{\partial\theta^{2\varepsilon}}+\frac{1}{R^{\varepsilon}}\frac{\partial^{\varepsilon}\Phi(R,\theta,\tau)}{\partial R^{\varepsilon}}\right)\\&=H(R,\theta,\tau),\end{aligned} \tag{1.281}$$

where ϖ is the thermal conductivity coefficient and $H(R,\theta,\tau)$ is the heat source.

The linear local homogeneous damped wave equation of fractal strings in 1 + 2 fractal dimensional space takes the form

$$\frac{\partial^{2\varepsilon}\Phi(R,\theta,\tau)}{\partial\tau^{2\varepsilon}}-\frac{\partial^{\varepsilon}\Phi(R,\theta,\tau)}{\partial\tau^{\varepsilon}}-\nabla^{2\varepsilon}\Phi(\mu,\sigma,\tau)=0. \tag{1.282}$$

The linear local homogeneous damped wave equation of fractal strings in Cantor-type circular coordinate system takes the form

$$\frac{\partial^{2\varepsilon}\Phi(\mu,\sigma,\tau)}{\partial\tau^{2\varepsilon}}-\frac{\partial^{\varepsilon}\Phi(\mu,\sigma,\tau)}{\partial\tau^{\varepsilon}} -\left(\frac{\partial^{2\varepsilon}\Phi(R,\theta,\tau)}{\partial R^{2\varepsilon}}+\frac{1}{R^{2\varepsilon}}\frac{\partial^{2\varepsilon}\Phi(R,\theta,\tau)}{\partial\theta^{2\varepsilon}}+\frac{1}{R^{\varepsilon}}\frac{\partial^{\varepsilon}\Phi(R,\theta,\tau)}{\partial R^{\varepsilon}}\right)=0. \tag{1.283}$$

The local fractional inhomogeneous dissipative wave equation of fractal strings in 1 + 2 fractal dimensional space takes the form

$$\frac{\partial^{2\varepsilon}\Phi(\mu,\sigma,\tau)}{\partial\tau^{2\varepsilon}}-\frac{\partial^{\varepsilon}\Phi(\mu,\sigma,\tau)}{\partial\tau^{\varepsilon}}-\nabla^{2\varepsilon}\Phi(\mu,\sigma,\tau)-\nabla^{\varepsilon}\Phi(\mu,\sigma,\tau)=0. \tag{1.284}$$

The local fractional inhomogeneous dissipative wave equation of fractal strings in Cantor-type circular coordinate system takes the form

$$\frac{\partial^{2\varepsilon}\Phi(R,\theta,\tau)}{\partial\tau^{2\varepsilon}}-\frac{\partial^{\varepsilon}\Phi(R,\theta,\tau)}{\partial\tau^{\varepsilon}} -\left(\frac{\partial^{2\varepsilon}\Phi(R,\theta,\tau)}{\partial R^{2\varepsilon}}+\frac{1}{R^{2\varepsilon}}\frac{\partial^{2\varepsilon}\Phi(R,\theta,\tau)}{\partial\theta^{2\varepsilon}}+\frac{1}{R^{\varepsilon}}\frac{\partial^{\varepsilon}\Phi(R,\theta,\tau)}{\partial R^{\varepsilon}}\right) -\left(e_R^{\varepsilon}\frac{\partial^{\varepsilon}\Phi(R,\theta,\tau)}{\partial R^{\varepsilon}}+e_\theta^{\varepsilon}\frac{1}{R^{\varepsilon}}\frac{\partial^{\varepsilon}\Phi(R,\theta,\tau)}{\partial\theta^{\varepsilon}}\right)=0. \tag{1.285}$$

The local fractional diffusion equation in 1 + 2 fractal dimensional space has the following form:

$$\frac{d^{\varepsilon}\Phi(\mu,\sigma,\tau)}{d\tau^{\varepsilon}}-D\nabla^{2\varepsilon}\Phi(\mu,\sigma,\tau)=0, \tag{1.286}$$

where D is the diffusive coefficient.

The local fractional diffusion equation in Cantor-type circular coordinate system takes the following form:

$$\frac{d^{\varepsilon}\Phi(R,\theta,\tau)}{d\tau^{\varepsilon}}-D\left(\frac{\partial^{2\varepsilon}\Phi(R,\theta,\tau)}{\partial R^{2\varepsilon}}+\frac{1}{R^{2\varepsilon}}\frac{\partial^{2\varepsilon}\Phi(R,\theta,\tau)}{\partial\theta^{2\varepsilon}}+\frac{1}{R^{\varepsilon}}\frac{\partial^{\varepsilon}\Phi(R,\theta,\tau)}{\partial R^{\varepsilon}}\right)=0, \tag{1.287}$$

where D is the diffusive coefficient.

Systems of local fractional Maxwell's equations in 1 + 2 fractal dimensional space take the following form:

$$\nabla^{\varepsilon}\cdot D(\mu,\sigma,\tau)=\rho(\mu,\sigma,\tau), \tag{1.288}$$

$$\nabla^{\varepsilon}\times H(\mu,\sigma,\tau)=\mathrm{J}_{\varepsilon}(\mu,\sigma,\tau)+\frac{\partial^{\varepsilon}D(\mu,\sigma,\tau)}{\partial\tau^{\varepsilon}}, \tag{1.289}$$

$$\nabla^{\varepsilon}\times E(\mu,\sigma,\tau)=-\frac{\partial^{\varepsilon}B(\mu,\sigma,\tau)}{\partial\tau^{\varepsilon}}, \tag{1.290}$$

$$\nabla^{\varepsilon}\cdot B(\mu,\sigma,\tau)=0, \tag{1.291}$$

where $\rho(\mu,\sigma,\tau)$ is the fractal electric charge density, $D(\mu,\sigma,\tau)$ is electric displacement in the fractal electric field, $H(\mu,\sigma,\tau)$ is the magnetic field strength in the fractal field, $E(\mu,\sigma,\tau)$ is the electric field strength in the fractal field, $J_\varepsilon(\mu,\sigma,\tau)$ is the conductive current, and $B(\mu,\sigma,\tau)$ is the magnetic induction in the fractal field, and the constitutive relationships in fractal electromagnetic can be written as follows:

$$D(\mu,\sigma,\tau) = \varepsilon_f E(\mu,\sigma,\tau) \tag{1.292}$$

and

$$H(\mu,\sigma,\tau) = \mu_f B(\mu,\sigma,\tau), \tag{1.293}$$

with the fractal dielectric permittivity ε_f and the fractal magnetic permeability μ_f.

We can write the systems of local fractional Maxwell's equations in Cantor-type circular coordinate system as follows:

$$\left(e_R^\varepsilon \frac{\partial^\varepsilon}{\partial R^\varepsilon} + e_\theta^\varepsilon \frac{1}{R^\varepsilon}\frac{\partial^\varepsilon}{\partial \theta^\varepsilon}\right) \cdot D(R,\theta,\tau) = \rho(R,\theta,\tau), \tag{1.294}$$

$$\left(e_R^\varepsilon \frac{\partial^\varepsilon}{\partial R^\varepsilon} + e_\theta^\varepsilon \frac{1}{R^\varepsilon}\frac{\partial^\varepsilon}{\partial \theta^\varepsilon}\right) \times H(R,\theta,\tau) = J_\varepsilon(R,\theta,\tau) + \frac{\partial^\varepsilon D(R,\theta,\tau)}{\partial \tau^\varepsilon}, \tag{1.295}$$

$$\left(e_R^\varepsilon \frac{\partial^\varepsilon}{\partial R^\varepsilon} + e_\theta^\varepsilon \frac{1}{R^\varepsilon}\frac{\partial^\varepsilon}{\partial \theta^\varepsilon}\right) \times E(R,\theta,\tau) = -\frac{\partial^\varepsilon B(R,\theta,\tau)}{\partial \tau^\varepsilon}, \tag{1.296}$$

$$\left(e_R^\varepsilon \frac{\partial^\varepsilon}{\partial R^\varepsilon} + e_\theta^\varepsilon \frac{1}{R^\varepsilon}\frac{\partial^\varepsilon}{\partial \theta^\varepsilon}\right) \cdot B(R,\theta,\tau) = 0, \tag{1.297}$$

where $\rho(R,\theta,\tau)$ is the fractal electric charge density, $D(R,\theta,\tau)$ is the electric displacement in the fractal electric field, $H(R,\theta,\tau)$ is the magnetic field strength in the fractal field, $E(R,\theta,\tau)$ is the electric field strength in the fractal field, $J_\varepsilon(R,\theta,\tau)$ is the conductive current, and $B(R,\theta,\tau)$ is the magnetic induction in the fractal field, and the constitutive relationships in fractal electromagnetic can be written as

$$D(R,\theta,\tau) = \varepsilon_f E(R,\theta,\tau) \tag{1.298}$$

and

$$H(R,\theta,\tau) = \mu_f B(R,\theta,\tau), \tag{1.299}$$

with the fractal dielectric permittivity ε_f and the fractal magnetic permeability μ_f.

The form of the linear local fractional wave equation for the velocity potential of fluid flow in Cantor-type cylindrical coordinate system is given by

$$\omega\left(\frac{\partial^{2\varepsilon}\Phi}{\partial R^{2\varepsilon}} + \frac{1}{R^{2\varepsilon}}\frac{\partial^{2\varepsilon}\Phi}{\partial \theta^{2\varepsilon}} + \frac{1}{R^\varepsilon}\frac{\partial^\varepsilon \Phi}{\partial R^\varepsilon} + \frac{\partial^{2\varepsilon}\Phi}{\partial \sigma^{2\varepsilon}}\right) - \frac{\partial^{2\varepsilon}\Phi}{\partial \tau^{2\varepsilon}} = 0, \tag{1.300}$$

where ω is a constant and $\Phi = \Phi(R,\theta,\sigma,\tau)$.

The linear local fractional wave equation for the velocity potential of fluid flow in Cantor-type spherical coordinate system takes the following form:

$$\frac{\partial^{2\varepsilon}\Phi}{\partial R^{2\varepsilon}} + \frac{1}{R^{2\varepsilon}\sin_{\varepsilon}(\vartheta^{\varepsilon})}\frac{\partial^{\varepsilon}}{\partial\vartheta^{\varepsilon}}\left(\sin_{\varepsilon}(\vartheta^{\varepsilon})\frac{\partial^{\varepsilon}\Phi}{\partial\vartheta^{\varepsilon}}\right) + \frac{2}{R^{\varepsilon}}\frac{\partial^{\varepsilon}\Phi}{\partial R^{\varepsilon}} + \frac{1}{R^{2\varepsilon}\sin_{\varepsilon}^{2}(\vartheta^{\varepsilon})}\frac{\partial^{2\varepsilon}\Phi}{\partial\theta^{2\varepsilon}} - \frac{\partial^{2\varepsilon}\Phi}{\partial\tau^{2\varepsilon}} = 0, \tag{1.301}$$

where ω is a constant and $\Phi = \Phi(\mu,\theta,\vartheta,\tau)$.

The local fractional Laplace equation in Cantor-type cylindrical coordinate system takes the following form:

$$\frac{\partial^{2\varepsilon}\Phi}{\partial R^{2\varepsilon}} + \frac{1}{R^{2\varepsilon}}\frac{\partial^{2\varepsilon}\Phi}{\partial\theta^{2\varepsilon}} + \frac{1}{R^{\varepsilon}}\frac{\partial^{\varepsilon}\Phi}{\partial R^{\varepsilon}} + \frac{\partial^{2\varepsilon}\Phi}{\partial\sigma^{2\varepsilon}} = 0, \tag{1.302}$$

where $\Phi = \Phi(R,\theta,\sigma,\tau)$.

The linear local fractional Laplace equation in Cantor-type spherical coordinate system takes the following form:

$$\frac{\partial^{2\varepsilon}\Phi}{\partial R^{2\varepsilon}} + \frac{1}{R^{2\varepsilon}\sin_{\varepsilon}(\vartheta^{\varepsilon})}\frac{\partial^{\varepsilon}}{\partial\vartheta^{\varepsilon}}\left(\sin_{\varepsilon}(\vartheta^{\varepsilon})\frac{\partial^{\varepsilon}\Phi}{\partial\vartheta^{\varepsilon}}\right) + \frac{2}{R^{\varepsilon}}\frac{\partial^{\varepsilon}\Phi}{\partial R^{\varepsilon}} + \frac{1}{R^{2\varepsilon}\sin_{\varepsilon}^{2}(\vartheta^{\varepsilon})}\frac{\partial^{2\varepsilon}\Phi}{\partial\theta^{2\varepsilon}} = 0, \tag{1.303}$$

where $\Phi = \Phi(\mu,\theta,\vartheta,\tau)$.

The local fractional Poisson equation in Cantor-type cylindrical coordinate system takes the following form:

$$\frac{\partial^{2\varepsilon}\Phi}{\partial R^{2\varepsilon}} + \frac{1}{R^{2\varepsilon}}\frac{\partial^{2\varepsilon}\Phi}{\partial\theta^{2\varepsilon}} + \frac{1}{R^{\varepsilon}}\frac{\partial^{\varepsilon}\Phi}{\partial R^{\varepsilon}} + \frac{\partial^{2\varepsilon}\Phi}{\partial\sigma^{2\varepsilon}} = \Phi(R,\theta,\sigma), \tag{1.304}$$

where $\Phi = \Phi(R,\theta,\sigma)$ and is a nondifferentiable function.

The linear local fractional Poisson equation in Cantor-type spherical coordinate system takes the following form:

$$\frac{\partial^{2\varepsilon}\Phi}{\partial R^{2\varepsilon}} + \frac{1}{R^{2\varepsilon}\sin_{\varepsilon}(\vartheta^{\varepsilon})}\frac{\partial^{\varepsilon}}{\partial\vartheta^{\varepsilon}}\left(\sin_{\varepsilon}(\vartheta^{\varepsilon})\frac{\partial^{\varepsilon}\Phi}{\partial\vartheta^{\varepsilon}}\right) + \frac{2}{R^{\varepsilon}}\frac{\partial^{\varepsilon}\Phi}{\partial R^{\varepsilon}} + \frac{1}{R^{2\varepsilon}\sin_{\varepsilon}^{2}(\vartheta^{\varepsilon})}\frac{\partial^{2\varepsilon}\Phi}{\partial\theta^{2\varepsilon}} = \Phi, \tag{1.305}$$

where $\Phi = \Phi(\mu,\theta,\vartheta)$.

The linear local fractional homogeneous Helmholtz equation in Cantor-type cylindrical coordinate system takes the following form:

$$\frac{\partial^{2\varepsilon}\Phi}{\partial R^{2\varepsilon}} + \frac{1}{R^{2\varepsilon}}\frac{\partial^{2\varepsilon}\Phi}{\partial\theta^{2\varepsilon}} + \frac{1}{R^{\varepsilon}}\frac{\partial^{\varepsilon}\Phi}{\partial R^{\varepsilon}} + \frac{\partial^{2\varepsilon}\Phi}{\partial\sigma^{2\varepsilon}} + \varpi\Phi = 0, \tag{1.306}$$

where ϖ is a constant and $\Phi = \Phi(R,\theta,\sigma)$.

The linear local fractional homogeneous Helmholtz equation in Cantor-type spherical coordinate system takes the following form:

$$\frac{\partial^{2\varepsilon}\Phi}{\partial R^{2\varepsilon}}+\frac{1}{R^{2\varepsilon}\sin_{\varepsilon}\left(\vartheta^{\varepsilon}\right)}\frac{\partial^{\varepsilon}}{\partial\vartheta^{\varepsilon}}\left(\sin_{\varepsilon}\left(\vartheta^{\varepsilon}\right)\frac{\partial^{\varepsilon}\Phi}{\partial\vartheta^{\varepsilon}}\right)+\frac{2}{R^{\varepsilon}}\frac{\partial^{\varepsilon}\Phi}{\partial R^{\varepsilon}}+\frac{1}{R^{2\varepsilon}\sin_{\varepsilon}^{2}\left(\vartheta^{\varepsilon}\right)}\frac{\partial^{2\varepsilon}\Phi}{\partial\theta^{2\varepsilon}}+\varpi\Phi=0, \tag{1.307}$$

where ϖ is a constant and $\Phi=\Phi\left(\mu,\theta,\vartheta\right)$.

We recall that the local fractional heat-conduction equation in Cantor-type cylindrical coordinate system takes the following form:

$$\frac{\partial^{\varepsilon}\Phi}{\partial\tau^{\varepsilon}}-\varpi\left(\frac{\partial^{2\varepsilon}\Phi}{\partial R^{2\varepsilon}}+\frac{1}{R^{2\varepsilon}}\frac{\partial^{2\varepsilon}\Phi}{\partial\theta^{2\varepsilon}}+\frac{1}{R^{\varepsilon}}\frac{\partial^{\varepsilon}\Phi}{\partial R^{\varepsilon}}+\frac{\partial^{2\varepsilon}\Phi}{\partial\sigma^{2\varepsilon}}\right)=H, \tag{1.308}$$

where ϖ is the thermal conductivity coefficient $\Phi=\Phi\left(R,\theta,\sigma,\tau\right)$ and $H\left(R,\theta,\sigma,\tau\right)$ is the heat source.

The local fractional heat-conduction equation in Cantor-type spherical coordinate system is given by

$$\frac{\partial^{\varepsilon}\Phi}{\partial\tau^{\varepsilon}}-\varpi\left(\frac{\partial^{2\varepsilon}\Phi}{\partial R^{2\varepsilon}}+\frac{1}{R^{2\varepsilon}\sin_{\varepsilon}\left(\vartheta^{\varepsilon}\right)}\frac{\partial^{\varepsilon}}{\partial\vartheta^{\varepsilon}}\left(\sin_{\varepsilon}\left(\vartheta^{\varepsilon}\right)\frac{\partial^{\varepsilon}\Phi}{\partial\vartheta^{\varepsilon}}\right)+\frac{2}{R^{\varepsilon}}\frac{\partial^{\varepsilon}\Phi}{\partial R^{\varepsilon}}+\frac{1}{R^{2\varepsilon}\sin_{\varepsilon}^{2}\left(\vartheta^{\varepsilon}\right)}\frac{\partial^{2\varepsilon}\Phi}{\partial\theta^{2\varepsilon}}\right)=H, \tag{1.309}$$

whereϖ is the thermal conductivity coefficient, $\Phi=\Phi\left(\mu,\theta,\vartheta,\tau\right)$, and $H\left(\mu,\theta,\vartheta,\tau\right)$ is the heat source.

The linear local homogeneous damped wave equation of fractal strings in Cantor-type cylindrical coordinate system can be written as follows:

$$\frac{\partial^{2\varepsilon}\Phi}{\partial\tau^{2\varepsilon}}-\frac{\partial^{\varepsilon}\Phi}{\partial\tau^{\varepsilon}}-\left(\frac{\partial^{2\varepsilon}\Phi}{\partial R^{2\varepsilon}}+\frac{1}{R^{2\varepsilon}}\frac{\partial^{2\varepsilon}\Phi}{\partial\theta^{2\varepsilon}}+\frac{1}{R^{\varepsilon}}\frac{\partial^{\varepsilon}\Phi}{\partial R^{\varepsilon}}+\frac{\partial^{2\varepsilon}\Phi}{\partial\sigma^{2\varepsilon}}\right)=0 \tag{1.310}$$

where $\Phi=\Phi\left(R,\theta,\sigma,\tau\right)$.

The linear local homogeneous damped wave equation of fractal strings in Cantor-type spherical coordinate system has the following form:

$$\frac{\partial^{2\varepsilon}\Phi}{\partial\tau^{2\varepsilon}}-\frac{\partial^{\varepsilon}\Phi}{\partial\tau^{\varepsilon}}-\left(\frac{\partial^{2\varepsilon}\Phi}{\partial R^{2\varepsilon}}+\frac{1}{R^{2\varepsilon}\sin_{\varepsilon}\left(\vartheta^{\varepsilon}\right)}\frac{\partial^{\varepsilon}}{\partial\vartheta^{\varepsilon}}\left(\sin_{\varepsilon}\left(\vartheta^{\varepsilon}\right)\frac{\partial^{\varepsilon}\Phi}{\partial\vartheta^{\varepsilon}}\right)+\frac{2}{R^{\varepsilon}}\frac{\partial^{\varepsilon}\Phi}{\partial R^{\varepsilon}}+\frac{1}{R^{2\varepsilon}\sin_{\varepsilon}^{2}\left(\vartheta^{\varepsilon}\right)}\frac{\partial^{2\varepsilon}\Phi}{\partial\theta^{2\varepsilon}}\right)=0, \tag{1.311}$$

where $\Phi=\Phi\left(\mu,\theta,\vartheta,\tau\right)$.

The local fractional inhomogeneous dissipative wave equation of fractal strings in Cantor-type cylindrical coordinate system takes the following form:

$$\frac{\partial^{2\varepsilon}\Phi}{\partial\tau^{2\varepsilon}} - \frac{\partial^{\varepsilon}\Phi}{\partial\tau^{\varepsilon}} - \left(\frac{\partial^{2\varepsilon}\Phi}{\partial R^{2\varepsilon}} + \frac{1}{R^{2\varepsilon}}\frac{\partial^{2\varepsilon}\Phi}{\partial\theta^{2\varepsilon}} + \frac{1}{R^{\varepsilon}}\frac{\partial^{\varepsilon}\Phi}{\partial R^{\varepsilon}} + \frac{\partial^{2\varepsilon}\Phi}{\partial\sigma^{2\varepsilon}}\right)$$
$$- \left(e_R^{\varepsilon}\frac{\partial^{\varepsilon}\Phi}{\partial R^{\varepsilon}} + e_\theta^{\varepsilon}\frac{1}{R^{\varepsilon}}\frac{\partial^{\varepsilon}\Phi}{\partial\theta^{\varepsilon}} + e_\sigma^{\varepsilon}\frac{\partial^{\varepsilon}\Phi}{\partial\sigma^{\varepsilon}}\right) = 0, \tag{1.312}$$

where $\Phi = \Phi(R,\theta,\sigma,\tau)$.

The local fractional inhomogeneous dissipative wave equation of fractal strings in Cantor-type spherical coordinate system takes the following form:

$$\frac{\partial^{2\varepsilon}\Phi}{\partial\tau^{2\varepsilon}} - \frac{\partial^{\varepsilon}\Phi}{\partial\tau^{\varepsilon}}$$
$$- \left(\frac{\partial^{2\varepsilon}\Phi}{\partial R^{2\varepsilon}} + \frac{1}{R^{2\varepsilon}\sin_{\varepsilon}(\vartheta^{\varepsilon})}\frac{\partial^{\varepsilon}}{\partial\vartheta^{\varepsilon}}\left(\sin_{\varepsilon}(\vartheta^{\varepsilon})\frac{\partial^{\varepsilon}\Phi}{\partial\vartheta^{\varepsilon}}\right) + \frac{2}{R^{\varepsilon}}\frac{\partial^{\varepsilon}\Phi}{\partial R^{\varepsilon}} + \frac{1}{R^{2\varepsilon}\sin_{\varepsilon}^{2}(\vartheta^{\varepsilon})}\frac{\partial^{2\varepsilon}\Phi}{\partial\theta^{2\varepsilon}}\right)$$
$$- \left(e_R^{\varepsilon}\frac{\partial^{\varepsilon}\Phi}{\partial R^{\varepsilon}} + e_\vartheta^{\varepsilon}\frac{1}{R^{\varepsilon}}\frac{\partial^{\varepsilon}\Phi}{\partial\vartheta^{\varepsilon}} + e_\theta^{\varepsilon}\frac{1}{R^{\varepsilon}}\frac{1}{\sin_{\varepsilon}(\vartheta^{\varepsilon})}\frac{\partial^{\varepsilon}\Phi}{\partial\theta^{\varepsilon}}\right) = 0, \tag{1.313}$$

where $\Phi = \Phi(\mu,\theta,\vartheta,\tau)$.

The local fractional diffusion equation in Cantor-type cylindrical coordinate system takes the following form:

$$\frac{d^{\varepsilon}\Phi}{d\tau^{\varepsilon}} - D\left(\frac{\partial^{2\varepsilon}\Phi}{\partial R^{2\varepsilon}} + \frac{1}{R^{2\varepsilon}}\frac{\partial^{2\varepsilon}\Phi}{\partial\theta^{2\varepsilon}} + \frac{1}{R^{\varepsilon}}\frac{\partial^{\varepsilon}\Phi}{\partial R^{\varepsilon}} + \frac{\partial^{2\varepsilon}\Phi}{\partial\sigma^{2\varepsilon}}\right) = 0, \tag{1.314}$$

where D is the diffusive coefficient and $\Phi = \Phi(R,\theta,\sigma,\tau)$.

The local fractional diffusion equation in Cantor-type spherical coordinate system has the following form:

$$\frac{d^{\varepsilon}\Phi}{d\tau^{\varepsilon}} - D\left(\frac{\partial^{2\varepsilon}\Phi}{\partial R^{2\varepsilon}} + \frac{1}{R^{2\varepsilon}\sin_{\varepsilon}(\vartheta^{\varepsilon})}\frac{\partial^{\varepsilon}}{\partial\vartheta^{\varepsilon}}\left(\sin_{\varepsilon}(\vartheta^{\varepsilon})\frac{\partial^{\varepsilon}\Phi}{\partial\vartheta^{\varepsilon}}\right)\right.$$
$$\left.+\frac{2}{R^{\varepsilon}}\frac{\partial^{\varepsilon}\Phi}{\partial R^{\varepsilon}} + \frac{1}{R^{2\varepsilon}\sin_{\varepsilon}^{2}(\vartheta^{\varepsilon})}\frac{\partial^{2\varepsilon}\Phi}{\partial\theta^{2\varepsilon}}\right) = 0, \tag{1.315}$$

where D is the diffusive coefficient and $\Phi = \Phi(\mu,\theta,\vartheta,\tau)$.

Systems of local fractional Maxwell's equations in Cantor-type cylindrical coordinate system is given by

$$\left(e_R^{\varepsilon}\frac{\partial^{\varepsilon}}{\partial R^{\varepsilon}} + e_\theta^{\varepsilon}\frac{1}{R^{\varepsilon}}\frac{\partial^{\varepsilon}}{\partial\theta^{\varepsilon}} + e_\sigma^{\varepsilon}\frac{\partial^{\varepsilon}}{\partial\sigma^{\varepsilon}}\right) \cdot D = \rho, \tag{1.316}$$

$$\left(e_R^{\varepsilon}\frac{\partial^{\varepsilon}}{\partial R^{\varepsilon}} + e_\theta^{\varepsilon}\frac{1}{R^{\varepsilon}}\frac{\partial^{\varepsilon}}{\partial\theta^{\varepsilon}} + e_\sigma^{\varepsilon}\frac{\partial^{\varepsilon}}{\partial\sigma^{\varepsilon}}\right) \times H = \mathrm{J}_{\varepsilon} + \frac{\partial^{\varepsilon}D}{\partial\tau^{\varepsilon}}, \tag{1.317}$$

$$\left(e_R^\varepsilon \frac{\partial^\varepsilon}{\partial R^\varepsilon} + e_\theta^\varepsilon \frac{1}{R^\varepsilon} \frac{\partial^\varepsilon}{\partial \theta^\varepsilon} + e_\sigma^\varepsilon \frac{\partial^\varepsilon}{\partial \sigma^\varepsilon}\right) \times E = -\frac{\partial^\varepsilon B}{\partial \tau^\varepsilon}, \tag{1.318}$$

$$\left(e_R^\varepsilon \frac{\partial^\varepsilon}{\partial R^\varepsilon} + e_\theta^\varepsilon \frac{1}{R^\varepsilon} \frac{\partial^\varepsilon}{\partial \theta^\varepsilon} + e_\sigma^\varepsilon \frac{\partial^\varepsilon}{\partial \sigma^\varepsilon}\right) \cdot B = 0, \tag{1.319}$$

where $\rho = \rho(\mu, \theta, \sigma, \tau)$ is the fractal electric charge density, $D = D(\mu, \theta, \sigma, \tau)$ is electric displacement in the fractal electric field, $H = H(\mu, \theta, \sigma, \tau)$ is the magnetic field strength in the fractal field, $E = E(\mu, \theta, \sigma, \tau)$ is the electric field strength in the fractal field, $\mathrm{J}_\varepsilon = \mathrm{J}_\varepsilon(\mu, \theta, \sigma, \tau)$ is the conductive current, and $B = B(\mu, \theta, \sigma, \tau)$ is the magnetic induction in the fractal field, and the constitutive relationships in fractal electromagnetic can be written as

$$D(\mu, \theta, \sigma, \tau) = \varepsilon_{\mathrm{f}} E(\mu, \theta, \sigma, \tau) \tag{1.320}$$

and

$$H(\mu, \theta, \sigma, \tau) = \mu_{\mathrm{f}} B(\mu, \theta, \sigma, \tau), \tag{1.321}$$

with the fractal dielectric permittivity ε_{f} and the fractal magnetic permeability μ_{f}.

Systems of local fractional Maxwell's equations in Cantor-type circular coordinate system take the following form:

$$\left(e_R^\varepsilon \frac{\partial^\varepsilon}{\partial R^\varepsilon} + e_\vartheta^\varepsilon \frac{1}{R^\varepsilon} \frac{\partial^\varepsilon}{\partial \vartheta^\varepsilon} + e_\theta^\varepsilon \frac{1}{R^\varepsilon} \frac{1}{\sin_\varepsilon(\vartheta^\varepsilon)} \frac{\partial^\varepsilon}{\partial \theta^\varepsilon}\right) \cdot D = \rho, \tag{1.322}$$

$$\left(e_R^\varepsilon \frac{\partial^\varepsilon}{\partial R^\varepsilon} + e_\vartheta^\varepsilon \frac{1}{R^\varepsilon} \frac{\partial^\varepsilon}{\partial \vartheta^\varepsilon} + e_\theta^\varepsilon \frac{1}{R^\varepsilon} \frac{1}{\sin_\varepsilon(\vartheta^\varepsilon)} \frac{\partial^\varepsilon}{\partial \theta^\varepsilon}\right) \times H = \mathrm{J}_\varepsilon + \frac{\partial^\varepsilon D}{\partial \tau^\varepsilon}, \tag{1.323}$$

$$\left(e_R^\varepsilon \frac{\partial^\varepsilon}{\partial R^\varepsilon} + e_\theta^\varepsilon \frac{1}{R^\varepsilon} \frac{\partial^\varepsilon}{\partial \theta^\varepsilon}\right) \times E(R, \theta, \tau) = -\frac{\partial^\varepsilon B(R, \theta, \tau)}{\partial \tau^\varepsilon}, \tag{1.324}$$

$$\left(e_R^\varepsilon \frac{\partial^\varepsilon}{\partial R^\varepsilon} + e_\theta^\varepsilon \frac{1}{R^\varepsilon} \frac{\partial^\varepsilon}{\partial \theta^\varepsilon}\right) \cdot B(R, \theta, \tau) = 0, \tag{1.325}$$

where $\rho = \rho(R, \theta, \vartheta, \tau)$ is the fractal electric charge density, $D = D(R, \theta, \vartheta, \tau)$ is electric displacement in the fractal electric field, $H = H(R, \theta, \vartheta, \tau)$ is the magnetic field strength in the fractal field, $E = E(R, \theta, \vartheta, \tau)$ is the electric field strength in the fractal field, $\mathrm{J}_\varepsilon = \mathrm{J}_\varepsilon(R, \theta, \vartheta, \tau)$ is the conductive current, and $B = B(R, \theta, \vartheta, \tau)$ is the magnetic induction in the fractal field, and the constitutive relationships in fractal electromagnetic can be written as

$$D(R, \theta, \vartheta, \tau) = \varepsilon_{\mathrm{f}} E(R, \theta, \vartheta, \tau) \tag{1.326}$$

and

$$H(R, \theta, \vartheta, \tau) = \mu_{\mathrm{f}} B(R, \theta, \vartheta, \tau), \tag{1.327}$$

with the fractal dielectric permittivity ε_{f} and the fractal magnetic permeability μ_{f}.

We present the wave equation, heat-conduction equation, damped wave equation, and diffusion equation in the Cantor-type cylindrical symmetry form and in the Cantor-type spherical symmetry form.

The linear local fractional wave equation for the velocity potential of fluid flow in the Cantor-type cylindrical symmetry form is presented as follows:

$$\omega\left(\frac{\partial^{2\varepsilon}\Phi(R,\tau)}{\partial R^{2\varepsilon}}+\frac{1}{R^{\varepsilon}}\frac{\partial^{\varepsilon}\Phi(R,\tau)}{\partial R^{\varepsilon}}\right)-\frac{\partial^{2\varepsilon}\Phi(R,\tau)}{\partial\tau^{2\varepsilon}}=0, \tag{1.328}$$

where ω is a constant.

The linear local fractional wave equation for the velocity potential of fluid flow in the Cantor-type spherical symmetry form is

$$\omega\left(\frac{\partial^{2\varepsilon}\Phi(R,\tau)}{\partial R^{2\varepsilon}}+\frac{2}{R^{\varepsilon}}\frac{\partial^{\varepsilon}\Phi(R,\tau)}{\partial R^{\varepsilon}}\right)-\frac{\partial^{2\varepsilon}\Phi(R,\tau)}{\partial\tau^{2\varepsilon}}=0, \tag{1.329}$$

where ω is a constant.

The local fractional heat-conduction equation in the Cantor-type cylindrical symmetry form can be written as

$$\frac{\partial^{\varepsilon}\Phi(R,\tau)}{\partial\tau^{\varepsilon}}-\varpi\left(\frac{\partial^{2\varepsilon}\Phi(R,\tau)}{\partial R^{2\varepsilon}}+\frac{1}{R^{\varepsilon}}\frac{\partial^{\varepsilon}\Phi(R,\tau)}{\partial R^{\varepsilon}}\right)=H(R,\tau), \tag{1.330}$$

where ϖ is the thermal conductivity coefficient and $H(R,\tau)$ is the heat source.

The local fractional heat-conduction equation in the Cantor-type spherical symmetry form has the following form:

$$\frac{\partial^{\varepsilon}\Phi(R,\tau)}{\partial\tau^{\varepsilon}}-\varpi\left(\frac{\partial^{2\varepsilon}\Phi(R,\tau)}{\partial R^{2\varepsilon}}+\frac{2}{R^{\varepsilon}}\frac{\partial^{\varepsilon}\Phi(R,\tau)}{\partial R^{\varepsilon}}\right)=H(R,\tau), \tag{1.331}$$

where ϖ is the thermal conductivity coefficient and $H(R,\tau)$ is the heat source.

The linear local homogeneous damped wave equation of fractal strings in the Cantor-type cylindrical symmetry form is given by

$$\frac{\partial^{2\varepsilon}\Phi(R,\tau)}{\partial\tau^{2\varepsilon}}-\frac{\partial^{\varepsilon}\Phi(R,\tau)}{\partial\tau^{\varepsilon}}-\left(\frac{\partial^{2\varepsilon}\Phi(R,\tau)}{\partial R^{2\varepsilon}}+\frac{1}{R^{\varepsilon}}\frac{\partial^{\varepsilon}\Phi(R,\tau)}{\partial R^{\varepsilon}}\right)=0. \tag{1.332}$$

The linear local homogeneous damped wave equation of fractal strings in the Cantor-type spherical symmetry form is written as

$$\frac{\partial^{2\varepsilon}\Phi(R,\tau)}{\partial\tau^{2\varepsilon}}-\frac{\partial^{\varepsilon}\Phi(R,\tau)}{\partial\tau^{\varepsilon}}-\left(\frac{\partial^{2\varepsilon}\Phi(R,\tau)}{\partial R^{2\varepsilon}}+\frac{2}{R^{\varepsilon}}\frac{\partial^{\varepsilon}\Phi(R,\tau)}{\partial R^{\varepsilon}}\right)=0. \tag{1.333}$$

Next, the local fractional diffusion equation in the Cantor-type cylindrical symmetry form is given by

$$\frac{d^{\varepsilon}\Phi(R,\tau)}{d\tau^{\varepsilon}}-D\left(\frac{\partial^{2\varepsilon}\Phi(R,\tau)}{\partial R^{2\varepsilon}}+\frac{1}{R^{\varepsilon}}\frac{\partial^{\varepsilon}\Phi(R,\tau)}{\partial R^{\varepsilon}}\right)=0, \tag{1.334}$$

where D is the diffusive coefficient.

The local fractional diffusion equation in the Cantor-type spherical symmetry form is given by

$$\frac{d^{\varepsilon}\Phi(R,\tau)}{d\tau^{\varepsilon}} - D\left(\frac{\partial^{2\varepsilon}\Phi(R,\tau)}{\partial R^{2\varepsilon}} + \frac{2}{R^{\varepsilon}}\frac{\partial^{\varepsilon}\Phi(R,\tau)}{\partial R^{\varepsilon}}\right) = 0, \tag{1.335}$$

where D is the diffusive coefficient.

Local fractional Fourier series

2

2.1 Introduction

In the Euclidean dimensional space, the sum of the special trigonometric functions is called the Fourier series in honor of the French mathematician, Jean Baptiste Joseph Fourier (1768–1830). The expansions of functions as trigonometric series play important roles in the analysis of periodic functions, which are studied in science and engineering [90–93].

In fractal dimensional space, there are the following special series via the Mittag–Leffler function defined on the fractal set (e.g., [16, 21, 94–98]):

$$\Theta(\mu) = \frac{\mathrm{A}_0}{2} + \sum_{k=1}^{\infty} \left(\mathrm{A}_k \cos_\varepsilon (k\tau)^\varepsilon + \mathrm{B}_k \sin_\varepsilon (k\tau)^\varepsilon \right) \tag{2.1}$$

and

$$\Theta(\mu) = \sum_{k=-\infty}^{\infty} \varphi_k E_\varepsilon \left(i^\varepsilon (k\mu)^\varepsilon \right), \tag{2.2}$$

where $\mathrm{A}_0, \mathrm{A}_k, \mathrm{B}_k$, and φ_k are local fractional Fourier coefficients of (2.1) and (2.2), respectively.

Just as in the classical mechanism of the Fourier series, we need to answer the following questions:

(a) In fractal dimensional space, how do we present the complex number defined on the fractal set $\Re$?

(b) Is there a generalized Hilbert space interpretation of local fractional Fourier series via nondifferentiable functions?

(c) How do we get the local fractional Fourier coefficients of the special series via the Mittag–Leffler function defined on the fractal set?

For $\mu, \eta \in \mathrm{R}$, and $0 < \varepsilon \leq 1$, the complex number defined on the fractal set $\Re$ is defined as follows [1, 16, 21, 94–104]:

$$Z^\varepsilon = \mu^\varepsilon + i^\varepsilon \eta^\varepsilon, \tag{2.3}$$

where $\mathrm{I}^\varepsilon \in \Re$.

The conjugate of (2.3) is defined by

$$\overline{Z}^\varepsilon = \mu^\varepsilon - i^\varepsilon \eta^\varepsilon, \tag{2.4}$$

where $\overline{Z}^\varepsilon \in \Re$.

Local Fractional Integral Transforms and Their Applications. http://dx.doi.org/10.1016/B978-0-12-804002-7.00002-4

The modulus of the complex number defined on the fractal set $\Re$ is given by [16, 21, 94]

$$\left|Z^{\varepsilon}\right| = \sqrt{\overline{Z}^{\varepsilon} Z^{\varepsilon}} = \sqrt{Z^{\varepsilon}\overline{Z}^{\varepsilon}} = \sqrt{\mu^{2\varepsilon} + \eta^{2\varepsilon}}, \tag{2.5}$$

where $E_{\varepsilon}\left(i^{\varepsilon}\left(2\pi\right)^{\varepsilon}\right) = 1$.

For $Z^{\varepsilon} \in \Re$ and $0 < \varepsilon \leq 1$, the complex Mittag–Leffler function defined on the fractal set $\Re$ is given by

$$E_{\varepsilon}\left(Z^{\varepsilon}\right) = \sum_{i=0}^{\infty} \frac{Z^{i\varepsilon}}{\Gamma\left(1 + i\varepsilon\right)}. \tag{2.6}$$

For $Z_1^{\varepsilon}, Z_2^{\varepsilon} \in \Re$, and $0 < \varepsilon \leq 1$, we present the following properties of the complex Mittag–Leffler function defined on the fractal set:

$$E_{\varepsilon}\left(Z_1^{\varepsilon} + Z_2^{\varepsilon}\right) = E_{\varepsilon}\left(Z_1^{\varepsilon}\right) E_{\varepsilon}\left(Z_2^{\varepsilon}\right), \tag{2.7}$$

$$E_{\varepsilon}\left(Z_1^{\varepsilon} - Z_2^{\varepsilon}\right) = E_{\varepsilon}\left(Z_1^{\varepsilon}\right) E_{\varepsilon}\left(-Z_2^{\varepsilon}\right), \tag{2.8}$$

and

$$E_{\varepsilon}\left(i^{\varepsilon} Z_1^{\varepsilon} + i^{\varepsilon} Z_2^{\varepsilon}\right) = E_{\varepsilon}\left(i^{\varepsilon} Z_1^{\varepsilon}\right) E_{\varepsilon}\left(i^{\varepsilon} Z_2^{\varepsilon}\right). \tag{2.9}$$

For $Z^{\varepsilon} \in \Re$ and $0 < \varepsilon \leq 1$, we have

$$E_{\varepsilon}\left(i^{\varepsilon} Z^{\varepsilon}\right) = \cos_{\varepsilon}\left(Z^{\varepsilon}\right) + i^{\varepsilon} \sin_{\varepsilon}\left(Z^{\varepsilon}\right), \tag{2.10}$$

where

$$\sin_{\varepsilon}\left(Z^{\varepsilon}\right) = \frac{E_{\varepsilon}\left(i^{\varepsilon} Z^{\varepsilon}\right) - E_{\varepsilon}\left(-i^{\varepsilon} Z^{\varepsilon}\right)}{2 i^{\varepsilon}} \tag{2.11}$$

and

$$\cos_{\varepsilon}\left(Z^{\varepsilon}\right) = \frac{E_{\varepsilon}\left(i^{\varepsilon} Z^{\varepsilon}\right) + E_{\varepsilon}\left(-i^{\varepsilon} Z^{\varepsilon}\right)}{2}. \tag{2.12}$$

We call (2.11) and (2.12) the analogues of trigonometric functions, which are derived from the complex Mittag–Leffler function defined on the fractal set.

Let us consider a set of functions given by [16–94]:

$$\Theta\left(E_{\varepsilon}\right) = \left\{E_{\varepsilon}\left(i^{\varepsilon}\left(2\pi\right)^{\varepsilon}\left(k\tau\right)^{\varepsilon}\right), k^{\varepsilon} \in \Re\right\} \tag{2.13}$$

is an orthonormal basis of the generalized Hilbert space of functions in the interval $[-\pi, \pi]$.

In the generalized Hilbert space, the dot product of $\varphi\left(\tau\right)$ and $\phi\left(\tau\right)$ with the 2π-period via local fractional integral is defined as follows [16, 21, 94]:

$$\langle \varphi, \phi \rangle_{\varepsilon} = \int_{-\pi}^{\pi} \varphi\left(\tau\right) \overline{\phi\left(\tau\right)} \left(d\tau\right)^{\varepsilon}. \tag{2.14}$$

For a given generalization of the Hilbert space H_{ε}, we have the following formula [16, 21, 94]:

$$\sum_{k=1}^{n} \left|\varphi_k^{\varepsilon}\right|^2 = \|\phi\|_{\varepsilon}^2, \tag{2.15}$$

where all functions $\phi \in \mathrm{H}_{\varepsilon}$,

$$\left|\varphi_k^{\varepsilon}\right|^2 = \left\langle \varphi_k^{\varepsilon}, \varphi_k^{\varepsilon} \right\rangle_{\varepsilon}, \tag{2.16}$$

$$\|\phi\|_{\varepsilon}^2 = \langle \phi, \phi \rangle_{\varepsilon}, \tag{2.17}$$

and

$$\varphi_k^{\varepsilon} = \left\langle \psi, e_k^{\varepsilon} \right\rangle_{\varepsilon}. \tag{2.18}$$

We also have

$$\phi = \sum_{k=1}^{n} \varphi_k^{\varepsilon} e_k^{\varepsilon} \tag{2.19}$$

with sum convergent in the generalized Hilbert space H_{ε} for all $\phi \in \mathrm{H}_{\varepsilon}$, where $\left\{e_k^{\varepsilon}\right\}$ is a basis of the generalized Hilbert space H_{ε}.

For $k \neq j$, the orthogonality condition of φ_k and φ_j is defined as follows:

$$\left\langle \varphi_k, \varphi_j \right\rangle_{\varepsilon} = \frac{1}{\pi^{\varepsilon}} \int_{-\pi}^{\pi} \varphi_k(\tau) \overline{\varphi_j(\tau)} (\mathrm{d}\tau)^{\varepsilon} = 0. \tag{2.20}$$

For $k \neq j$, the normalized condition of φ_k is defined as follows:

$$\langle \varphi_k, \varphi_k \rangle_{\varepsilon} = \frac{1}{\pi^{\varepsilon}} \int_{-\pi}^{\pi} \varphi_k^2(\tau) (\mathrm{d}\tau)^{\varepsilon} = 1. \tag{2.21}$$

Based upon the space given by

$$\begin{aligned} H_{\varepsilon} &= \operatorname{span}\left\{e_1^{\varepsilon}, \ldots, e_n^{\varepsilon}\right\} \\ &= \operatorname{span}\left\{1, E_{\varepsilon}\left(\frac{\pi^{\varepsilon} i^{\varepsilon} \tau^{\varepsilon}}{L^{\varepsilon}}\right), E_{\varepsilon}\left(\frac{\pi^{\varepsilon} i^{\varepsilon} (2\tau)^{\varepsilon}}{L^{\varepsilon}}\right), \ldots, E_{\varepsilon}\left(\frac{\pi^{\varepsilon} i^{\varepsilon} (n\tau)^{\varepsilon}}{L^{\varepsilon}}\right)\right\}, \end{aligned} \tag{2.22}$$

we have

$$\varphi_0^{\varepsilon} = \langle \phi, 1 \rangle_{\varepsilon} \tag{2.23}$$

and

$$\varphi_k^{\varepsilon} = \left\langle \phi, E_{\varepsilon}\left(i^{\varepsilon} (k\tau)^{\varepsilon}\right) \right\rangle_{\varepsilon}, \tag{2.24}$$

where $k \in \mathrm{Z}$.

One also gets the inverse relations as follows:

$$\phi = \sum_{k=1}^{n} \varphi_k^{\varepsilon} e_k^{\varepsilon}, \tag{2.25}$$

where $k \in \mathrm{N}$.

Hence, the local fractional Fourier series from the generalized Hilbert space H_ε is presented as follows:

$$\phi = \sum_{k=1}^{n} \left\langle \phi, E_\varepsilon\left(i^\varepsilon (k\tau)^\varepsilon\right)\right\rangle_\varepsilon e_k^\varepsilon, \tag{2.26}$$

where $\{e_k^\varepsilon\}_{k=1}^{\infty}$ is a complete orthonormal set of functions.

Let $\phi(\tau)$ be $2L$-periodic. For $k \in Z$, the local fractional Fourier series of the nondifferentiable function $\phi(\tau)$ is defined as

$$\phi(\tau) = \sum_{k=-\infty}^{\infty} \varphi_k^\varepsilon E_\varepsilon\left(\frac{\pi^\varepsilon i^\varepsilon (k\tau)^\varepsilon}{L^\varepsilon}\right), \tag{2.27}$$

where the local fractional Fourier coefficients are written as follows:

$$\varphi_k^\varepsilon = \frac{1}{(2L)^\varepsilon} \int_{-L}^{L} \phi(\tau) E_\varepsilon\left(-\frac{\pi^\varepsilon i^\varepsilon (k\tau)^\varepsilon}{L^\varepsilon}\right)(d\tau)^\varepsilon. \tag{2.28}$$

Therefore, for $L = \pi$, we get the following pair:

$$\phi(\tau) = \sum_{k=1}^{n} \varphi_k^\varepsilon E_\varepsilon\left(i^\varepsilon k^\varepsilon \tau^\varepsilon\right) \tag{2.29}$$

and

$$\varphi_k^\varepsilon = \frac{1}{(2\pi)^\varepsilon} \int_{-\pi}^{\pi} \phi(\tau) E_\varepsilon\left(-i^\varepsilon k^\varepsilon \tau^\varepsilon\right)(d\tau)^\varepsilon. \tag{2.30}$$

For $k \in N$, a set of functions given by

$$\frac{1}{2}, \sin_\varepsilon(\tau)^\varepsilon, \cos_\varepsilon(\tau)^\varepsilon, \ldots, \sin_\varepsilon(k\tau)^\varepsilon, \cos_\varepsilon(k\tau)^\varepsilon \tag{2.31}$$

are orthogonal. In this case, we have

$$\Theta(\tau) = \frac{A_0}{2} + \sum_{k=1}^{\infty}\left(A_k \cos_\varepsilon(k\tau)^\varepsilon + B_k \sin_\varepsilon(k\tau)^\varepsilon\right), \tag{2.32}$$

where

$$A_0 = \frac{1}{\pi^\varepsilon} \int_{-\pi}^{\pi} \phi(\tau)(d\tau)^\varepsilon, \tag{2.33}$$

$$A_k = \frac{1}{\pi^\varepsilon} \int_{-\pi}^{\pi} \phi(\tau) \cos_\varepsilon(k\tau)^\varepsilon (d\tau)^\varepsilon, \tag{2.34}$$

and

$$B_k = \frac{1}{\pi^\varepsilon} \int_{-\pi}^{\pi} \phi(\tau) \sin_\varepsilon(k\tau)^\varepsilon (d\tau)^\varepsilon. \tag{2.35}$$

Hence, the local fractional Fourier series from the generalized Hilbert space H_ε is presented as follows:

$$\phi = \sum_{k=1}^{n} \left\langle \phi, e_k^\varepsilon \right\rangle_\varepsilon e_k^\varepsilon, \tag{2.36}$$

where $\left\{e_k^\varepsilon\right\}_{k=1}^{\infty}$ is a complete orthonormal set of functions.

In different bases, we can obtain a class of different Fourier series from the generalized Hilbert space. In this case, we find the local fractional Fourier coefficients of the special series via the Mittag–Leffler function defined on the fractal set.

2.2 Definitions and properties

2.2.1 Analogous trigonometric form of local fractional Fourier series

Let $\phi(\tau)$ be $2L$-periodic. For $k \in \mathrm{N}$, a generalized local fractional Fourier series of $\psi(\tau)$ is defined as follows:

$$\psi(\tau) \sim \frac{\mathrm{A}(0,\varepsilon)}{2} + \sum_{k=1}^{\infty} \left(\mathrm{A}(k,\varepsilon) \cos_\varepsilon \left(\frac{\pi k \tau}{L} \right)^\varepsilon + \mathrm{B}(k,\varepsilon) \sin_\varepsilon \left(\frac{\pi k \tau}{L} \right)^\varepsilon \right), \tag{2.37}$$

where

$$\mathrm{A}(0,\varepsilon) = \frac{1}{L^\varepsilon} \int_{-L}^{L} \phi(\tau) (\mathrm{d}\tau)^\varepsilon, \tag{2.38}$$

$$\mathrm{A}(k,\varepsilon) = \frac{1}{L^\varepsilon} \int_{-L}^{L} \psi(\tau) \cos_\varepsilon \left(\frac{\pi k \tau}{L} \right)^\varepsilon (\mathrm{d}\tau)^\varepsilon, \tag{2.39}$$

and

$$\mathrm{B}(k,\varepsilon) = \frac{1}{L^\varepsilon} \int_{-L}^{L} \psi(\tau) \sin_\varepsilon \left(\frac{\pi k \tau}{L} \right)^\varepsilon (\mathrm{d}\tau)^\varepsilon \tag{2.40}$$

are the local fractional Fourier coefficients of a generalized local fractional Fourier series of $\psi(\tau)$.

Let $\phi(\tau)$ be 2π-periodic. For $k \in \mathrm{N}$, a local fractional Fourier series of $\phi(\tau)$ is defined as follows:

$$\phi(\tau) \sim \frac{\mathrm{A}(0,\varepsilon)}{2} + \sum_{k-1}^{\infty} \left(\mathrm{A}(k,\varepsilon) \cos_\varepsilon (k\tau)^\varepsilon + \mathrm{B}(k,\varepsilon) \sin_\varepsilon (k\tau)^\varepsilon \right), \tag{2.41}$$

where

$$\mathrm{A}(0,\varepsilon)=\frac{1}{\pi^{\varepsilon}}\int_{-\pi}^{\pi}\phi(\tau)(\mathrm{d}\tau)^{\varepsilon}, \tag{2.42}$$

$$\mathrm{A}(k,\varepsilon)=\frac{1}{\pi^{\varepsilon}}\int_{-\pi}^{\pi}\phi(\tau)\cos_{\varepsilon}(k\tau)^{\varepsilon}(\mathrm{d}\tau)^{\varepsilon}, \tag{2.43}$$

and

$$\mathrm{B}(k,\varepsilon)=\frac{1}{\pi^{\varepsilon}}\int_{-\pi}^{\pi}\phi(\tau)\sin_{\varepsilon}(k\tau)^{\varepsilon}(\mathrm{d}\tau)^{\varepsilon} \tag{2.44}$$

are the local fractional Fourier coefficients of a local fractional Fourier series of $\phi(\tau)$ in (2.39).

2.2.2 Complex Mittag–Leffler form of local fractional Fourier series

Let $\phi(\tau)$ be $2L$-periodic. For $k \in \mathrm{Z}$, a generalized local fractional Fourier series of $\psi(\tau)$ is defined as follows:

$$\psi(\tau)\sim\sum_{k=-\infty}^{\infty}\varphi(k,\varepsilon)E_{\varepsilon}\left(\frac{\pi^{\varepsilon}i^{\varepsilon}(k\tau)^{\varepsilon}}{L^{\varepsilon}}\right), \tag{2.45}$$

where

$$\varphi(k,\varepsilon)=\frac{1}{(2L)^{\varepsilon}}\int_{-L}^{L}\psi(\tau)E_{\varepsilon}\left(-\frac{\pi^{\varepsilon}i^{\varepsilon}(k\tau)^{\varepsilon}}{L^{\varepsilon}}\right)(\mathrm{d}\tau)^{\varepsilon} \tag{2.46}$$

is the local fractional Fourier coefficient of a generalized local fractional Fourier series of $\psi(\tau)$.

Let $\phi(\tau)$ be 2π-periodic. For $k \in \mathrm{Z}$, a local fractional Fourier series of $\phi(\tau)$ is defined as follows:

$$\phi(\tau)\sim\sum_{k=-\infty}^{\infty}\varphi(k,\varepsilon)E_{\varepsilon}\left(i^{\varepsilon}(k\tau)^{\varepsilon}\right), \tag{2.47}$$

where

$$\varphi(k,\varepsilon)=\frac{1}{(2L)^{\varepsilon}}\int_{-L}^{L}\phi(\tau)E_{\varepsilon}\left(-i^{\varepsilon}(k\tau)^{\varepsilon}\right)(\mathrm{d}\tau)^{\varepsilon} \tag{2.48}$$

is the local fractional Fourier coefficient of a local fractional Fourier series of $\phi(\tau)$.

Adopting (2.11) and (2.12), we easily obtain the following relationships:

$$\varphi(0,\varepsilon)=\frac{\mathrm{A}(0,\varepsilon)}{2^{\varepsilon}}, \tag{2.49}$$

$$\varphi(k,\varepsilon)=\frac{\mathrm{A}(k,\varepsilon)-i^{\varepsilon}\mathrm{B}(k,\varepsilon)}{2^{\varepsilon}}, \tag{2.50}$$

and

$$\varphi(-k,\varepsilon) = \frac{\mathrm{A}(k,\varepsilon) + i^{\varepsilon}\mathrm{B}(k,\varepsilon)}{2^{\varepsilon}}. \tag{2.51}$$

In this case, we also write transformation pairs in the following forms:

$$\psi(\tau) \leftrightarrow \varphi(k,\varepsilon). \tag{2.52}$$

Other forms of local fractional Fourier series were presented in the literature (e.g., [1, 16, 21, 94–104]).

2.2.3 Properties of local fractional Fourier series

Property 3 (Linearity of local fractional Fourier series). *Suppose that*

$$\psi_1(\tau) \leftrightarrow \varphi_1(k,\varepsilon)$$

and

$$\psi_2(\tau) \leftrightarrow \varphi_2(k,\varepsilon).$$

Then, for two constants a and b, we have

$$a\psi_1(\tau) + b\psi_2(\tau) \leftrightarrow a\varphi_1(k,\varepsilon) + b\varphi_2(k,\varepsilon). \tag{2.53}$$

Proof. The proof of this property is a straightforward application of the linearity property of integration. □

Property 4 (Conjugation of local fractional Fourier series). *Suppose that $\varphi(k,\varepsilon)$ is the local fractional Fourier coefficient of $\psi(\tau)$. Then, we get*

$$\overline{\psi(\tau)} \leftrightarrow \overline{\varphi(-k,\varepsilon)}. \tag{2.54}$$

Proof. Since

$$\overline{E_{\varepsilon}\left(-i^{\varepsilon}(k\tau)^{\varepsilon}\right)} = E_{\varepsilon}\left(i^{\varepsilon}(k\tau)^{\varepsilon}\right), \tag{2.55}$$

it follows by direct calculation of the local fractional Fourier coefficient of $\psi(\tau)$ that

$$\begin{aligned}
\frac{1}{(2\pi)^{\varepsilon}}\int_{-\pi}^{\pi}\overline{\psi(\tau)}E_{\varepsilon}\left(-i^{\varepsilon}(k\tau)^{\varepsilon}\right)(\mathrm{d}\tau)^{\varepsilon} &= \overline{\frac{1}{(2\pi)^{\varepsilon}}\int_{-\pi}^{\pi}\psi(\tau)E_{\varepsilon}\left(i^{\varepsilon}(k\tau)^{\varepsilon}\right)(\mathrm{d}\tau)^{\varepsilon}} \\
&= \overline{\frac{1}{(2\pi)^{\varepsilon}}\int_{-\pi}^{\pi}\psi(\tau)E_{\varepsilon}\left(-i^{\varepsilon}(-k\tau)^{\varepsilon}\right)(\mathrm{d}\tau)^{\varepsilon}} \\
&= \overline{\varphi(-k,\varepsilon)}.
\end{aligned} \tag{2.56}$$

The proof of this property is thus completed. □

Property 5 (Shift in fractal time of local fractional Fourier series). *Suppose that $\varphi(k,\varepsilon)$ is the local fractional Fourier coefficient of $\psi(\tau)$. Then, we have*

$$\psi(\tau - \tau_0) \leftrightarrow E_\varepsilon\left(-i^\varepsilon (k\tau_0)^\varepsilon\right)\varphi(k,\varepsilon). \tag{2.57}$$

Proof. Adopting the definition of local fractional Fourier series, we find that

$$\begin{aligned}
&\frac{1}{(2\pi)^\varepsilon}\int_{-\pi}^{\pi}\psi(\tau-\tau_0)\,E_\varepsilon\left(-i^\varepsilon (k\tau)^\varepsilon\right)(\mathrm{d}\tau)^\varepsilon \\
&= E_\varepsilon\left(-i^\varepsilon (k\tau_0)^\varepsilon\right)\frac{1}{(2\pi)^\varepsilon}\int_{-\pi}^{\pi}\psi(\tau-\tau_0)\,E_\varepsilon\left(-i^\varepsilon [k(\tau-\tau_0)]^\varepsilon\right)(\mathrm{d}\tau)^\varepsilon \\
&= E_\varepsilon\left(-i^\varepsilon (k\tau_0)^\varepsilon\right)\frac{1}{(2\pi)^\varepsilon}\int_{-\pi}^{\pi}\psi(\tau)\,E_\varepsilon\left(-i^\varepsilon (k\tau)^\varepsilon\right)(\mathrm{d}\tau)^\varepsilon \\
&= E_\varepsilon\left(-i^\varepsilon (k\tau_0)^\varepsilon\right)\varphi(k,\varepsilon).
\end{aligned} \tag{2.58}$$

The proof of this property is evidently completed. □

Property 6 (Fractal time reversal of local fractional Fourier series). *Suppose that $\varphi(k,\varepsilon)$ is the local fractional Fourier coefficient of $\psi(\tau)$. Then, we obtain*

$$\psi(-\tau) \leftrightarrow \varphi(-k,\varepsilon). \tag{2.59}$$

Proof. With the help of the definition of local fractional Fourier series, we have

$$\begin{aligned}
\frac{1}{(2\pi)^\varepsilon}\int_{-\pi}^{\pi}\psi(-\tau)\,E_\varepsilon\left(-i^\varepsilon (k\tau)^\varepsilon\right)(\mathrm{d}\tau)^\varepsilon &= \frac{1}{(2\pi)^\varepsilon}\int_{-\pi}^{\pi}\psi(-\tau)\,E_\varepsilon\left(i^\varepsilon (-k\tau)^\varepsilon\right)(\mathrm{d}\tau)^\varepsilon \\
&= \frac{1}{(2\pi)^\varepsilon}\int_{-\pi}^{\pi}\psi(\tau)\,E_\varepsilon\left(i^\varepsilon (k\tau)^\varepsilon\right)(\mathrm{d}\tau)^\varepsilon \\
&= \frac{1}{(2\pi)^\varepsilon}\int_{-\pi}^{\pi}\psi(\tau)\,E_\varepsilon\left(-i^\varepsilon (-k\tau)^\varepsilon\right)(\mathrm{d}\tau)^\varepsilon \\
&= \varphi(-k,\varepsilon).
\end{aligned} \tag{2.60}$$

The proof of this property is thus completed. □

2.2.4 Theorems of local fractional Fourier series

Theorem 2.1 (Bessel inequality for local fractional Fourier series). *Suppose that $\psi(\tau)$ is 2π-periodic, bounded, and locally fractional integrable on $[-\pi,\pi]$. Then, the following inequality holds true:*

$$\frac{A^2(0,\varepsilon)}{2} + \sum_{k=1}^{n}\left(A^2(k,\varepsilon) + B^2(k,\varepsilon)\right) \le \frac{1}{\pi^\varepsilon}\int_{-\pi}^{\pi}\psi^2(\tau)\,(d\tau)^\varepsilon, \tag{2.61}$$

provided that $A(0,\varepsilon)$, $A(k,\varepsilon)$, and $B(k,\varepsilon)$ are the local fractional Fourier coefficients of $\psi(\tau)$.

Proof. Let us consider the sum of the local fractional Fourier series, namely,

$$S_{n,\varepsilon}(\tau) = \frac{\mathrm{A}(0,\varepsilon)}{2} + \sum_{k=1}^{n} \left(\mathrm{A}(k,\varepsilon)\cos_{\varepsilon}(k\tau)^{\varepsilon} + \mathrm{B}(k,\varepsilon)\sin_{\varepsilon}(k\tau)^{\varepsilon}\right). \tag{2.62}$$

Following (2.62), we calculate

$$\frac{1}{\pi^{\varepsilon}}\int_{-\pi}^{\pi}\left[\psi(\tau) - S_{n,\varepsilon}(\tau)\right]^{2}(\mathrm{d}\tau)^{\varepsilon} = \frac{1}{\pi^{\varepsilon}}\int_{-\pi}^{\pi}\Big[\psi^{2}(\tau) - 2\psi(\tau)S_{n,\varepsilon}(\tau) + S_{n,\varepsilon}^{2}(\tau)\Big](\mathrm{d}\tau)^{\varepsilon}, \tag{2.63}$$

where

$$\begin{aligned}
&\frac{1}{\pi^{\varepsilon}}\int_{-\pi}^{\pi} S_{n,\varepsilon}^{2}(\tau)(\mathrm{d}\tau)^{\varepsilon} \\
&= \frac{1}{\pi^{\varepsilon}}\int_{-\pi}^{\pi}\left[\sum_{k=0}^{\infty}\left(\mathrm{A}(k,\varepsilon)\cos_{\varepsilon}(k\tau)^{\varepsilon} + \mathrm{B}(k,\varepsilon)\sin_{\varepsilon}(k\tau)^{\varepsilon}\right)\right]^{2}(\mathrm{d}\tau)^{\varepsilon} \\
&= \frac{\mathrm{A}^{2}(0,\varepsilon)}{2} + \frac{1}{\pi^{\varepsilon}}\int_{-\pi}^{\pi}\left[\sum_{k=1}^{\infty}\left(\mathrm{A}(k,\varepsilon)\cos_{\varepsilon}(k\tau)^{\varepsilon} + \mathrm{B}(k,\varepsilon)\sin_{\varepsilon}(k\tau)^{\varepsilon}\right)\right]^{2}(\mathrm{d}\tau)^{\varepsilon} \\
&= \frac{\mathrm{A}^{2}(0,\varepsilon)}{2} + \sum_{k=1}^{n}\left(\mathrm{A}^{2}(k,\varepsilon) + \mathrm{B}^{2}(k,\varepsilon)\right)
\end{aligned} \tag{2.64}$$

and

$$\begin{aligned}
&\frac{1}{\pi^{\varepsilon}}\int_{-\pi}^{\pi}\left[\psi(\tau)S_{n,\varepsilon}(\tau)\right](\mathrm{d}\tau)^{\varepsilon} \\
&= \frac{1}{\pi^{\varepsilon}}\int_{-\pi}^{\pi}\left\{\left[\mathrm{A}(0,\varepsilon) + \sum_{k=1}^{n}\left(\mathrm{A}(k,\varepsilon)\cos_{\varepsilon}(k\tau)^{\varepsilon} + \mathrm{B}(k,\varepsilon)\sin_{\varepsilon}(k\tau)^{\varepsilon}\right)\right] S_{k,\varepsilon}(\tau)\right\}(\mathrm{d}\tau)^{\varepsilon} \\
&= \frac{\mathrm{A}^{2}(0,\varepsilon)}{2} + \sum_{k=1}^{n}\left(\mathrm{A}^{2}(k,\varepsilon) + \mathrm{B}^{2}(k,\varepsilon)\right) \\
&= \frac{\mathrm{A}^{2}(0,\varepsilon)}{2} + \sum_{k=1}^{n}\left(\mathrm{A}^{2}(k,\varepsilon) + \mathrm{B}^{2}(k,\varepsilon)\right)
\end{aligned} \tag{2.65}$$

with

$$\psi(\tau) \sim \mathrm{A}(0,\varepsilon) + \sum_{k=1}^{\infty}\left(\mathrm{A}(k,\varepsilon)\cos_{\varepsilon}(k\tau)^{\varepsilon} + \mathrm{B}(k,\varepsilon)\sin_{\varepsilon}(k\tau)^{\varepsilon}\right). \tag{2.66}$$

Therefore, this theorem is proved. □

Theorem 2.2 (Riemann-Lebesgue theorem for local fractional Fourier series). *Suppose that $\psi(\tau)$ is 2π-periodic, bounded, and locally fractional integrable on $[-\pi,\pi]$. Then,*

$$\lim_{k\to\infty}\frac{1}{\pi^{\varepsilon}}\int_{-\pi}^{\pi}\psi(\tau)\sin_{\varepsilon}(k\tau)^{\varepsilon}(d\tau)^{\varepsilon}=0 \tag{2.67}$$

and

$$\lim_{k\to\infty}\frac{1}{\pi^{\varepsilon}}\int_{-\pi}^{\pi}\psi(\tau)\cos_{\varepsilon}(k\tau)^{\varepsilon}(d\tau)^{\varepsilon}=0. \tag{2.68}$$

Proof. Considering the integration

$$\Theta(k)=\frac{1}{\pi^{\varepsilon}}\int_{-\pi}^{\pi}\psi(\tau)\sin_{\varepsilon}(k\tau)^{\varepsilon}(d\tau)^{\varepsilon} \tag{2.69}$$

and changing the variable in (2.66) with

$$\tau=t+\frac{\pi}{k}, \tag{2.70}$$

we find that

$$\sin_{\varepsilon}(k\tau)^{\varepsilon}=-\sin_{\varepsilon}(kt)^{\varepsilon} \tag{2.71}$$

and

$$\Theta(k)=-\frac{1}{\pi^{\varepsilon}}\int_{-\pi\left(1-\frac{1}{k}\right)}^{\pi\left(1-\frac{1}{k}\right)}\psi\left(t+\frac{\pi}{k}\right)\sin_{\varepsilon}(kt)^{\varepsilon}(dt)^{\varepsilon}. \tag{2.72}$$

Therefore, from (2.66) and (2.69), we have

$$\begin{aligned}
2\Theta(k)&=\frac{1}{\pi^{\varepsilon}}\int_{-\pi}^{\pi}\psi(t)\sin_{\varepsilon}(kt)^{\varepsilon}(dt)^{\varepsilon}\\
&\quad-\frac{1}{\pi^{\varepsilon}}\int_{-\pi\left(1-\frac{1}{k}\right)}^{\pi\left(1-\frac{1}{k}\right)}\psi\left(t+\frac{\pi}{k}\right)\sin_{\varepsilon}(kt)^{\varepsilon}(dt)^{\varepsilon}\\
&=-\frac{1}{\pi^{\varepsilon}}\int_{-\pi\left(1-\frac{1}{k}\right)}^{-\pi}\psi\left(t+\frac{\pi}{k}\right)\sin_{\varepsilon}(kt)^{\varepsilon}(dt)^{\varepsilon}\\
&\quad+\frac{1}{\pi^{\varepsilon}}\int_{\pi\left(1-\frac{1}{k}\right)}^{\pi}\psi\left(t+\frac{\pi}{k}\right)\sin_{\varepsilon}(kt)^{\varepsilon}(dt)^{\varepsilon}\\
&\quad+\frac{1}{\pi^{\varepsilon}}\int_{-\pi}^{\pi}\left[\psi(t)-\psi\left(t+\frac{\pi}{k}\right)\right]\sin_{\varepsilon}(kt)^{\varepsilon}(dt)^{\varepsilon}.
\end{aligned} \tag{2.73}$$

Since $\psi(\tau)$ is 2π-periodic, bounded, and locally fractional integrable on $[-\pi,\pi]$, there exists M such that

$$|\psi(t)|\leq \mathrm{M} \tag{2.74}$$

$\forall t\in[-\pi,\pi]$.

In this case, we get

$$\left|\frac{1}{\pi^{\varepsilon}}\int_{-\pi\left(1-\frac{1}{k}\right)}^{-\pi}\psi\left(t+\frac{\pi}{k}\right)\sin_{\varepsilon}(kt)^{\varepsilon}(\mathrm{d}t)^{\varepsilon}\right| \leq \frac{1}{\pi^{\varepsilon}}\int_{-\pi\left(1-\frac{1}{k}\right)}^{-\pi}\left|\psi\left(t+\frac{\pi}{k}\right)\sin_{\varepsilon}(kt)^{\varepsilon}\right|(\mathrm{d}t)^{\varepsilon}$$
$$\leq \frac{1}{\pi^{\varepsilon}}\int_{-\pi\left(1-\frac{1}{k}\right)}^{-\pi}\left|\psi\left(t+\frac{\pi}{k}\right)\right|(\mathrm{d}t)^{\varepsilon}$$
$$\leq \frac{\mathrm{M}}{\pi^{\varepsilon}}\left(\frac{\pi}{k}\right)^{\varepsilon} \tag{2.75}$$

and

$$\left|\frac{1}{\pi^{\varepsilon}}\int_{\pi\left(1-\frac{1}{k}\right)}^{\pi}\psi\left(t+\frac{\pi}{k}\right)\sin_{\varepsilon}(kt)^{\varepsilon}(\mathrm{d}t)^{\varepsilon}\right| \leq \frac{1}{\pi^{\varepsilon}}\int_{\pi\left(1-\frac{1}{k}\right)}^{\pi}\left|\psi\left(t+\frac{\pi}{k}\right)\sin_{\varepsilon}(kt)^{\varepsilon}\right|(\mathrm{d}t)^{\varepsilon}$$
$$\leq \frac{1}{\pi^{\varepsilon}}\int_{\pi\left(1-\frac{1}{k}\right)}^{\pi}\left|\psi\left(t+\frac{\pi}{k}\right)\right|(\mathrm{d}t)^{\varepsilon}$$
$$\leq \frac{\mathrm{M}}{\pi^{\varepsilon}}\left(\frac{\pi}{k}\right)^{\varepsilon}. \tag{2.76}$$

It then follows that

$$|2\Theta(k)| \leq \frac{\mathrm{M}}{\pi^{\varepsilon}}\left(\frac{\pi}{k}\right)^{\varepsilon}+\frac{\mathrm{M}}{\pi^{\varepsilon}}\left(\frac{\pi}{k}\right)^{\varepsilon}+\left|\frac{1}{\pi^{\varepsilon}}\int_{-\pi}^{\pi}\left[\psi(t)-\psi\left(t+\frac{\pi}{k}\right)\right]\sin_{\varepsilon}(kt)^{\varepsilon}(\mathrm{d}t)^{\varepsilon}\right|. \tag{2.77}$$

Therefore, for $\rho > 0$, we can find K such that (for $k > K$)

$$\left|\psi(t)-\psi\left(t+\frac{\pi}{k}\right)\right| \leq \left(\frac{\rho}{2}\right)^{\varepsilon}. \tag{2.78}$$

In this case, we also choose K large enough such that (2.74) can be written as follows:

$$\frac{\mathrm{M}}{\pi^{\varepsilon}}\left(\frac{\pi}{k}\right)^{\varepsilon} \leq \left(\frac{\rho}{4}\right)^{\varepsilon}. \tag{2.79}$$

Then, for $k > K$, we get

$$|\Theta(k)| \leq \left(\frac{\rho}{4}\right)^{\varepsilon}+\left(\frac{\rho}{4}\right)^{\varepsilon}+\left(\frac{\rho}{2}\right)^{\varepsilon}=\rho^{\varepsilon}, \tag{2.80}$$

which implies (2.67).

In a similar manner, we also obtain (2.68). Therefore, this theorem is proved. □

We also generalize (2.67) and (2.68). In this case, the following theorem holds true:

Theorem 2.3. *Suppose that $\psi(\tau)$ is 2π-periodic, bounded, and locally fractional integrable on $[-\pi,\pi]$. Then,*

$$\lim_{k\to\infty}\frac{1}{\pi^{\varepsilon}}\int_{0}^{\pi}\psi(\tau)\sin_{\varepsilon}\left(\frac{2k+1}{2}\tau\right)^{\varepsilon}(d\tau)^{\varepsilon}=0, \tag{2.81}$$

$$\lim_{k\to\infty}\frac{1}{\pi^{\varepsilon}}\int_{-\pi}^{0}\psi(\tau)\sin_{\varepsilon}\left(\frac{2k+1}{2}\tau\right)^{\varepsilon}(d\tau)^{\varepsilon}=0, \tag{2.82}$$

$$\lim_{k\to\infty}\frac{1}{\pi^{\varepsilon}}\int_{0}^{\pi}\psi(\tau)\cos_{\varepsilon}\left(\frac{2k+1}{2}\tau\right)^{\varepsilon}(d\tau)^{\varepsilon}=0, \tag{2.83}$$

and

$$\lim_{k\to\infty}\frac{1}{\pi^{\varepsilon}}\int_{-\pi}^{0}\psi(\tau)\cos_{\varepsilon}\left(\frac{2k+1}{2}\tau\right)^{\varepsilon}(d\tau)^{\varepsilon}=0. \tag{2.84}$$

We next state the following theorem:

Theorem 2.4. *Suppose that*

$$S_{n,\varepsilon}(\tau)\sim\frac{A(0,\varepsilon)}{2}+\sum_{k=1}^{n}\left(A(k,\varepsilon)\cos_{\varepsilon}(k\tau)^{\varepsilon}+B(k,\varepsilon)\sin_{\varepsilon}(k\tau)^{\varepsilon}\right).$$

Then,

$$S_{n,\varepsilon}(\mu)=\frac{1}{\pi^{\varepsilon}}\int_{-\pi}^{\pi}S_{n,\varepsilon}(\mu+\tau)D_{n,\varepsilon}(\tau)(d\tau)^{\varepsilon}, \tag{2.85}$$

where

$$D_{n,\varepsilon}(t)=\frac{1}{2}+\sum_{k=1}^{n}\cos_{\varepsilon}(k\tau)^{\varepsilon}. \tag{2.86}$$

Proof. We can expand $S_{n,\varepsilon}(\mu+\tau)$ in the following form:

$$\begin{aligned}
S_{n,\varepsilon}(\mu+\tau)&=\frac{\mathrm{A}(0,\varepsilon)}{2}+\sum_{k=1}^{n}\left(\mathrm{A}(k,\varepsilon)\cos_{\varepsilon}[k(\mu+\tau)]^{\varepsilon}+\mathrm{B}(k,\varepsilon)\sin_{\varepsilon}[k(\mu+\tau)]^{\varepsilon}\right)\\
&=\frac{\mathrm{A}(0,\varepsilon)}{2}+\sum_{k=1}^{n}\mathrm{A}(k,\varepsilon)\left[\cos_{\varepsilon}(k\mu)^{\varepsilon}\cos_{\varepsilon}(k\tau)^{\varepsilon}-\sin_{\varepsilon}(k\mu)^{\varepsilon}\sin_{\varepsilon}(k\tau)^{\varepsilon}\right]\\
&\quad+\sum_{k=1}^{n}\mathrm{B}(k,\varepsilon)\left[\sin_{\varepsilon}(k\mu)^{\varepsilon}\cos_{\varepsilon}(k\tau)^{\varepsilon}+\sin_{\varepsilon}(k\tau)^{\varepsilon}\cos_{\varepsilon}(k\mu)^{\varepsilon}\right]\\
&=\frac{\mathrm{A}(0,\varepsilon)}{2}+\sum_{k=0}^{n}\mathrm{A}(k,\varepsilon)\cos_{\varepsilon}(k\mu)^{\varepsilon}\cos_{\varepsilon}(k\tau)^{\varepsilon}\\
&\quad-\sum_{k=1}^{n}\mathrm{A}(k,\varepsilon)\sin_{\varepsilon}(k\mu)^{\varepsilon}\sin_{\varepsilon}(k\tau)^{\varepsilon}\\
&\quad+\sum_{k=1}^{n}\mathrm{B}(k,\varepsilon)\sin_{\varepsilon}(k\mu)^{\varepsilon}\cos_{\varepsilon}(k\tau)^{\varepsilon}\\
&\quad+\sum_{k=1}^{n}\mathrm{B}(k,\varepsilon)\cos_{\varepsilon}(k\mu)^{\varepsilon}\sin_{\varepsilon}(k\tau)^{\varepsilon}.
\end{aligned} \tag{2.87}$$

In this case, we consider the following integral:

$$\frac{1}{\pi^{\varepsilon}}\int_{-\pi}^{\pi} S_{n,\varepsilon}(\mu+\tau) D_{n,\varepsilon}(\tau)(\mathrm{d}\tau)^{\varepsilon}$$

$$=\frac{1}{\pi^{\varepsilon}}\int_{-\pi}^{\pi}\left\{\left[\sum_{k=1}^{n}\mathrm{A}(k,\varepsilon)\cos_{\varepsilon}(k\mu)^{\varepsilon}\cos_{\varepsilon}(k\tau)^{\varepsilon}\right]\left(\frac{1}{2}+\sum_{k=1}^{n}\cos_{\varepsilon}(k\tau)^{\varepsilon}\right)\right\}(\mathrm{d}\tau)^{\varepsilon}$$

$$-\frac{1}{\pi^{\varepsilon}}\int_{-\pi}^{\pi}\left\{\left[\sum_{k=1}^{n}\mathrm{A}(k,\varepsilon)\sin_{\varepsilon}(k\mu)^{\varepsilon}\sin_{\varepsilon}(k\tau)^{\varepsilon}\right]\left(\frac{1}{2}+\sum_{k=1}^{n}\cos_{\varepsilon}(k\tau)^{\varepsilon}\right)\right\}(\mathrm{d}\tau)^{\varepsilon}$$

$$+\frac{1}{\pi^{\varepsilon}}\int_{-\pi}^{\pi}\left\{\left[\sum_{k=1}^{n}\mathrm{B}(k,\varepsilon)\sin_{\varepsilon}(k\mu)^{\varepsilon}\cos_{\varepsilon}(k\tau)^{\varepsilon}\right]\left(\frac{1}{2}+\sum_{k=1}^{n}\cos_{\varepsilon}(k\tau)^{\varepsilon}\right)\right\}(\mathrm{d}\tau)^{\varepsilon}$$

$$+\frac{1}{\pi^{\varepsilon}}\int_{-\pi}^{\pi}\left\{\left[\sum_{k=1}^{n}\mathrm{B}(k,\varepsilon)\cos_{\varepsilon}(k\mu)^{\varepsilon}\sin_{\varepsilon}(k\tau)^{\varepsilon}\right]\left(\frac{1}{2}+\sum_{k=1}^{n}\cos_{\varepsilon}(k\tau)^{\varepsilon}\right)\right\}(\mathrm{d}\tau)^{\varepsilon}$$

$$+\frac{1}{\pi^{\varepsilon}}\int_{-\pi}^{\pi}\left\{\frac{\mathrm{A}(0,\varepsilon)}{2}\left(\frac{1}{2}+\sum_{k=1}^{n}\cos_{\varepsilon}(k\tau)^{\varepsilon}\right)\right\}(\mathrm{d}\tau)^{\varepsilon}$$

$$=\frac{\mathrm{A}(0,\varepsilon)}{2}+\sum_{k=0}^{n}\mathrm{A}(k,\varepsilon)\cos_{\varepsilon}(k\mu)^{\varepsilon}+\sum_{k=0}^{n}\mathrm{B}(k,\varepsilon)\sin_{\varepsilon}(k\mu)^{\varepsilon}, \tag{2.88}$$

where

$$\sum_{k=0}^{n}\mathrm{A}(k,\varepsilon)\cos_{\varepsilon}(k\mu)^{\varepsilon}$$

$$=\int_{-\pi}^{\pi}\left\{\left[\sum_{k=0}^{n}\mathrm{A}(k,\varepsilon)\cos_{\varepsilon}(k\mu)^{\varepsilon}\cos_{\varepsilon}(k\tau)^{\varepsilon}\right]\left(\frac{1}{2}+\sum_{k=1}^{n}\cos_{\varepsilon}(k\tau)^{\varepsilon}\right)\right\}(\mathrm{d}\tau)^{\varepsilon}, \tag{2.89}$$

$$\frac{1}{\pi^{\varepsilon}}\int_{-\pi}^{\pi}\left\{\left[\sum_{k=0}^{n}\mathrm{A}(k,\varepsilon)\sin_{\varepsilon}(k\mu)^{\varepsilon}\sin_{\varepsilon}(k\tau)^{\varepsilon}\right]\left(\frac{1}{2}+\sum_{k=1}^{n}\cos_{\varepsilon}(k\tau)^{\varepsilon}\right)\right\}(\mathrm{d}\tau)^{\varepsilon}=0, \tag{2.90}$$

$$\sum_{k=0}^{n}\mathrm{B}(k,\varepsilon)\sin_{\varepsilon}(k\mu)^{\varepsilon}$$

$$=\frac{1}{\pi^{\varepsilon}}\int_{-\pi}^{\pi}\left\{\left[\sum_{k=0}^{n}\mathrm{B}(k,\varepsilon)\sin_{\varepsilon}(k\mu)^{\varepsilon}\cos_{\varepsilon}(k\tau)^{\varepsilon}\right]\left(\frac{1}{2}+\sum_{k=1}^{n}\cos_{\varepsilon}(k\tau)^{\varepsilon}\right)\right\}(\mathrm{d}\tau)^{\varepsilon}, \tag{2.91}$$

and

$$\frac{1}{\pi^{\varepsilon}}\int_{-\pi}^{\pi}\left\{\left[\sum_{k=0}^{n}\mathrm{B}(k,\varepsilon)\cos_{\varepsilon}(k\mu)^{\varepsilon}\sin_{\varepsilon}(k\tau)^{\varepsilon}\right]\left(\frac{1}{2}+\sum_{k=1}^{n}\cos_{\varepsilon}(k\tau)^{\varepsilon}\right)\right\}(\mathrm{d}\tau)^{\varepsilon}=0. \tag{2.92}$$

Therefore, we obtain

$$\begin{aligned}\frac{1}{\pi^{\varepsilon}}\int_{-\pi}^{\pi}S_{n,\varepsilon}(\mu+\tau)D_{n,\varepsilon}(\tau)(\mathrm{d}\tau)^{\varepsilon}&=\sum_{k=0}^{n}\mathrm{A}(k,\varepsilon)\cos_{\varepsilon}(k\mu)^{\varepsilon}+\sum_{k=0}^{n}\mathrm{B}(k,\varepsilon)\sin_{\varepsilon}(k\mu)^{\varepsilon}\\&=\sum_{k=0}^{n}\left(\mathrm{A}(k,\varepsilon)\cos_{\varepsilon}(k\mu)^{\varepsilon}+\mathrm{B}(k,\varepsilon)\sin_{\varepsilon}(k\mu)^{\varepsilon}\right).\end{aligned} \tag{2.93}$$

Clearly, this theorem is proved. □

Theorem 2.5. *Suppose that $\psi(\tau)$ is 2π-periodic, bounded, and locally fractional integrable on $[-\pi,\pi]$. Then,*

$$\frac{1}{\pi^{\varepsilon}}\int_{-\pi}^{\pi}\psi^{2}(\tau)(d\tau)^{\varepsilon}=\frac{A^{2}(0,\varepsilon)}{2}+\sum_{k=1}^{\infty}\left(A^{2}(k,\varepsilon)+B^{2}(k,\varepsilon)\right), \tag{2.94}$$

provided that

$$\psi(\tau)\sim\frac{A(0,\varepsilon)}{2}+\sum_{k=1}^{n}\left(A(k,\varepsilon)\cos_{\varepsilon}(k\tau)^{\varepsilon}+B(k,\varepsilon)\sin_{\varepsilon}(k\tau)^{\varepsilon}\right), \tag{2.95}$$

where

$$A(0,\varepsilon)=\frac{1}{\pi^{\varepsilon}}\int_{-\pi}^{\pi}\phi(\tau)(d\tau)^{\varepsilon}, \tag{2.96}$$

$$A(k,\varepsilon)=\frac{1}{\pi^{\varepsilon}}\int_{-\pi}^{\pi}\phi(\tau)\cos_{\varepsilon}(k\tau)^{\varepsilon}(d\tau)^{\varepsilon}, \tag{2.97}$$

and

$$B(k,\varepsilon)=\frac{1}{\pi^{\varepsilon}}\int_{-\pi}^{\pi}\phi(\tau)\sin_{\varepsilon}(k\tau)^{\varepsilon}(d\tau)^{\varepsilon}. \tag{2.98}$$

Proof. Following (2.94), we consider following integral:

$$\begin{aligned}&\frac{1}{\pi^{\varepsilon}}\int_{-\pi}^{\pi}\psi^{2}(\tau)(\mathrm{d}\tau)^{\varepsilon}\\&=\frac{1}{\pi^{\varepsilon}}\int_{-\pi}^{\pi}\left\{\psi(\tau)\left(\frac{\mathrm{A}(0,\varepsilon)}{2}+\sum_{k=1}^{n}\left(\mathrm{A}(k,\varepsilon)\cos_{\varepsilon}(k\tau)^{\varepsilon}+\mathrm{B}(k,\varepsilon)\sin_{\varepsilon}(k\tau)^{\varepsilon}\right)\right)\right\}(\mathrm{d}\tau)^{\varepsilon}\end{aligned}$$

$$= \frac{1}{\pi^{\varepsilon}} \int_{-\pi}^{\pi} \left[\psi(\tau) \frac{\mathrm{A}(0,\varepsilon)}{2} \right] (\mathrm{d}\tau)^{\varepsilon}$$
$$+ \frac{1}{\pi^{\varepsilon}} \int_{-\pi}^{\pi} \left\{ \psi(\tau) \left[\sum_{k=1}^{n} \left(\mathrm{A}(k,\varepsilon) \cos_{\varepsilon}(k\tau)^{\varepsilon} + \mathrm{B}(k,\varepsilon) \sin_{\varepsilon}(k\tau)^{\varepsilon} \right) \right] \right\} (\mathrm{d}\tau)^{\varepsilon}, \quad (2.99)$$

as well as the following relations:

$$\frac{1}{\pi^{\varepsilon}} \int_{-\pi}^{\pi} [\psi(\tau) \mathrm{A}(0,\varepsilon)] (\mathrm{d}\tau)^{\varepsilon}$$
$$= \frac{1}{\pi^{c}} \int_{-\pi}^{\pi} \left\{ \left[\frac{\mathrm{A}(0,\varepsilon)}{2} + \sum_{k=1}^{n} \left(\mathrm{A}(k,\varepsilon) \cos_{\varepsilon}(k\tau)^{\varepsilon} + \mathrm{B}(k,\varepsilon) \sin_{\varepsilon}(k\tau)^{\varepsilon} \right) \right] \frac{\mathrm{A}(0,\varepsilon)}{2} \right\} (\mathrm{d}\tau)^{\varepsilon}$$
$$= \frac{\mathrm{A}^{2}(0,\varepsilon)}{2} \quad (2.100)$$

and

$$\sum_{k=1}^{\infty} \left(\mathrm{A}^{2}(k,\varepsilon) + \mathrm{B}^{2}(k,\varepsilon) \right) = \frac{1}{\pi^{\varepsilon}} \int_{-\pi}^{\pi} \left\{ \psi(\tau) \left[\sum_{k=1}^{n} \left(\mathrm{A}(k,\varepsilon) \cos_{\varepsilon}(k\tau)^{\varepsilon} \right. \right. \right.$$
$$\left. \left. \left. + \mathrm{B}(k,\varepsilon) \sin_{\varepsilon}(k\tau)^{\varepsilon} \right) \right] \right\} (\mathrm{d}\tau)^{\varepsilon}. \quad (2.101)$$

Making use of (2.99), we obtain the result asserted by this theorem. Therefore, this theorem is proved. □

Theorem 2.6 (Convergence theorem for local fractional Fourier series)**.** *Suppose that $\psi(\tau)$ is 2π-periodic, bounded, and locally fractional integrable on $[-\pi,\pi]$. Then, the local fractional Fourier series of $\psi(\tau)$ converges to $\psi(\tau)$ at $\tau \in [-\pi,\pi]$ and*

$$\frac{\psi(\tau+0)+\psi(\tau-0)}{2} = \frac{A(0,\varepsilon)}{2} + \sum_{k=1}^{\infty} \left(A(k,\varepsilon) \cos_{\varepsilon}(k\tau)^{\varepsilon} + B(k,\varepsilon) \sin_{\varepsilon}(k\tau)^{\varepsilon} \right), \quad (2.102)$$

where

$$A(0,\varepsilon) = \frac{1}{\pi^{\varepsilon}} \int_{-\pi}^{\pi} \phi(\tau) (d\tau)^{\varepsilon}, \quad (2.103)$$

$$A(k,\varepsilon) = \frac{1}{\pi^{\varepsilon}} \int_{-\pi}^{\pi} \phi(\tau) \cos_{\varepsilon}(k\tau)^{\varepsilon} (d\tau)^{\varepsilon}, \quad (2.104)$$

and

$$B(k,\varepsilon) = \frac{1}{\pi^{\varepsilon}} \int_{-\pi}^{\pi} \phi(\tau) \sin_{c}(k\tau)^{\varepsilon} (d\tau)^{\varepsilon}. \quad (2.105)$$

Proof. Let us define the sum of the local fractional Fourier series in the form

$$S_{n,\varepsilon}(\tau) = \frac{\mathrm{A}(0,\varepsilon)}{2} + \sum_{k=1}^{n}\left(\mathrm{A}(k,\varepsilon)\cos_{\varepsilon}(k\tau)^{\varepsilon} + \mathrm{B}(k,\varepsilon)\sin_{\varepsilon}(k\tau)^{\varepsilon}\right). \tag{2.106}$$

In this case, we transform (2.95) into the following equation:

$$\lim_{n\to\infty}\left\{\left(\frac{\psi(\tau+0)+\psi(\tau-0)}{2}\right) - S_{n,\varepsilon}(\tau)\right\} = 0. \tag{2.107}$$

Adopting the formula (2.85), we find that

$$S_{n,\varepsilon}(\tau) = \frac{1}{\pi^{\varepsilon}}\int_{-\pi}^{\pi}\psi(\tau+t)\,D_{n,\varepsilon}(t)\,(\mathrm{d}t)^{\varepsilon}, \tag{2.108}$$

where

$$D_{n,\varepsilon}(t) = \frac{1}{2} + \sum_{k=1}^{n}\cos_{\varepsilon}(kt)^{\varepsilon} = \frac{\sin_{\varepsilon}\left[\left(\left(k+\frac{1}{2}\right)t\right)^{\varepsilon}\right]}{2\sin_{\varepsilon}\left[\left(\frac{t}{2}\right)^{\varepsilon}\right]}. \tag{2.109}$$

In this case, starting from (2.109), we are led to the following formulas:

$$\frac{1}{\pi^{\varepsilon}}\int_{0}^{\pi}\psi(\tau+0)\,D_{n,\varepsilon}(t)\,(\mathrm{d}t)^{\varepsilon} = \psi(\tau+0) \tag{2.110}$$

and

$$\frac{1}{\pi^{\varepsilon}}\int_{-\pi}^{0}\psi(\tau-0)\,D_{n,\varepsilon}(t)\,(\mathrm{d}t)^{\varepsilon} = \psi(\tau-0)\,. \tag{2.111}$$

From (2.109) to (2.111), we expand (2.107) as follows:

$$\begin{aligned}
&\left(\frac{\psi(\tau+0)+\psi(\tau-0)}{2}\right) - S_{n,\varepsilon}(\tau)\\
&= \left(\frac{\psi(\tau+0)+\psi(\tau-0)}{2}\right) - \frac{1}{\pi^{\varepsilon}}\int_{-\pi}^{\pi}\psi(\tau+t)\,D_{n,\varepsilon}(t)\,(\mathrm{d}t)^{\varepsilon}\\
&= \left[\frac{\psi(\tau+0)}{2} - \frac{1}{\pi^{\varepsilon}}\int_{0}^{\pi}\psi(\tau+t)\,D_{n,\varepsilon}(t)\,(\mathrm{d}t)^{\varepsilon}\right]\\
&\quad + \left[\frac{\psi(\tau-0)}{2} - \frac{1}{\pi^{\varepsilon}}\int_{-\pi}^{0}\psi(\tau-t)\,D_{n,\varepsilon}(t)\,(\mathrm{d}t)^{\varepsilon}\right]\\
&= \frac{1}{\pi^{\varepsilon}}\int_{0}^{\pi}\left(\psi(\tau+0)-\psi(\tau+t)\right)D_{n,\varepsilon}(t)\,(\mathrm{d}t)^{\varepsilon}\\
&\quad + \frac{1}{\pi^{\varepsilon}}\int_{-\pi}^{0}\left(\psi(\tau-0)-\psi(\tau-t)\right)D_{n,\varepsilon}(t)\,(\mathrm{d}t)^{\varepsilon}\\
&= \frac{1}{\pi^{\varepsilon}}\int_{0}^{\pi}\left(\psi(\tau+0)-\psi(\tau+t)\right)\frac{\sin_{\varepsilon}\left[\left(\left(k+\frac{1}{2}\right)t\right)^{\varepsilon}\right]}{2\sin_{\varepsilon}\left[\left(\frac{t}{2}\right)^{\varepsilon}\right]}(\mathrm{d}t)^{\varepsilon}
\end{aligned}$$

$$+\frac{1}{\pi^{\varepsilon}}\int_{-\pi}^{0}\left(\psi\left(\tau-0\right)-\psi\left(\tau-t\right)\right)\frac{\sin_{\varepsilon}\left[\left(\left(k+\frac{1}{2}\right)t\right)^{\varepsilon}\right]}{2\sin_{\varepsilon}\left[\left(\frac{t}{2}\right)^{\varepsilon}\right]}(\mathrm{d}t)^{\varepsilon}$$

$$=\frac{1}{\pi^{\varepsilon}}\int_{0}^{\pi}\left[\frac{\psi\left(\tau+0\right)-\psi\left(\tau+t\right)}{\frac{t^{\varepsilon}}{\Gamma(1+\varepsilon)}}\right]\left\{\frac{\frac{t^{\varepsilon}}{\Gamma(1+\varepsilon)}}{2\sin_{\varepsilon}\left[\left(\frac{t}{2}\right)^{\varepsilon}\right]}\right\}\sin_{\varepsilon}\left[\left(\left(k+\frac{1}{2}\right)t\right)^{\varepsilon}\right](\mathrm{d}t)^{\varepsilon}$$

$$+\frac{1}{\pi^{\varepsilon}}\int_{-\pi}^{0}\left[\frac{\psi\left(\tau-0\right)-\psi\left(\tau-t\right)}{\frac{t^{\varepsilon}}{\Gamma(1+\varepsilon)}}\right]\left\{\frac{\frac{t^{\varepsilon}}{\Gamma(1+\varepsilon)}}{2\sin_{\varepsilon}\left[\left(\frac{t}{2}\right)^{\varepsilon}\right]}\right\}\sin_{\varepsilon}\left[\left(\left(k+\frac{1}{2}\right)t\right)^{\varepsilon}\right](\mathrm{d}t)^{\varepsilon}. \tag{2.112}$$

For $\tau\in[0,1]$, we simulate the following formula:

$$\lim_{t\to 0}\left[\frac{\psi\left(\tau+0\right)-\psi\left(\tau+t\right)}{\frac{t^{\varepsilon}}{\Gamma(1+\varepsilon)}}\right]\left\{\frac{\frac{t^{\varepsilon}}{\Gamma(1+\varepsilon)}}{2\sin_{\varepsilon}\left[\left(\frac{t}{2}\right)^{\varepsilon}\right]}\right\}=2^{\varepsilon-1}\psi^{(\varepsilon)}\left(\tau+0\right), \tag{2.113}$$

where

$$\lim_{t\to 0}\left[\frac{\psi\left(\tau+0\right)-\psi\left(\tau+t\right)}{\frac{t^{\varepsilon}}{\Gamma(1+\varepsilon)}}\right]=\psi^{(\varepsilon)}\left(\tau+0\right) \tag{2.114}$$

and

$$\lim_{t\to 0}\frac{\frac{t^{\varepsilon}}{\Gamma(1+\varepsilon)}}{2\sin_{\varepsilon}\left[\left(\frac{t}{2}\right)^{\varepsilon}\right]}=\frac{\frac{\partial^{\varepsilon}}{\partial t^{\varepsilon}}\left[\frac{t^{\varepsilon}}{\Gamma(1+\varepsilon)}\right]}{\frac{\partial^{\varepsilon}}{\partial t^{\varepsilon}}\left\{2\sin_{\varepsilon}\left[\left(\frac{t}{2}\right)^{\varepsilon}\right]\right\}}=\lim_{t\to 0}\frac{1}{2^{1-\varepsilon}\cos_{\varepsilon}\left[\left(\frac{t}{2}\right)^{\varepsilon}\right]}=2^{\varepsilon-1}. \tag{2.115}$$

In a similar way, we have

$$\lim_{t\to 0}\left[\frac{\psi\left(\tau-0\right)-\psi\left(\tau-t\right)}{\frac{t^{\varepsilon}}{\Gamma(1+\varepsilon)}}\right]\left\{\frac{\frac{t^{\varepsilon}}{\Gamma(1+\varepsilon)}}{2\sin_{\varepsilon}\left[\left(\frac{t}{2}\right)^{\varepsilon}\right]}\right\}=2^{\varepsilon-1}\psi^{(\varepsilon)}\left(\tau-0\right), \tag{2.116}$$

where

$$\lim_{t\to 0}\left[\frac{\psi\left(\tau-0\right)-\psi\left(\tau-t\right)}{\frac{t^{\varepsilon}}{\Gamma(1+\varepsilon)}}\right]=\psi^{(\varepsilon)}\left(\tau-0\right) \tag{2.117}$$

and

$$\lim_{t\to 0}\frac{\frac{t^{\varepsilon}}{\Gamma(1+\varepsilon)}}{2\sin_{\varepsilon}\left[\left(\frac{t}{2}\right)^{\varepsilon}\right]}=\frac{\frac{\partial^{\varepsilon}}{\partial t^{\varepsilon}}\left[\frac{t^{\varepsilon}}{\Gamma(1+\varepsilon)}\right]}{\frac{\partial^{\varepsilon}}{\partial t^{\varepsilon}}\left\{2\sin_{\varepsilon}\left[\left(\frac{t}{2}\right)^{\varepsilon}\right]\right\}}=\lim_{t\to 0}\frac{1}{2^{1-\varepsilon}\cos_{\varepsilon}\left[\left(\frac{t}{2}\right)^{\varepsilon}\right]}=2^{\varepsilon-1}. \tag{2.118}$$

Making use of the formulas (2.115) and (2.118), we obtain

$$\begin{aligned}
&\lim_{n\to\infty}\left[\left(\frac{\psi(\tau+0)+\psi(\tau-0)}{2}\right)-S_{n,\varepsilon}(\tau)\right]\\
&=\frac{1}{\pi^{\varepsilon}}\int_{0}^{\pi}2^{\varepsilon-1}\psi^{(\varepsilon)}(\tau+0)\sin_{\varepsilon}\left[\left(\left(k+\frac{1}{2}\right)t\right)^{\varepsilon}\right](dt)^{\varepsilon}\\
&+\frac{1}{\pi^{\varepsilon}}\int_{-\pi}^{0}2^{\varepsilon-1}\psi^{(\varepsilon)}(\tau-0)\sin_{\varepsilon}\left[\left(\left(k+\frac{1}{2}\right)t\right)^{\varepsilon}\right](dt)^{\varepsilon}.
\end{aligned} \tag{2.119}$$

In this case, using the Riemann–Lebesgue theorem for local fractional Fourier series, we have

$$\frac{1}{\pi^{\varepsilon}}\int_{0}^{\pi}\left\{\left[\frac{\psi(\tau+0)-\psi(\tau+t)}{\frac{t^{\varepsilon}}{\Gamma(1+\varepsilon)}}\right]\frac{\frac{t^{\varepsilon}}{\Gamma(1+\varepsilon)}}{2\sin_{\varepsilon}\left[\left(\frac{t}{2}\right)^{\varepsilon}\right]}\right\}\sin_{\varepsilon}\left[\left(\left(k+\frac{1}{2}\right)t\right)^{\varepsilon}\right](dt)^{\varepsilon}=0 \tag{2.120}$$

and

$$\frac{1}{\pi^{\varepsilon}}\int_{0}^{\pi}\left\{\left[\frac{\psi(\tau-0)-\psi(\tau-t)}{\frac{t^{\varepsilon}}{\Gamma(1+\varepsilon)}}\right]\frac{\frac{t^{\varepsilon}}{\Gamma(1+\varepsilon)}}{2\sin_{\varepsilon}\left[\left(\frac{t}{2}\right)^{\varepsilon}\right]}\right\}\sin_{\varepsilon}\left[\left(\left(k+\frac{1}{2}\right)t\right)^{\varepsilon}\right](dt)^{\varepsilon}=0. \tag{2.121}$$

Considering the above case, we obtain

$$\lim_{n\to\infty}\left[\left(\frac{\psi(\tau+0)+\psi(\tau-0)}{2}\right)-S_{n,\varepsilon}(\tau)\right]=0. \tag{2.122}$$

Therefore, the proof of this theorem is completed. □

As a direct application, we present the following result:

Theorem 2.7. *Suppose that $\psi(\tau)$ is 2π-periodic, locally fractional continuous, and locally fractional integrable on $[-\pi,\pi]$. Then,*

$$\psi(\tau)=\frac{A(0,\varepsilon)}{2}+\sum_{k=1}^{\infty}\left(A(k,\varepsilon)\cos_{\varepsilon}(k\tau)^{\varepsilon}+B(k,\varepsilon)\sin_{\varepsilon}(k\tau)^{\varepsilon}\right), \tag{2.123}$$

where

$$A(0,\varepsilon)=\frac{1}{\pi^{\varepsilon}}\int_{-\pi}^{\pi}\phi(\tau)(d\tau)^{\varepsilon}, \tag{2.124}$$

$$A(k,\varepsilon)=\frac{1}{\pi^{\varepsilon}}\int_{-\pi}^{\pi}\phi(\tau)\cos_{\varepsilon}(k\tau)^{\varepsilon}(d\tau)^{\varepsilon}, \tag{2.125}$$

and

$$B(k,\varepsilon) = \frac{1}{\pi^{\varepsilon}} \int_{-\pi}^{\pi} \phi(\tau) \sin_{\varepsilon}(k\tau)^{\varepsilon} (d\tau)^{\varepsilon}. \tag{2.126}$$

Proof. Since $\psi(\tau)$ is locally fractional continuous on $[-\pi, \pi]$, we have

$$\frac{\psi(\tau+0) + \psi(\tau-0)}{2} = \psi(\tau). \tag{2.127}$$

Now, by using (2.102) and (2.127), we obtain the result asserted by the theorem. Hence, we have completed the proof of this theorem. □

Theorem 2.8. *Suppose that $\psi(\tau)$ is 2π-periodic, bounded, and locally fractional integrable on $[-\pi, \pi]$. Then,*

$$\psi(\tau) = \frac{A(0,\varepsilon)}{2} + \sum_{k=1}^{\infty} A(k,\varepsilon) \cos_{\varepsilon}(k\tau)^{\varepsilon}, \tag{2.128}$$

where

$$A(0,\varepsilon) = \frac{1}{\pi^{\varepsilon}} \int_{-\pi}^{\pi} \phi(\tau) (d\tau)^{\varepsilon} \tag{2.129}$$

and

$$A(k,\varepsilon) = \frac{1}{\pi^{\varepsilon}} \int_{-\pi}^{\pi} \phi(\tau) \cos_{\varepsilon}(k\tau)^{\varepsilon} (d\tau)^{\varepsilon}, \tag{2.130}$$

provided that

$$\psi(\tau) = \psi(-\tau). \tag{2.131}$$

Proof. Since $\psi(\tau) = \psi(-\tau)$, we obtain

$$\sum_{k=1}^{\infty} \mathrm{B}(k,\varepsilon) \sin_{\varepsilon}(k\tau)^{\varepsilon} = 0. \tag{2.132}$$

Therefore, we obtain the asserted result, and it completes the proof of this theorem. □

Theorem 2.9. *Suppose that $\psi(\tau)$ is 2π-periodic, bounded, and locally fractional integrable on $[-\pi, \pi]$. Then,*

$$\psi(\tau) = \sum_{k=1}^{\infty} B(k,\varepsilon) \sin_{\varepsilon}(k\tau)^{\varepsilon}, \tag{2.133}$$

where

$$B(k,\varepsilon) = \frac{1}{\pi^{\varepsilon}} \int_{-\pi}^{\pi} \phi(\tau) \sin_{\varepsilon}(k\tau)^{\varepsilon} (d\tau)^{\varepsilon}, \tag{2.134}$$

provided that

$$\psi(\tau) = -\psi(\tau). \tag{2.135}$$

Proof. Since $\psi(\tau) = -\psi(\tau)$, we obtain

$$\frac{\mathrm{A}(0,\varepsilon)}{2} + \sum_{k=1}^{\infty} \mathrm{A}(k,\varepsilon)\cos_{\varepsilon}(k\tau)^{\varepsilon} = 0. \tag{2.136}$$

Therefore, we obtain the asserted result, and the proof of this theorem is completed. □

Theorem 2.10. *Suppose that $\psi(\tau)$ is 2π-periodic, locally fractional continuous, and locally fractional integrable on $[-L, L]$. Then,*

$$\psi(\tau) = \frac{A(0,\varepsilon)}{2} + \sum_{k=1}^{\infty}\left(A(k,\varepsilon)\cos_{\varepsilon}\left(\frac{\pi k\tau}{L}\right)^{\varepsilon} + B(k,\varepsilon)\sin_{\varepsilon}\left(\frac{\pi k\tau}{L}\right)^{\varepsilon}\right), \tag{2.137}$$

where

$$A(0,\varepsilon) = \frac{1}{L^{\varepsilon}}\int_{-L}^{L}\phi(\tau)(d\tau)^{\varepsilon}, \tag{2.138}$$

$$A(k,\varepsilon) = \frac{1}{L^{\varepsilon}}\int_{-L}^{L}\psi(\tau)\cos_{\varepsilon}\left(\frac{\pi k\tau}{L}\right)^{\varepsilon}(d\tau)^{\varepsilon}, \tag{2.139}$$

and

$$B(k,\varepsilon) = \frac{1}{L^{\varepsilon}}\int_{-L}^{L}\psi(\tau)\sin_{\varepsilon}\left(\frac{\pi k\tau}{L}\right)^{\varepsilon}(d\tau)^{\varepsilon}. \tag{2.140}$$

Proof. Suppose that $\Theta(\eta)$ is 2π-periodic and locally fractional continuous on the interval $[-\pi, \pi]$. Then, by defining the variable η by

$$\eta = \left(\frac{\pi}{L}\right)\tau, \tag{2.141}$$

we get

$$\Theta(\eta) = \Theta\left(\left(\frac{\pi}{L}\right)\tau\right) = \psi(\tau). \tag{2.142}$$

Using (2.102), we obtain the asserted result. Therefore, the proof of this theorem is completed. □

Theorem 2.11. *Suppose that $\psi(\tau)$ is 2π-periodic and locally fractional continuous on the interval $[-\pi, \pi]$. Then,*

$$\psi(\tau) = \frac{A(0,\varepsilon)}{2} + \sum_{k=1}^{\infty} A(k,\varepsilon)\cos_{\varepsilon}\left(\frac{\pi k \tau}{L}\right)^{\varepsilon}, \tag{2.143}$$

where

$$A(0,\varepsilon) = \frac{1}{L^{\varepsilon}} \int_{-L}^{L} \phi(\tau)(d\tau)^{\varepsilon} \tag{2.144}$$

and

$$A(k,\varepsilon) = \frac{1}{L^{\varepsilon}} \int_{-L}^{L} \psi(\tau)\cos_{\varepsilon}\left(\frac{\pi k \tau}{L}\right)^{\varepsilon}(d\tau)^{\varepsilon}, \tag{2.145}$$

provided that

$$\psi(\tau) = \psi(-\tau). \tag{2.146}$$

Proof. Since $\psi(\tau) = \psi(-\tau)$, we obtain

$$\sum_{k=1}^{\infty} \mathrm{B}(k,\varepsilon)\sin_{\varepsilon}\left(\frac{\pi k \tau}{L}\right)^{\varepsilon} = 0. \tag{2.147}$$

Therefore, we obtain the asserted result and complete the proof of this theorem. □

Theorem 2.12. *Suppose that $\psi(\tau)$ is 2π-periodic, bounded, and locally fractional integrable on $[-\pi, \pi]$. Then,*

$$\psi(\tau) = \sum_{k=1}^{\infty} B(k,\varepsilon)\sin_{\varepsilon}\left(\frac{\pi k \tau}{L}\right)^{\varepsilon}, \tag{2.148}$$

where

$$B(k,\varepsilon) = \frac{1}{L^{\varepsilon}} \int_{-L}^{L} \psi(\tau)\sin_{\varepsilon}\left(\frac{\pi k \tau}{L}\right)^{\varepsilon}(d\tau)^{\varepsilon}, \tag{2.149}$$

provided that

$$\psi(\tau) = -\psi(\tau). \tag{2.150}$$

Proof. Since $\psi(\tau) = -\psi(\tau)$, we obtain

$$\frac{\mathrm{A}(0,\varepsilon)}{2} + \sum_{k=1}^{\infty} \mathrm{A}(k,\varepsilon)\cos_{\varepsilon}\left(\frac{\pi k \tau}{L}\right)^{\varepsilon} = 0. \tag{2.151}$$

Therefore, we obtain the asserted result and complete the proof of this theorem. □

2.3 Applications to signal analysis

In this section, we consider some applications of local fractional Fourier series in signal analysis defined on Cantor sets. The aim of this part is to investigate the fractal signal processes with help of the local fractional Fourier series. The technique is a powerful tool to process fractal signals to applied scientists and engineers. We will give some examples (see also [94–98]).

We consider the $2L$-periodic fractal signal given by

$$\psi(\tau) = C \tag{2.152}$$

on the interval $0 \le \tau \le L$ in local fractional Fourier series, where $\varepsilon = \ln 2/\ln 3$ and C is a constant.

The local fractional Fourier coefficients take the following form:

$$\begin{aligned} \mathrm{A}(0,\varepsilon) &= \frac{1}{L^{\varepsilon}} \int_{-L}^{L} \psi(\tau)(\mathrm{d}\tau)^{\varepsilon} \\ &= \frac{1}{L^{\varepsilon}} \int_{-L}^{L} C(\mathrm{d}\tau)^{\varepsilon} \\ &= \frac{1}{L^{\varepsilon}} \int_{0}^{L} C(\mathrm{d}\tau)^{\varepsilon} \\ &= C, \end{aligned} \tag{2.153}$$

$$\begin{aligned} \mathrm{A}(k,\varepsilon) &= \frac{1}{L^{\varepsilon}} \int_{-L}^{L} \psi(\tau)\cos_{\varepsilon}\left(\frac{\pi k \tau}{L}\right)^{\varepsilon} (\mathrm{d}\tau)^{\varepsilon} \\ &= \frac{1}{L^{\varepsilon}} \int_{0}^{L} C\cos_{\varepsilon}\left(\frac{\pi k \tau}{L}\right)^{\varepsilon} (\mathrm{d}\tau)^{\varepsilon} \\ &= \frac{\Gamma(1+\varepsilon)\sin_{\varepsilon}\left(\frac{\pi k L}{L}\right)^{\varepsilon}}{(\pi k)^{\varepsilon}} \\ &= 0, \end{aligned} \tag{2.154}$$

and

$$\begin{aligned} \mathrm{B}(k,\varepsilon) &= \frac{1}{L^{\varepsilon}} \int_{-L}^{L} \psi(\tau)\sin_{\varepsilon}\left(\frac{\pi k \tau}{L}\right)^{\varepsilon} (\mathrm{d}\tau)^{\varepsilon} \\ &= \frac{1}{L^{\varepsilon}} \int_{0}^{L} C\sin_{\varepsilon}\left(\frac{\pi k \tau}{L}\right)^{\varepsilon} (\mathrm{d}\tau)^{\varepsilon} \\ &= \frac{2C\Gamma(1+\varepsilon)\left(1-(-1)^{k}\right)}{(\pi k)^{\varepsilon}}. \end{aligned} \tag{2.155}$$

For $0 \le \tau \le L$, the fractal signal $\psi(\tau)$ is represented as follows:

$$\psi(\tau) = \frac{C}{2} + \sum_{k=1}^{\infty} \left(\frac{2C\Gamma(1+\varepsilon)\left(1-(-1)^{k}\right)}{(\pi k)^{\varepsilon}} \right) \sin_{\varepsilon}\left(\frac{\pi k \tau}{L}\right)^{\varepsilon}. \tag{2.156}$$

When $k=0$, the fractal signal $\psi(\tau)$ is expanded as follows:

$$\psi(\tau)=\frac{C}{2}. \tag{2.157}$$

When $k=1$, the fractal signal $\psi(\tau)$ is expanded as follows:

$$\psi(\tau)=\frac{C}{2}+\frac{4C\Gamma(1+\varepsilon)}{\pi^{\varepsilon}}\sin_{\varepsilon}\left(\frac{\pi\tau}{L}\right)^{\varepsilon}. \tag{2.158}$$

When $k=2$, we have

$$\frac{2C\Gamma(1+\varepsilon)\left(1-(-1)^{2}\right)}{(2\pi)^{\varepsilon}}\sin_{\varepsilon}\left(\frac{2\pi\tau}{L}\right)^{\varepsilon}=0. \tag{2.159}$$

When $k=3$, we expand the fractal signal $\psi(\tau)$ as given below:

$$\psi(\tau)=\frac{C}{2}+4C\Gamma(1+\varepsilon)\left[\frac{1}{\pi^{\varepsilon}}\sin_{\varepsilon}\left(\frac{\pi\tau}{L}\right)^{\varepsilon}+\frac{1}{(3\pi)^{\varepsilon}}\sin_{\varepsilon}\left(\frac{3\pi\tau}{L}\right)^{\varepsilon}\right]. \tag{2.160}$$

When $k=4$, we have

$$\frac{2C\Gamma(1+\varepsilon)\left(1-(-1)^{4}\right)}{(\pi k)^{\varepsilon}}\sin_{\varepsilon}\left(\frac{4\pi\tau}{L}\right)^{\varepsilon}=0. \tag{2.161}$$

When $k=5$, the fractal signal $\psi(\tau)$ in local fractional Fourier series can be written as follows:

$$\begin{aligned}\psi(\tau)=\frac{C}{2}+4C\Gamma(1+\varepsilon)&\left[\frac{1}{\pi^{\varepsilon}}\sin_{\varepsilon}\left(\frac{\pi\tau}{L}\right)^{\varepsilon}+\frac{1}{(3\pi)^{\varepsilon}}\sin_{\varepsilon}\left(\frac{3\pi\tau}{L}\right)^{\varepsilon}\right.\\&\left.+\frac{1}{(5\pi)^{\varepsilon}}\sin_{\varepsilon}\left(\frac{5\pi\tau}{L}\right)^{\varepsilon}\right].\end{aligned} \tag{2.162}$$

The local fractional Fourier series of the fractal signal $\psi(\tau)$ when $k=0$, $k=1$, $k=3$, and $k=5$ is shown in Figure 2.1.

In this way, we can expand the fractal signal $\psi(\tau)$ into the local fractional Fourier series representation.

Expand the $2L$-periodic fractal signal

$$\psi(\tau)=\frac{\tau^{\varepsilon}}{\Gamma(1+\varepsilon)} \tag{2.163}$$

on the interval $-L<\tau\leq L$ in local fractional Fourier series. In this case, we present the local fractional Fourier coefficients as follows:

$$\begin{aligned}\mathrm{A}(0,\varepsilon)&=\frac{1}{L^{\varepsilon}}\int_{-L}^{L}\psi(\tau)(d\tau)^{\varepsilon}\\&=\frac{1}{L^{\varepsilon}}\int_{-L}^{L}\frac{\tau^{\varepsilon}}{\Gamma(1+\varepsilon)}(d\tau)^{\varepsilon}\\&=0,\end{aligned} \tag{2.164}$$

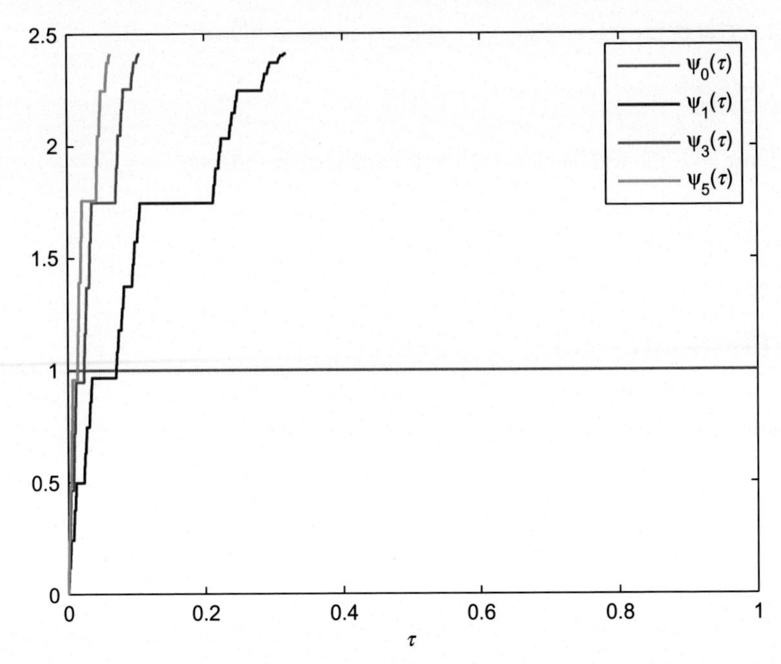

Figure 2.1 The local fractional Fourier series representation of fractal signal $\psi(\tau)$ when $\varepsilon = \ln 2/\ln 3$, $k = 0$, $k = 1$, $k = 3$, and $k = 5$.

$$\begin{aligned}
\mathrm{A}(k,\varepsilon) &= \frac{1}{L^{\varepsilon}} \int_{-L}^{L} \psi(\tau) \cos_{\varepsilon}\left(\frac{\pi k \tau}{L}\right)^{\varepsilon} (d\tau)^{\varepsilon} \\
&= \frac{1}{L^{\varepsilon}} \int_{-L}^{L} \frac{\tau^{\varepsilon}}{\Gamma(1+\varepsilon)} \cos_{\varepsilon}\left(\frac{\pi k \tau}{L}\right)^{\varepsilon} (d\tau)^{\varepsilon} \\
&= 0,
\end{aligned} \tag{2.165}$$

and

$$\begin{aligned}
\mathrm{B}(k,\varepsilon) &= \frac{1}{L^{\varepsilon}} \int_{-L}^{L} \psi(\tau) \sin_{\varepsilon}\left(\frac{\pi k \tau}{L}\right)^{\varepsilon} (d\tau)^{\varepsilon} \\
&= \frac{1}{L^{\varepsilon}} \int_{-L}^{L} \frac{\tau^{\varepsilon}}{\Gamma(1+\varepsilon)} \sin_{\varepsilon}\left(\frac{\pi k \tau}{L}\right)^{\varepsilon} (d\tau)^{\varepsilon} \\
&= \frac{2}{L^{\varepsilon}} \int_{0}^{L} \frac{\tau^{\varepsilon}}{\Gamma(1+\varepsilon)} \sin_{\varepsilon}\left(\frac{\pi k \tau}{L}\right)^{\varepsilon} (d\tau)^{\varepsilon} \\
&= \frac{2L^{\varepsilon}(-1)^{k+1}}{(k\pi)^{\varepsilon}}.
\end{aligned} \tag{2.166}$$

For $-L < \tau \leq L$, the local fractional Fourier series representation of fractal signal finally can be written as follows:

$$\psi(\tau) = \sum_{k=1}^{\infty} \left[\frac{2L^{\varepsilon}(-1)^{k+1}}{(k\pi)^{\varepsilon}} \sin_{\varepsilon}\left(\frac{\pi k \tau}{L}\right)^{\varepsilon} \right]. \tag{2.167}$$

When $k = 1$, the fractal signal $\psi(\tau)$ is expanded as follows:

$$\psi(\tau) = \frac{2L^{\varepsilon}}{\pi^{\varepsilon}} \sin_{\varepsilon}\left(\frac{\pi\tau}{L}\right)^{\varepsilon}. \tag{2.168}$$

When $k = 2$, we expand the fractal signal $\psi(\tau)$ as given below:

$$\psi(\tau) = \frac{2L^{\varepsilon}}{\pi^{\varepsilon}} \sin_{\varepsilon}\left(\frac{\pi\tau}{L}\right)^{\varepsilon} - \frac{2L^{\varepsilon}}{(2\pi)^{\varepsilon}} \sin_{\varepsilon}\left(\frac{2\pi\tau}{L}\right)^{\varepsilon}. \tag{2.169}$$

When $k = 3$, we expand the fractal signal $\psi(\tau)$ as follows:

$$\psi(\tau) = \frac{2L^{\varepsilon}}{\pi^{\varepsilon}} \sin_{\varepsilon}\left(\frac{\pi\tau}{L}\right)^{\varepsilon} - \frac{2L^{\varepsilon}}{(2\pi)^{\varepsilon}} \sin_{\varepsilon}\left(\frac{2\pi\tau}{L}\right)^{\varepsilon} + \frac{2L^{\varepsilon}}{(3\pi)^{\varepsilon}} \sin_{\varepsilon}\left(\frac{3\pi\tau}{L}\right)^{\varepsilon}. \tag{2.170}$$

When $k = 4$, we have

$$\begin{aligned}\psi(\tau) =& \frac{2L^{\varepsilon}}{\pi^{\varepsilon}} \sin_{\varepsilon}\left(\frac{\pi\tau}{L}\right)^{\varepsilon} - \frac{2L^{\varepsilon}}{(2\pi)^{\varepsilon}} \sin_{\varepsilon}\left(\frac{2\pi\tau}{L}\right)^{\varepsilon} \\ &+ \frac{2L^{\varepsilon}}{(3\pi)^{\varepsilon}} \sin_{\varepsilon}\left(\frac{3\pi\tau}{L}\right)^{\varepsilon} - \frac{2L^{\varepsilon}}{(4\pi)^{\varepsilon}} \sin_{\varepsilon}\left(\frac{4\pi\tau}{L}\right)^{\varepsilon}.\end{aligned} \tag{2.171}$$

When $k = 5$, the fractal signal $\psi(\tau)$ in local fractional Fourier series can be written as follows:

$$\begin{aligned}\psi(\tau) =& \frac{2L^{\varepsilon}}{\pi^{\varepsilon}} \sin_{\varepsilon}\left(\frac{\pi\tau}{L}\right)^{\varepsilon} - \frac{2L^{\varepsilon}}{(2\pi)^{\varepsilon}} \sin_{\varepsilon}\left(\frac{2\pi\tau}{L}\right)^{\varepsilon} \\ &+ \frac{2L^{\varepsilon}}{(3\pi)^{\varepsilon}} \sin_{\varepsilon}\left(\frac{3\pi\tau}{L}\right)^{\varepsilon} - \frac{2L^{\varepsilon}}{(4\pi)^{\varepsilon}} \sin_{\varepsilon}\left(\frac{4\pi\tau}{L}\right)^{\varepsilon} + \frac{2L^{\varepsilon}}{(5\pi)^{\varepsilon}} \sin_{\varepsilon}\left(\frac{5\pi\tau}{L}\right)^{\varepsilon}.\end{aligned} \tag{2.172}$$

The local fractional Fourier series of the fractal signal $\psi(\tau)$ when $k = 1$, $k = 2$, $k = 3$, $k = 4$, and $k = 5$ is shown in Figure 2.2.

We notice that the expression (2.148) is also applied to find the local fractional Fourier series representation of the fractal signal $\psi(\tau)$.

Expand the 2π-periodic fractal signal,

$$\psi(\tau) = \frac{\tau^{2\varepsilon}}{\Gamma(1+2\varepsilon)}, \tag{2.173}$$

on the interval $-\pi < \tau \leq \pi$ in local fractional Fourier series, and its graph is shown in Figure 2.3.

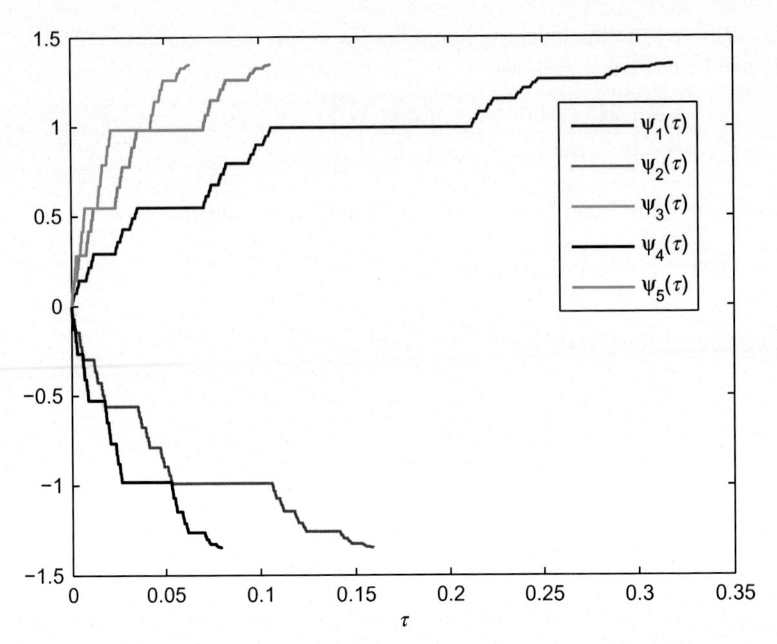

Figure 2.2 The local fractional Fourier series representation of fractal signal $\psi(\tau)$ when $\varepsilon = \ln 2/\ln 3$, $k = 1$, $k = 2$, $k = 3$, $k = 4$, and $k = 5$.

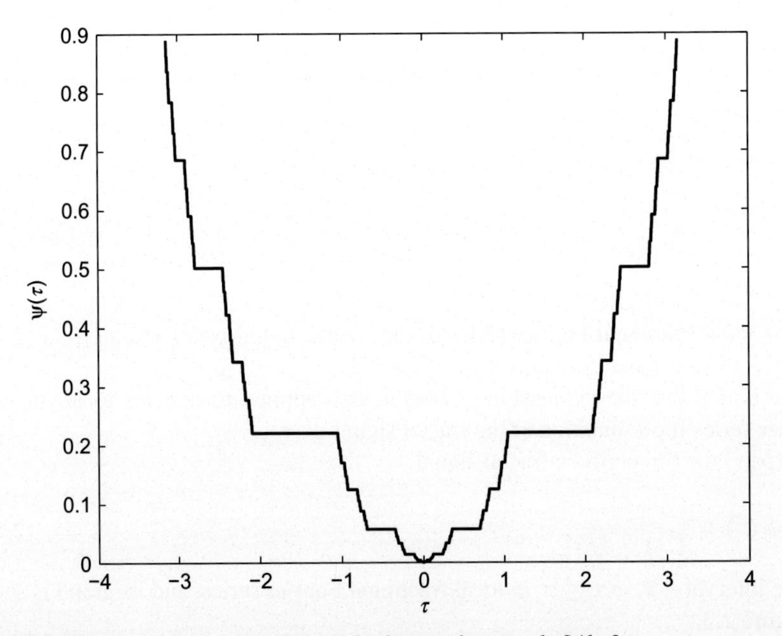

Figure 2.3 The plot of fractal signal $\psi(\tau)$ is shown when $\varepsilon = \ln 2/\ln 3$.

Making use of the expression in (2.128) to find the local fractional Fourier series representation of fractal signal $\psi(\tau)$, we find that

$$\begin{aligned} A(0,\varepsilon) &= \frac{1}{\pi^{\varepsilon}} \int_{-\pi}^{\pi} \psi(\tau)(d\tau)^{\varepsilon} \\ &= \frac{1}{\pi^{\varepsilon}} \int_{-\pi}^{\pi} \frac{\tau^{2\varepsilon}}{\Gamma(1+2\varepsilon)}(d\tau)^{\varepsilon} \\ &= \frac{2\pi^{2\varepsilon}\Gamma(1+\varepsilon)}{\Gamma(1+3\varepsilon)} \end{aligned} \tag{2.174}$$

and

$$\begin{aligned} A(k,\varepsilon) &= \frac{1}{\pi^{\varepsilon}} \int_{-\pi}^{\pi} \psi(\tau)\cos_{\varepsilon}(k\tau)^{\varepsilon}(d\tau)^{\varepsilon} \\ &= \frac{1}{\pi^{\varepsilon}} \int_{-\pi}^{\pi} \frac{\tau^{2\varepsilon}}{\Gamma(1+2\varepsilon)}\cos_{\varepsilon}(k\tau)^{\varepsilon}(d\tau)^{\varepsilon} \\ &= \frac{2}{\pi^{\varepsilon}} \int_{0}^{\pi} \frac{\tau^{2\varepsilon}}{\Gamma(1+2\varepsilon)}\cos_{\varepsilon}(k\tau)^{\varepsilon}(d\tau)^{\varepsilon} \\ &= -\frac{2}{k^{2\alpha}}. \end{aligned} \tag{2.175}$$

Therefore, the 2π-periodic fractal signal $\psi(\tau)$ on the interval $0 < \tau \leq 2\pi$ is expressed as follows:

$$\psi(\tau) = \frac{2\pi^{2\varepsilon}\Gamma(1+\varepsilon)}{\Gamma(1+3\varepsilon)} - \sum_{k=1}^{\infty}\left[\frac{2}{k^{2\varepsilon}}\cos_{\varepsilon}(k\tau)^{\varepsilon}\right]. \tag{2.176}$$

Let us consider the 2π-periodic fractal signal

$$\psi(\tau) = \frac{\tau^{3\varepsilon}}{\Gamma(1+3\varepsilon)} \tag{2.177}$$

on the interval $-\pi < \tau \leq \pi$ in local fractional Fourier series, and its graph is shown in Figure 2.4.

Adopting the expression in (2.133) to find the local fractional Fourier series representation of the fractal signal $\psi(\tau)$, we present the following local fractional Fourier series coefficient:

$$\begin{aligned} B(k,\varepsilon) &= \frac{1}{\pi^{\varepsilon}} \int_{-\pi}^{\pi} \psi(\tau)\sin_{\varepsilon}(k\tau)^{\varepsilon}(d\tau)^{\varepsilon} \\ &= \frac{1}{\pi^{\varepsilon}} \int_{-\pi}^{\pi} \left[\frac{\tau^{3\varepsilon}}{\Gamma(1+3\varepsilon)}\sin_{\varepsilon}(k\tau)^{\varepsilon}\right](d\tau)^{\varepsilon} \\ &= \frac{2}{\pi^{\varepsilon}} \int_{0}^{\pi} \left[\frac{\tau^{3\varepsilon}}{\Gamma(1+3\varepsilon)}\sin_{\varepsilon}(k\tau)^{\varepsilon}\right](d\tau)^{\varepsilon} \\ &= \frac{1}{k^{\varepsilon}}\frac{\pi^{3\varepsilon}}{\Gamma(1+\varepsilon\alpha)} - \frac{1}{k^{3\varepsilon}}\frac{\pi^{\varepsilon}}{\Gamma(1+\varepsilon)}. \end{aligned} \tag{2.178}$$

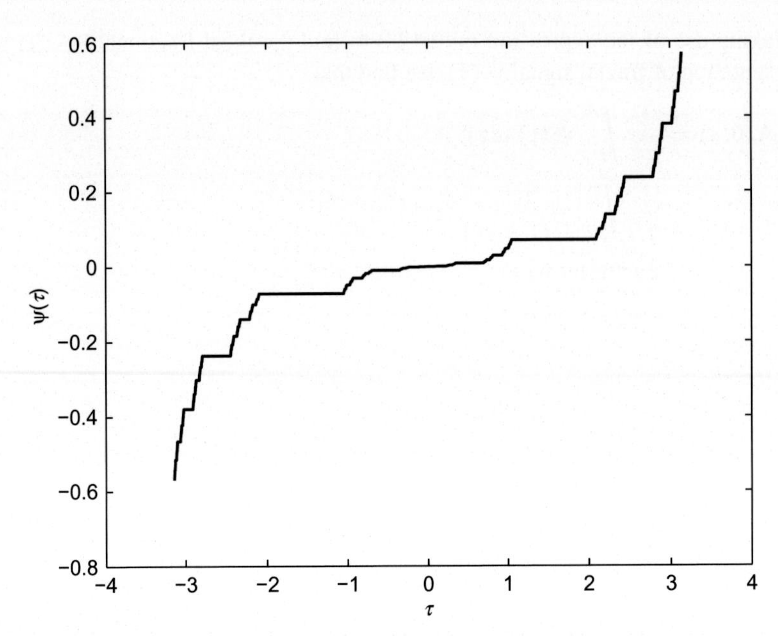

Figure 2.4 The plot of fractal signal $\psi(\tau)$ is shown when $\varepsilon = \ln 2/\ln 3$.

Therefore, the nondifferentiable signal is expressed as follows:

$$\psi(\tau) = \sum_{k=1}^{\infty} \left(\frac{1}{k^{\varepsilon}} \frac{\pi^{3\varepsilon}}{\Gamma(1+\varepsilon\alpha)} - \frac{1}{k^{3\varepsilon}} \frac{\pi^{\varepsilon}}{\Gamma(1+\varepsilon)} \right) \sin_{\varepsilon}(k\tau)^{\varepsilon}. \tag{2.179}$$

2.4 Solving local fractional differential equations

We now consider some applications of local fractional Fourier series to handle the local fractional differential equations (see [99–104]).

2.4.1 Applications of local fractional ordinary differential equations

We now consider the following local fractional ordinary differential equation

$$\frac{d^{2\varepsilon}\psi(\tau)}{d\tau^{2\varepsilon}} + \psi(\tau) = \frac{\tau^{\varepsilon}}{\Gamma(1+\varepsilon)}, \quad \tau \in (0,\pi) \tag{2.180}$$

subject to initial value conditions given by

$$\psi(0) = 0, \quad \psi(\pi) = 0. \tag{2.181}$$

We can write the function

$$\tau^{\varepsilon}/\Gamma(1+\varepsilon)$$

as follows:

$$\frac{\tau^{\varepsilon}}{\Gamma(1+\varepsilon)} = \sum_{k=1}^{\infty} B(k,\varepsilon)\sin_{\varepsilon}(k\tau)^{\varepsilon}, \quad \tau \in (0,\pi), \tag{2.182}$$

where

$$B(k,\varepsilon) = \frac{1}{\pi^{\varepsilon}}\int_0^{2\pi} \frac{\tau^{\varepsilon}}{\Gamma(1+\varepsilon)}\left[\sin_{\varepsilon}(kx)^{\varepsilon}\right](\mathrm{d}x)^{\varepsilon}$$
$$= -\frac{\left(\frac{2\pi}{k}\right)^{\varepsilon}}{\Gamma(1+\varepsilon)}. \tag{2.183}$$

In this case, we can rewrite $\psi(\tau)$ as follows:

$$\psi(\tau) = \sum_{k=1}^{\infty} B(k,\varepsilon)\sin_{\varepsilon}(k\tau)^{\varepsilon}, \quad \tau \in (0,\pi), \tag{2.184}$$

which leads us to

$$\frac{\mathrm{d}^{2\varepsilon}\psi(\tau)}{\mathrm{d}\tau^{2\varepsilon}} = -\sum_{k=1}^{\infty} B(k,\varepsilon)k^{2\varepsilon}\sin_{\varepsilon}(k\tau)^{\varepsilon}, \quad \tau \in (0,\pi). \tag{2.185}$$

Using (2.180), (2.184), and (2.185), we obtain

$$-\sum_{n=1}^{\infty} B(k,\varepsilon)k^{2\varepsilon}\sin_{\varepsilon}(k\tau)^{\varepsilon} + \sum_{n=1}^{\infty} B(k,\varepsilon)\sin_{\varepsilon}(k\tau)^{\varepsilon} = -\sum_{k=1}^{\infty} -\frac{\left(\frac{2\pi}{k}\right)^{\varepsilon}}{\Gamma(1+\varepsilon)}\sin_{\varepsilon}(k\tau)^{\varepsilon}. \tag{2.186}$$

Matching the coefficients for the same terms in the two series, we find from (2.186) that

$$\left(1-k^{2\varepsilon}\right)B(k,\varepsilon) = -\frac{\left(\frac{2\pi}{k}\right)^{\varepsilon}}{\Gamma(1+\varepsilon)}, \quad k \in \mathrm{N}, \tag{2.187}$$

which leads us to

$$B(k,\varepsilon) = \frac{\left(\frac{2\pi}{k}\right)^{\varepsilon}}{\left(k^{2\varepsilon}-1\right)\Gamma(1+\varepsilon)}. \tag{2.188}$$

Therefore, the local fractional Fourier solution for the local fractional differential equation is given as follows:

$$\psi(\tau) = \sum_{k=1}^{\infty} \frac{\left(\frac{2\pi}{k}\right)^{\varepsilon}}{\left(k^{2\varepsilon}-1\right)\Gamma(1+\varepsilon)}\sin_{\varepsilon}(k\tau)^{\varepsilon}. \tag{2.189}$$

2.4.2 Applications of local fractional partial differential equations

We now present some examples for solving the local fractional PDEs in mathematical physics using the local fractional Fourier series.

We consider the following homogeneous local fractional heat equation in the nondimensional case:

$$\frac{\partial^{\varepsilon}\psi(\mu,\tau)}{\partial\tau^{\varepsilon}}-\frac{\partial^{2\varepsilon}\psi(\mu,\tau)}{\partial\mu^{2\varepsilon}}=0, \tag{2.190}$$

subject to initial value conditions given by

$$\psi(0,\tau)=0, \tag{2.191}$$

$$\psi(L,\tau)=0, \tag{2.192}$$

and

$$\psi(\mu,0)=\vartheta(\mu). \tag{2.193}$$

Suppose that

$$\psi(\mu,\tau)=\Theta(\mu)\Phi(\tau). \tag{2.194}$$

In this case, we transform (2.190) into the following formula:

$$\gamma=\frac{\Phi^{(\varepsilon)}(\tau)}{\Phi(\tau)}=\frac{\Theta^{(2\varepsilon)}(\mu)}{\Theta(\mu)}. \tag{2.195}$$

We set $\gamma=-\lambda^{2\varepsilon}$ such that

$$\Phi^{(\varepsilon)}(\tau)+\lambda^{2\varepsilon}\Phi(\tau)=0 \tag{2.196}$$

and

$$\Theta^{(2\varepsilon)}(\mu)+\lambda^{2\varepsilon}\Theta(\mu)=0. \tag{2.197}$$

We now present the following terms:

$$\Phi(\tau)=C(\lambda,\varepsilon)E_{\varepsilon}\left(-\lambda^{2\varepsilon}\tau^{\varepsilon}\right) \tag{2.198}$$

and

$$\Theta(\mu)=A(\lambda,\varepsilon)\cos_{\varepsilon}\left(\lambda^{\varepsilon}\mu^{\varepsilon}\right)+B(\lambda,\varepsilon)\sin_{\varepsilon}\left(\lambda^{\varepsilon}\mu^{\varepsilon}\right), \tag{2.199}$$

where $A(\lambda,\varepsilon), B(\lambda,\varepsilon)$, and $C(\lambda,\varepsilon)$ are constants to be determined.

In this case, we give the general solution as follows:

$$\psi(\mu,\tau)=C(\lambda,\varepsilon)E_{\varepsilon}\left(-\lambda^{2\varepsilon}\tau^{\varepsilon}\right)\left[A(\lambda,\varepsilon)\cos_{\varepsilon}\left(\lambda^{\varepsilon}\mu^{\varepsilon}\right)+B(\lambda,\varepsilon)\sin_{\varepsilon}\left(\lambda^{\varepsilon}\mu^{\varepsilon}\right)\right]. \tag{2.200}$$

Using (2.191), (2.200) can be rewritten in the form

$$\psi(\mu,\tau)=A(\lambda,\varepsilon)\,C(\lambda,\varepsilon)\,E_{\varepsilon}\left(-\lambda^{2\varepsilon}\tau^{\varepsilon}\right)\sin_{\varepsilon}\left(\lambda^{\varepsilon}\mu^{\varepsilon}\right) \tag{2.201}$$

or, equivalently,

$$B(\lambda,\varepsilon)=0. \tag{2.202}$$

Thus, by using (3.192), we have

$$A(\lambda,\varepsilon)=0 \tag{2.203}$$

such that

$$\lambda=\frac{k\pi}{L},\quad k\in \mathrm{N}_0. \tag{2.204}$$

Therefore, we rewrite (2.201) as follows:

$$\psi(\mu,\tau)=A\left(\frac{k\pi}{L},\varepsilon\right)C\left(\frac{k\pi}{L},\varepsilon\right)E_{\varepsilon}\left(-\left(\frac{k\pi}{L}\right)^{2\varepsilon}\tau^{\varepsilon}\right)\sin_{\varepsilon}\left(\left(\frac{k\pi}{L}\right)^{\varepsilon}\mu^{\varepsilon}\right). \tag{2.205}$$

In this case, we set the local fractional Fourier series as given below:

$$\psi(\mu,\tau)=\frac{\theta(0,\varepsilon)}{2}+\sum_{k=0}^{\infty}\theta(k,\varepsilon)E_{\varepsilon}\left(-\left(\frac{k\pi}{L}\right)^{2\varepsilon}\tau^{\varepsilon}\right)\sin_{\varepsilon}\left(\left(\frac{k\pi}{L}\right)^{\varepsilon}\mu^{\varepsilon}\right), \tag{2.206}$$

which leads us to

$$\psi(\mu,0)=\frac{\theta(0,\varepsilon)}{2}+\sum_{k=0}^{\infty}\theta(k,\varepsilon)\sin_{\varepsilon}\left(\left(\frac{k\pi}{L}\right)^{\varepsilon}\mu^{\varepsilon}\right), \tag{2.207}$$

where the local fractional Fourier coefficients are confirmed by

$$\theta(k,\varepsilon)=\frac{1}{L^{\varepsilon}}\int_{-L}^{L}\left[\vartheta(\mu)\sin_{\varepsilon}\left(\left(\frac{k\pi}{L}\right)^{\varepsilon}\mu^{\varepsilon}\right)\right](d\mu)^{\varepsilon}. \tag{2.208}$$

Therefore, we obtain the local fractional Fourier solution in the form

$$\psi(\mu,\tau)=\frac{\theta(0,\varepsilon)}{2}+\sum_{k=0}^{\infty}\theta(k,\varepsilon)E_{\varepsilon}\left(-\left(\frac{k\pi}{L}\right)^{2\varepsilon}\tau^{\varepsilon}\right)\sin_{\varepsilon}\left(\left(\frac{k\pi}{L}\right)^{\varepsilon}\mu^{\varepsilon}\right), \tag{2.209}$$

where the local fractional Fourier series can be written in the form

$$\theta(k,\varepsilon)=\frac{1}{L^{\varepsilon}}\int_{0}^{L}\left[\vartheta(\mu)\sin_{\varepsilon}\left(\left(\frac{k\pi}{L}\right)^{\varepsilon}\mu^{\varepsilon}\right)\right](d\mu)^{\varepsilon},\quad k\in \mathrm{N}_0. \tag{2.210}$$

We solve the following nonhomogeneous local fractional heat equation in the nondimensional case:

$$\frac{d^{\varepsilon}\psi(\mu,\tau)}{d\tau^{\varepsilon}} - \frac{d^{2\varepsilon}\psi(\mu,\tau)}{d\mu^{2\varepsilon}} = G(\mu,\tau), \tag{2.211}$$

subject to the initial value conditions given by

$$\psi(0,\tau) = 0, \tag{2.212}$$

$$\psi(L,\tau) = 0, \tag{2.213}$$

and

$$\psi(\mu,0) = \vartheta(\mu). \tag{2.214}$$

We present the local fractional Fourier solution given by

$$\psi(\mu,\tau) = \sum_{k=1}^{\infty} \psi_k(\tau) \sin_{\varepsilon}\left(\frac{k\pi\mu}{L}\right)^{\varepsilon}. \tag{2.215}$$

We suppose that there are $\{G_k(\tau)\}_{k=1}^{\infty}$ and $\{\psi_k(0)\}_{k=1}^{\infty}$ such that

$$G(\mu,\tau) = \sum_{k=1}^{\infty} G_k(\tau) \sin_{\varepsilon}\left(\frac{k\pi\mu}{L}\right)^{\varepsilon} \tag{2.216}$$

and

$$\psi(\mu,0) = \sum_{k=1}^{\infty} \psi_k(0) \sin_{\varepsilon}\left(\frac{k\pi\mu}{L}\right)^{\varepsilon}, \tag{2.217}$$

where

$$G_k(\tau) = \frac{1}{L^{\varepsilon}} \int_0^L \left[G(\mu,\tau) \sin_{\varepsilon}\left(\left(\frac{k\pi}{L}\right)^{\varepsilon} \mu^{\varepsilon}\right)\right] (d\mu)^{\varepsilon} \tag{2.218}$$

and

$$\psi_k(0) = \frac{1}{L^{\varepsilon}} \int_0^L \left[\psi(\mu,0) \sin_{\varepsilon}\left(\left(\frac{k\pi}{L}\right)^{\varepsilon} \mu^{\varepsilon}\right)\right] (d\mu)^{\varepsilon}. \tag{2.219}$$

Hence, by adopting (2.215) and (2.216), (2.211) becomes

$$\sum_{k=1}^{\infty} \left[\frac{d^{\varepsilon}\psi_k(\tau)}{d\tau^{\varepsilon}} - \left(\frac{\pi k}{L}\right)^{\varepsilon} \psi_k(\tau) - G_k(\tau)\right] \sin_{\varepsilon}\left(\frac{k\pi\mu}{L}\right)^{\varepsilon} = 0. \tag{2.220}$$

For any value $\sin_{\varepsilon}\left(\frac{k\pi\mu}{L}\right)^{\varepsilon}$, we present

$$\frac{d^{\varepsilon}\psi_k(\tau)}{d\tau^{\varepsilon}} - \left(\frac{\pi k}{L}\right)^{\varepsilon} \psi_k(\tau) - G_k(\tau) = 0, \tag{2.221}$$

where

$$\psi_k(0) = \frac{1}{L^\varepsilon}\int_0^L \left[\psi(\mu, 0)\sin_\varepsilon\left(\left(\frac{k\pi}{L}\right)^\varepsilon \mu^\varepsilon\right)\right](\mathrm{d}\mu)^\varepsilon$$
$$= \frac{1}{L^\varepsilon}\int_0^L \left[\vartheta(\mu)\sin_\varepsilon\left(\left(\frac{k\pi}{L}\right)^\varepsilon \mu^\varepsilon\right)\right](\mathrm{d}\mu)^\varepsilon. \quad (2.222)$$

Hence, we obtain

$$\psi_k(\tau) = \psi_k(0)\, E_\varepsilon\left(\frac{k\pi\tau}{L}\right)^\varepsilon + \frac{1}{\Gamma(1+\varepsilon)}\int_0^\tau E_\varepsilon\left(\frac{k\pi(\tau-\chi)}{L}\right)^\varepsilon G_k(\chi)(\mathrm{d}\chi)^\varepsilon. \quad (2.223)$$

From (2.217) and (2.223), we have the local fractional Fourier series solution given by

$$\psi(\mu,\tau) = \sum_{k=1}^\infty \psi_k(\tau)\sin_\varepsilon\left(\frac{k\pi\mu}{L}\right)^\varepsilon$$
$$= \sum_{k=1}^\infty\left[\psi_k(0)\, E_\varepsilon\left(\frac{k\pi\tau}{L}\right)^\varepsilon\right]\sin_\varepsilon\left(\frac{k\pi\mu}{L}\right)^\varepsilon$$
$$+ \sum_{k=1}^\infty\left[\frac{1}{\Gamma(1+\varepsilon)}\int_0^\tau E_\varepsilon\left(\frac{k\pi(\tau-\chi)}{L}\right)^\varepsilon G_k(\chi)(\mathrm{d}\chi)^\varepsilon\right]\sin_\varepsilon\left(\frac{k\pi\mu}{L}\right)^\varepsilon. \quad (2.224)$$

Let us now consider the local fractional Laplace equation in the form

$$\frac{\partial^{2\varepsilon}\psi(\mu,\eta)}{\partial\mu^{2\varepsilon}} + \frac{\partial^{2\varepsilon}\psi(\mu,\eta)}{\partial\eta^{2\varepsilon}} = 0, \quad (2.225)$$

subject to initial value conditions given by

$$\psi(\mu, 0) = \vartheta(\mu), \quad (2.226)$$
$$\psi(\mu, \rho) = 0, \quad (2.227)$$
$$\psi(0, \eta) = 0, \quad (2.228)$$

and

$$\psi(L, \eta) = 0. \quad (2.229)$$

Suppose that we construct a special solution in the form

$$\psi(\mu,\eta) = \Theta(\mu)\,\Phi(\eta). \quad (2.230)$$

In this case, we transform (2.225) into the following formula:

$$\gamma = \frac{\Theta^{(2\varepsilon)}(\mu)}{\Theta(\mu)} = -\frac{\Phi^{(2\varepsilon)}(\eta)}{\Phi(\eta)}. \quad (2.231)$$

Then, by separating the variables, we obtain

$$\gamma = -\lambda^{2\varepsilon} \quad (2.232)$$

such that

$$\Theta^{(2\varepsilon)}(\mu) + \lambda^{2\varepsilon}\Theta(\mu) = 0 \tag{2.233}$$

and

$$\Phi^{(2\varepsilon)}(\eta) - \lambda^{2\varepsilon}\Phi(\eta) = 0. \tag{2.234}$$

Using (2.227) and (2.228), we find that

$$\Theta(0) = \Theta(L) = 0. \tag{2.235}$$

From (3.234), we get the general solution in the form

$$\Theta(\mu) = A(\lambda,\varepsilon)\cos_\varepsilon\left(\lambda^\varepsilon\mu^\varepsilon\right) + B(\lambda,\varepsilon)\sin_\varepsilon\left(\lambda^\varepsilon\mu^\varepsilon\right), \tag{2.236}$$

where $A(\lambda,\varepsilon)$ and $B(\lambda,\varepsilon)$ are the coefficients involved.

Using (2.235), (2.236) becomes

$$\Theta(\mu) = B(\lambda,\varepsilon)\sin_\varepsilon\left(\lambda^\varepsilon\mu^\varepsilon\right), \tag{2.237}$$

$$A(\lambda,\varepsilon) = 0, \tag{2.238}$$

and

$$\sin_\varepsilon\left(\lambda^\varepsilon L^\varepsilon\right) = 0. \tag{2.239}$$

Hence, we find from (2.239) that

$$\lambda^\varepsilon = \left(\frac{k\pi}{L}\right)^\varepsilon, \quad k \in \mathrm{N}_0, \tag{2.240}$$

such that

$$\Theta(\mu) = B(\lambda,\varepsilon)\sin_\varepsilon\left(\left(\frac{k\pi}{L}\right)^\varepsilon\mu^\varepsilon\right). \tag{2.241}$$

From (2.240) and (2.241), we have

$$\Phi(\eta) = C(\lambda,\varepsilon)\cosh_\varepsilon\left(\left(\frac{k\pi}{L}\right)^\varepsilon(\eta-\rho)^\varepsilon\right) + D(\lambda,\varepsilon)\sinh_\varepsilon\left(\left(\frac{k\pi}{L}\right)^\varepsilon(\eta-\rho)^\varepsilon\right). \tag{2.242}$$

Adopting (2.226), we get

$$C(\lambda,\varepsilon) = 0 \tag{2.243}$$

such that (2.242) becomes

$$\Phi(\eta) = D(\lambda,\varepsilon)\sinh_\varepsilon\left(\left(\frac{k\pi}{L}\right)^\varepsilon(\eta-\rho)^\varepsilon\right). \tag{2.244}$$

Therefore, we have

$$\psi(\mu,\eta) = D(\lambda,\varepsilon)B(\lambda,\varepsilon)\sin_\varepsilon\left(\left(\frac{k\pi}{L}\right)^\varepsilon\mu^\varepsilon\right)\sinh_\varepsilon\left(\left(\frac{k\pi}{L}\right)^\varepsilon(\eta-\rho)^\varepsilon\right). \tag{2.245}$$

The local fractional Fourier solution reads as follows:

$$\psi(\mu,\eta)=\sum_{k=1}^{\infty}\left\{D(\lambda,\varepsilon)B(\lambda,\varepsilon)\sin_{\varepsilon}\left(\left(\frac{k\pi}{L}\right)^{\varepsilon}\mu^{\varepsilon}\right)\sinh_{\varepsilon}\left(\left(\frac{k\pi}{L}\right)^{\varepsilon}(\eta-\rho)^{\varepsilon}\right)\right\}. \tag{2.246}$$

With the help of (2.246), we get

$$\begin{aligned}\psi(\mu,0)&=\sum_{k=1}^{\infty}\left\{D(\lambda,\varepsilon)B(\lambda,\varepsilon)\sinh_{\varepsilon}\left(\left(\frac{k\pi}{L}\right)^{\varepsilon}(0-\rho)^{\varepsilon}\right)\sin_{\varepsilon}\left(\left(\frac{k\pi}{L}\right)^{\varepsilon}\mu^{\varepsilon}\right)\right\}\\&=\sum_{k=1}^{\infty}\left\{\psi_k(\mu,0)\sin_{\varepsilon}\left(\left(\frac{k\pi}{L}\right)^{\varepsilon}\mu^{\varepsilon}\right)\right\}\\&=\vartheta(\mu),\end{aligned} \tag{2.247}$$

where

$$\psi_k(\mu,0)=-D(\lambda,\varepsilon)B(\lambda,\varepsilon)\sinh_{\varepsilon}\left(\left(\frac{k\pi}{L}\right)^{\varepsilon}\rho^{\varepsilon}\right). \tag{2.248}$$

Consequently, the local fractional Fourier coefficients are given by

$$\begin{aligned}\psi_k(\mu,0)&=\frac{1}{L^{\varepsilon}}\int_0^L\left[\psi(\mu,0)\sin_{\varepsilon}\left(\left(\frac{k\pi}{L}\right)^{\varepsilon}\mu^{\varepsilon}\right)\right](\mathrm{d}\mu)^{\varepsilon}\\&=\frac{1}{L^{\varepsilon}}\int_0^L\left[\vartheta(\mu)\sin_{\varepsilon}\left(\left(\frac{k\pi}{L}\right)^{\varepsilon}\mu^{\varepsilon}\right)\right](\mathrm{d}\mu)^{\varepsilon}.\end{aligned} \tag{2.249}$$

In this case, making use of (2.248), we obtain

$$D(\lambda,\varepsilon)B(\lambda,\varepsilon)=-\frac{\frac{1}{L^{\varepsilon}}\int_0^L\left[\vartheta(\mu)\sin_{\varepsilon}\left(\left(\frac{k\pi}{L}\right)^{\varepsilon}\mu^{\varepsilon}\right)\right](\mathrm{d}\mu)^{\varepsilon}}{\sinh_{\varepsilon}\left(\left(\frac{k\pi}{L}\right)^{\varepsilon}\rho^{\varepsilon}\right)}. \tag{2.250}$$

Hence, the local fractional Fourier solution of the local fractional Laplace equation is given by

$$\begin{aligned}\psi(\mu,\eta)&=\sum_{k=1}^{\infty}\left\{D\left(\frac{k\pi}{L},\varepsilon\right)B\left(\frac{k\pi}{L},\varepsilon\right)\sin_{\varepsilon}\left(\left(\frac{k\pi}{L}\right)^{\varepsilon}\mu^{\varepsilon}\right)\sinh_{\varepsilon}\left(\left(\frac{k\pi}{L}\right)^{\varepsilon}(\eta-\rho)^{\varepsilon}\right)\right\}\\&=\sum_{k=1}^{\infty}\left\{-\frac{\frac{1}{L^{\varepsilon}}\int_0^L\left[\vartheta(\mu)\sin_{\varepsilon}\left(\left(\frac{k\pi}{L}\right)^{\varepsilon}\mu^{\varepsilon}\right)\right](\mathrm{d}\mu)^{\varepsilon}}{\sinh_{\varepsilon}\left(\left(\frac{k\pi}{L}\right)^{\varepsilon}\rho^{\varepsilon}\right)}\sin_{\varepsilon}\left(\left(\frac{k\pi}{L}\right)^{\varepsilon}\mu^{\varepsilon}\right)\right.\\&\left.\sinh_{\varepsilon}\left(\left(\frac{k\pi}{L}\right)^{\varepsilon}(\eta-\rho)^{\varepsilon}\right)\right\}.\end{aligned} \tag{2.251}$$

The local fractional wave equation in the nondimensional case takes the following form:

$$\frac{\partial^{2\varepsilon}\psi(\mu,\tau)}{\partial\mu^{2\varepsilon}}-\frac{\partial^{2\varepsilon}\psi(\mu,\tau)}{\partial\tau^{2\varepsilon}}=0, \tag{2.252}$$

subject to initial value conditions given by

$$\psi(\mu,0)=\vartheta(\mu), \tag{2.253}$$

$$\psi(\mu,\rho)=0, \tag{2.254}$$

$$\psi(0,\tau)=0, \tag{2.255}$$

and

$$\psi(L,\tau)=0. \tag{2.256}$$

Suppose now that we construct a special solution in the form

$$\psi(\mu,\eta)=\Theta(\mu)\Phi(\tau). \tag{2.257}$$

In this case, we transform (2.252) into the following formula:

$$\gamma=\frac{\Theta^{(2\varepsilon)}(\mu)}{\Theta(\mu)}=\frac{\Phi^{(2\varepsilon)}(\tau)}{\Phi(\tau)}. \tag{2.258}$$

Thus, by separating the variables, we have

$$\gamma=-\lambda^{2\varepsilon} \tag{2.259}$$

such that

$$\Theta^{(2\varepsilon)}(\mu)+\lambda^{2\varepsilon}\Theta(\mu)=0 \tag{2.260}$$

and

$$\Phi^{(2\varepsilon)}(\tau)+\lambda^{2\varepsilon}\Phi(\tau)=0. \tag{2.261}$$

Using (2.255) and (2.256), we obtain

$$\Theta(0)=\Theta(L)=0. \tag{2.262}$$

From (2.260), we get the general solution in the form

$$\Theta(\mu)=A(\lambda,\varepsilon)\cos_\varepsilon\left(\lambda^\varepsilon\mu^\varepsilon\right)+B(\lambda,\varepsilon)\sin_\varepsilon\left(\lambda^\varepsilon\mu^\varepsilon\right), \tag{2.263}$$

where $A(\lambda,\varepsilon)$ and $B(\lambda,\varepsilon)$ are coefficients.

From (2.261), the general solution reads as follows:

$$\Phi(\tau)=C(\lambda,\varepsilon)\cos_\varepsilon\left(\lambda^\varepsilon(\tau-\rho)^\varepsilon\right)+D(\lambda,\varepsilon)\sin_\varepsilon\left(\lambda^\varepsilon(\tau-\rho)^\varepsilon\right), \tag{2.264}$$

where $C(\lambda,\varepsilon)$ and $D(\lambda,\varepsilon)$ are coefficients.

Using (2.262), (2.263) can be written as follows:

$$\Theta(\mu)=B(\lambda,\varepsilon)\sin_\varepsilon\left(\lambda^\varepsilon\mu^\varepsilon\right), \tag{2.265}$$

$$A(\lambda, \varepsilon) = 0, \tag{2.266}$$

and

$$\sin_\varepsilon\left(\lambda^\varepsilon L^\varepsilon\right) = 0. \tag{2.267}$$

Hence, with the help of (2.267), we get

$$\lambda^\varepsilon = \left(\frac{k\pi}{L}\right)^\varepsilon, \quad k \in \mathrm{N}_0, \tag{2.268}$$

such that

$$\Theta(\mu) = B(\lambda, \varepsilon)\sin_\varepsilon\left(\left(\frac{k\pi}{L}\right)^\varepsilon \mu^\varepsilon\right). \tag{2.269}$$

In view of (2.264), (2.265), and (2.268), we refer to the general solution in the form

$$\begin{aligned}\psi(\mu, \tau) &= \Theta(\mu)\Phi(\tau) \\ &= B(\lambda, \varepsilon)\sin_\varepsilon\left(\lambda^\varepsilon \mu^\varepsilon\right)\left[C(\lambda, \varepsilon)\cos_\varepsilon\left(\lambda^\varepsilon(\tau - \rho)^\varepsilon\right)\right. \\ &\quad \left.+ D(\lambda, \varepsilon)\sin_\varepsilon\left(\lambda^\varepsilon(\tau - \rho)^\varepsilon\right)\right].\end{aligned} \tag{2.270}$$

Hence, we have

$$\begin{aligned}\psi(\mu, \tau) = \sum_{k=1}^{\infty} B(\lambda, \varepsilon)\sin_\varepsilon\left(\lambda^\varepsilon \mu^\varepsilon\right)&\left[C(\lambda, \varepsilon)\cos_\varepsilon\left(\lambda^\varepsilon(\tau - \rho)^\varepsilon\right)\right. \\ &\left.+ D(\lambda, \varepsilon)\sin_\varepsilon\left(\lambda^\varepsilon(\tau - \rho)^\varepsilon\right)\right],\end{aligned} \tag{2.271}$$

which, in light of (2.254), leads us to

$$C(\lambda, \varepsilon) = 0 \tag{2.272}$$

and

$$\psi(\mu, \tau) = \sum_{k=1}^{\infty} B(\lambda, \varepsilon) D(\lambda, \varepsilon)\sin_\varepsilon\left(\lambda^\varepsilon(\tau - \rho)^\varepsilon\right)\sin_\varepsilon\left(\lambda^\varepsilon \mu^\varepsilon\right). \tag{2.273}$$

Utilizing (2.253) and (2.268), we get

$$\begin{aligned}\psi(\mu, 0) &= \sum_{k=1}^{\infty} B(\lambda, \varepsilon) D(\lambda, \varepsilon)\sin_\varepsilon\left(\lambda^\varepsilon(0 - \rho)^\varepsilon\right)\sin_\varepsilon\left(\lambda^\varepsilon \mu^\varepsilon\right) \\ &= -\sum_{k=1}^{\infty} B\left(\frac{k\pi}{L}, \varepsilon\right) D\left(\frac{k\pi}{L}, \varepsilon\right)\sin_\varepsilon\left(\left(\frac{k\pi}{L}\right)^\varepsilon \rho^\varepsilon\right)\sin_\varepsilon\left(\left(\frac{k\pi}{L}\right)^\varepsilon \mu^\varepsilon\right) \\ &= \sum_{k=1}^{\infty} \psi_k(\mu, 0)\sin_\varepsilon\left(\left(\frac{k\pi}{L}\right)^\varepsilon \mu^\varepsilon\right) \\ &= \vartheta(\mu),\end{aligned} \tag{2.274}$$

where

$$\psi_k(\mu,0) = -B\left(\frac{k\pi}{L},\varepsilon\right)D\left(\frac{k\pi}{L},\varepsilon\right)\sin_\varepsilon\left(\left(\frac{k\pi}{L}\right)^\varepsilon\rho^\varepsilon\right) \tag{2.275}$$

and

$$\begin{aligned}\psi_k(\mu,0) &= \frac{1}{L^\varepsilon}\int_0^L\left[\psi(\mu,0)\sin_\varepsilon\left(\left(\frac{k\pi}{L}\right)^\varepsilon\mu^\varepsilon\right)\right](\mathrm{d}\mu)^\varepsilon \\ &= \frac{1}{L^\varepsilon}\int_0^L\left[\vartheta(\mu)\sin_\varepsilon\left(\left(\frac{k\pi}{L}\right)^\varepsilon\mu^\varepsilon\right)\right](\mathrm{d}\mu)^\varepsilon.\end{aligned} \tag{2.276}$$

From (2.275) and (2.276), we easily obtain

$$B\left(\frac{k\pi}{L},\varepsilon\right)D\left(\frac{k\pi}{L},\varepsilon\right) = -\frac{\frac{1}{L^\varepsilon}\int_0^L\left[\vartheta(\mu)\sin_\varepsilon\left(\left(\frac{k\pi}{L}\right)^\varepsilon\mu^\varepsilon\right)\right](\mathrm{d}\mu)^\varepsilon}{\sin_\varepsilon\left(\left(\frac{k\pi}{L}\right)^\varepsilon\rho^\varepsilon\right)}. \tag{2.277}$$

Therefore, the local fractional Fourier solution of the local fractional wave equation in the nondimensional case has the form

$$\begin{aligned}\psi(\mu,\eta) &= \sum_{k=1}^{\infty}B\left(\frac{k\pi}{L},\varepsilon\right)D\left(\frac{k\pi}{L},\varepsilon\right)\sin_\varepsilon\left(\left(\frac{k\pi}{L}\right)^\varepsilon(\tau-\rho)^\varepsilon\right)\sin_\varepsilon\left(\left(\frac{k\pi}{L}\right)^\varepsilon\mu^\varepsilon\right) \\ &= \sum_{k=1}^{\infty}\left\{-\frac{\frac{1}{L^\varepsilon}\int_0^L\left[\vartheta(\mu)\sin_\varepsilon\left(\left(\frac{k\pi}{L}\right)^\varepsilon\mu^\varepsilon\right)\right](\mathrm{d}\mu)^\varepsilon}{\sin_\varepsilon\left(\left(\frac{k\pi}{L}\right)^\varepsilon\rho^\varepsilon\right)}\sin_\varepsilon\left(\left(\frac{k\pi}{L}\right)^\varepsilon\mu^\varepsilon\right)\right. \\ &\quad \left.\sin_\varepsilon\left(\left(\frac{k\pi}{L}\right)^\varepsilon(\tau-\rho)^\varepsilon\right)\right\}.\end{aligned} \tag{2.278}$$

Local fractional Fourier transform and applications

3

3.1 Introduction

Fourier transforms play an important role in the theoretical analysis of mathematical models for the problems that appear in mathematical physics, engineering applications, and theoretical and applied physics, as in quantum mechanics, signal analysis, control theory, and both pure and applied mathematics. Especially, the Fourier theory is utilized to analyze the nonperiodic phenomena of heat conduction.

Recently, it was found that a nondifferentiable function in fractal time is decomposed into the local fractional Fourier series via the Mittag–Leffler functions or the analogous sine functions and the analogous cosine functions defined on the fractal set because there are analogous trigonometric functions defined on Cantor sets. We expand the idea of the Fourier transform operator to the local fractional integral transform in the case of the Mittag–Leffler functions defined on Cantor sets. In order to understand the concept, the following problems should be considered first:

Problem 1. What mechanism does the local fractional Fourier transform operator possess?

Problem 2. Does the local fractional Fourier transform preserve the energy of the original quantity of nondifferentiable signals?

We give the answers to the above problems in this chapter. We start by presenting the basic mechanism of local fractional Fourier transform operator.

3.2 Definitions and properties

3.2.1 *Mathematical mechanism is the local fractional Fourier transform operator*

Suppose that $\phi(\tau)$ is $2L$-periodic and locally fractional continuous on the interval $[-L, L]$. Then, $\phi(\tau)$ can be decomposed into the local fractional Fourier series in the form

$$\phi(\tau) = \sum_{k=-\infty}^{\infty} \varphi_k^{\varepsilon} E_{\varepsilon}\left(\frac{\pi^{\varepsilon} i^{\varepsilon} (k\tau)^{\varepsilon}}{L^{\varepsilon}}\right), \quad k \in Z, \tag{3.1}$$

where the local fractional Fourier coefficients can determined by

$$\varphi_k^{\varepsilon} = \frac{1}{(2L)^{\varepsilon}} \int_{-L}^{L} \phi(\tau) E_{\varepsilon}\left(-\frac{\pi^{\varepsilon} i^{\varepsilon} (k\tau)^{\varepsilon}}{L^{\varepsilon}}\right) (d\tau)^{\varepsilon}. \tag{3.2}$$

Local Fractional Integral Transforms and Their Applications. http://dx.doi.org/10.1016/B978-0-12-804002-7.00003-6

Now, let us define the local fractional Fourier coefficients as follows:

$$\varphi_k^{\varepsilon} = \frac{\Gamma(1+\varepsilon)}{(2L)^{\varepsilon}} \Phi_k^{\varepsilon}, \quad k \in \mathrm{Z}. \tag{3.3}$$

Therefore, (3.1) and (3.2) can be rewritten in the form

$$\phi(\tau) = \frac{1}{(2L)^{\varepsilon}} \sum_{k=-\infty}^{\infty} \Phi_k^{\varepsilon} E_{\varepsilon}\left(\frac{\pi^{\varepsilon} \mathrm{i}^{\varepsilon} (k\tau)^{\varepsilon}}{L^{\varepsilon}}\right) \tag{3.4}$$

and

$$\Phi_k^{\varepsilon} = \frac{1}{\Gamma(1+\varepsilon)} \int_{-L}^{L} \phi(\tau) E_{\varepsilon}\left(-\frac{\pi^{\varepsilon} \mathrm{i}^{\varepsilon} (k\tau)^{\varepsilon}}{L^{\varepsilon}}\right) (\mathrm{d}\tau)^{\varepsilon}, \tag{3.5}$$

where $k \in \mathrm{Z}$.

Defining

$$\theta_k^{\varepsilon} = \left(\frac{\pi k}{L}\right)^{\varepsilon}, \tag{3.6}$$

we have

$$(\Delta\theta_k)^{\varepsilon} = (\theta_{k+1} - \theta_k)^{\varepsilon} = \left(\frac{\pi}{L}\right)^{\varepsilon}. \tag{3.7}$$

Therefore, (3.4) can be rewritten as follows:

$$\begin{aligned} \phi(\tau) &= \frac{1}{(2L)^{\varepsilon}} \sum_{k=-\infty}^{\infty} \Phi_k^{\varepsilon} E_{\varepsilon}\left(\mathrm{i}^{\varepsilon} \tau^{\varepsilon} \left(\frac{\pi k}{L}\right)^{\varepsilon}\right) \\ &= \frac{1}{(2\pi)^{\varepsilon}} \sum_{k=-\infty}^{\infty} \Phi_k^{\varepsilon} E_{\varepsilon}\left(\mathrm{i}^{\varepsilon} \tau^{\varepsilon} \theta_k^{\varepsilon}\right) (\Delta\theta_k)^{\varepsilon}, \end{aligned} \tag{3.8}$$

which leads us to

$$\begin{aligned} \lim_{L\to\infty} \phi(\tau) &= \lim_{L\to\infty} \left[\frac{1}{(2L)^{\varepsilon}} \sum_{k=-\infty}^{\infty} \Phi_k^{\varepsilon} E_{\varepsilon}\left(\mathrm{i}^{\varepsilon} \tau^{\varepsilon} \left(\frac{\pi k}{L}\right)^{\varepsilon}\right)\right] \\ &= \lim_{L\to\infty} \left[\frac{1}{(2\pi)^{\varepsilon}} \sum_{k=-\infty}^{\infty} \Phi_k^{\varepsilon} E_{\varepsilon}\left(\mathrm{i}^{\varepsilon} \tau^{\varepsilon} \theta_k^{\varepsilon}\right) (\Delta\theta_k)^{\varepsilon}\right] \\ &= \frac{1}{(2\pi)^{\varepsilon}} \int_{-\infty}^{\infty} \Phi_k^{\varepsilon} E_{\varepsilon}\left(\mathrm{i}^{\varepsilon} \tau^{\varepsilon} \theta_k^{\varepsilon}\right) (\mathrm{d}\theta_k)^{\varepsilon} \end{aligned} \tag{3.9}$$

and

$$\begin{aligned} \lim_{L\to\infty} \Phi_k^{\varepsilon} &= \lim_{L\to\infty} \left[\frac{1}{\Gamma(1+\varepsilon)} \int_{-L}^{L} \phi(\tau) E_{\varepsilon}\left(-\mathrm{i}^{\varepsilon} \tau^{\varepsilon} \left(\frac{\pi k}{L}\right)^{\varepsilon}\right) (\mathrm{d}\tau)^{\varepsilon}\right] \\ &= \lim_{L\to\infty} \left[\frac{1}{\Gamma(1+\varepsilon)} \int_{-\infty}^{\infty} \phi(\tau) E_{\varepsilon}\left(\mathrm{i}^{\varepsilon} \tau^{\varepsilon} \theta_k^{\varepsilon}\right) (\mathrm{d}\tau)^{\varepsilon}\right], \end{aligned} \tag{3.10}$$

where $k \in \mathrm{Z}$.

Thus, (3.9) and (3.10) can be written as follows:

$$\phi(\tau) = \frac{1}{(2\pi)^{\varepsilon}} \int_{-\infty}^{\infty} \Phi(\omega) E_{\varepsilon}\left(\mathrm{i}^{\varepsilon}\tau^{\varepsilon}\omega^{\varepsilon}\right)(\mathrm{d}\omega)^{\varepsilon}$$
$$= \frac{\Gamma(1+\varepsilon)}{(2\pi)^{\varepsilon}} {}_{-\infty}I_{\infty}^{(\varepsilon)}\left[\Phi(\omega) E_{\varepsilon}\left(\mathrm{i}^{\varepsilon}\tau^{\varepsilon}\omega^{\varepsilon}\right)\right] \tag{3.11}$$

and

$$\Phi(\omega) = \frac{1}{\Gamma(1+\varepsilon)} \int_{-\infty}^{\infty} \phi(\tau) E_{\varepsilon}\left(-\mathrm{i}^{\varepsilon}\tau^{\varepsilon}\omega^{\varepsilon}\right)(\mathrm{d}\tau)^{\varepsilon}$$
$$= {}_{-\infty}I_{\infty}^{(\varepsilon)}\left[\phi(\tau) E_{\varepsilon}\left(\ \mathrm{i}^{\varepsilon}\tau^{\varepsilon}\omega^{\varepsilon}\right)\right], \tag{3.12}$$

where $\tau, \varpi \in \mathrm{R}$,

$$\omega^{\varepsilon} = \lim_{L\to\infty}\left(\frac{\pi k}{L}\right)^{\varepsilon},$$

$$(\mathrm{d}\omega)^{\varepsilon} = \lim_{L\to\infty}\left(\frac{\pi}{L}\right)^{\varepsilon},$$

and

$$\Phi(\omega) = \lim_{L\to\infty} \Phi_k^{\varepsilon}$$

and they converge.

Let us define the variable ω by

$$\omega = 2\pi\varpi. \tag{3.13}$$

Then, we rewrite (3.11) and (3.12) in the form

$$\phi(\tau) = \int_{-\infty}^{\infty} \Phi(\varpi) E_{\varepsilon}\left[(2\pi\ \mathrm{i})^{\varepsilon}\,\tau^{\varepsilon}\varpi^{\varepsilon}\right](\mathrm{d}\varpi)^{\varepsilon}$$
$$= \Gamma(1+\varepsilon)\, {}_{-\infty}I_{\infty}^{(\varepsilon)}\Phi(\varpi) E_{\varepsilon}\left[(2\pi\ \mathrm{i})^{\varepsilon}\,\tau^{\varepsilon}\varpi^{\varepsilon}\right] \tag{3.14}$$

and

$$\Phi(\varpi) = \frac{1}{\Gamma(1+\varepsilon)} \int_{-\infty}^{\infty} \phi(\tau) E_{\varepsilon}\left[-(2\pi\ \mathrm{i})^{\varepsilon}\,\tau^{\varepsilon}\varpi^{\varepsilon}\right](\mathrm{d}\tau)^{\varepsilon}$$
$$= {}_{-\infty}I_{\infty}^{(\varepsilon)}\left\{\phi(\tau) E_{\varepsilon}\left[-(2\pi\ \mathrm{i})^{\varepsilon}\,\tau^{\varepsilon}\varpi^{\varepsilon}\right]\right\}, \tag{3.15}$$

where $\tau, \varpi \in \mathrm{R}$.

The next step is to define the variable ω^{ε} by

$$\omega^{\varepsilon} = \sigma^{\varepsilon} h_0, \tag{3.16}$$

where

$$h_0 = \frac{(2\pi)^{\varepsilon}}{\Gamma(1+\varepsilon)},$$

and (3.11) and (3.12) are restructured as follows:

$$\phi(\tau) = \frac{1}{\Gamma(1+\varepsilon)} \int_{-\infty}^{\infty} \left[\Phi(\sigma) E_\varepsilon\left(\mathrm{i}^\varepsilon h_0 \tau^\varepsilon \sigma^\varepsilon\right)\right] (\mathrm{d}\sigma)^\varepsilon$$
$$= {}_{-\infty}I_{\infty}^{(\varepsilon)} \left[\Phi(\sigma) E_\varepsilon\left(\mathrm{i}^\varepsilon h_0 \tau^\varepsilon \sigma^\varepsilon\right)\right] \tag{3.17}$$

and

$$\Phi(\sigma) = \frac{1}{\Gamma(1+\varepsilon)} \int_{-\infty}^{\infty} \left[\phi(\tau) E_\varepsilon\left(-\mathrm{i}^\varepsilon h_0 \tau^\varepsilon \sigma^\varepsilon\right)\right] (\mathrm{d}\tau)^\varepsilon$$
$$= {}_{-\infty}I_{\infty}^{(\varepsilon)} \left[\phi(\tau) E_\varepsilon\left(-\mathrm{i}^\varepsilon h_0 \tau^\varepsilon \sigma^\varepsilon\right)\right], \tag{3.18}$$

where $\tau, \sigma \in \mathrm{R}$.

The alternative forms of (3.17) and (3.18) are written as follows:

$$\phi(\tau) = \frac{1}{\Gamma(1+\varepsilon)} \int_{-\infty}^{\infty} \left\{\Phi(\sigma) E_\varepsilon\left[(2\pi\ \mathrm{i})^\varepsilon \zeta(\varepsilon) \tau^\varepsilon \sigma^\varepsilon\right]\right\} (\mathrm{d}\sigma)^\varepsilon$$
$$= {}_{-\infty}I_{\infty}^{(\varepsilon)} \left[\Phi(\sigma) E_\varepsilon\left[(2\pi\ \mathrm{i})^\varepsilon \zeta(\varepsilon) \tau^\varepsilon \sigma^\varepsilon\right]\right] \tag{3.19}$$

and

$$\Phi(\sigma) = \frac{1}{\Gamma(1+\varepsilon)} \int_{-\infty}^{\infty} \phi(\tau) E_\varepsilon\left(-(2\pi\ \mathrm{i})^\varepsilon \zeta(\varepsilon) \tau^\varepsilon \sigma^\varepsilon\right) (\mathrm{d}\tau)^\varepsilon$$
$$= {}_{-\infty}I_{\infty}^{(\varepsilon)} \left[\phi(\tau) E_\varepsilon\left(-(2\pi\ \mathrm{i})^\varepsilon \zeta(\varepsilon) \tau^\varepsilon \sigma^\varepsilon\right)\right], \tag{3.20}$$

where $\tau, \sigma \in \mathrm{R}$ and

$$\zeta(\varepsilon) = \frac{1}{\Gamma(1+\varepsilon)}. \tag{3.21}$$

In a similar manner, we have the following result.

Suppose that $\phi(\tau)$ is $2L$-periodic, bounded, and locally fractional integrable on the interval $[-L, L]$. In a similar way, we conclude that

$$(\mathrm{I}\phi)(\tau) \equiv \frac{\phi(\tau+0)+\phi(\tau-0)}{2} = \frac{\phi(\tau+0)+\phi(\tau-0)}{2}$$
$$= \frac{1}{(2\pi)^\varepsilon} \int_{-\infty}^{\infty} \Phi(\omega) E_\varepsilon\left(\mathrm{i}^\varepsilon \tau^\varepsilon \omega^\varepsilon\right) (\mathrm{d}\omega)^\varepsilon \tag{3.22}$$

and

$$\Phi(\omega) = \frac{1}{\Gamma(1+\varepsilon)} \int_{-\infty}^{\infty} (\mathrm{I}\phi)(\tau) E_\varepsilon\left(-\mathrm{i}^\varepsilon \tau^\varepsilon \omega^\varepsilon\right) (\mathrm{d}\tau)^\varepsilon$$
$$= {}_{-\infty}I_{\infty}^{(\varepsilon)} \left[(\mathrm{I}\phi)(\tau) E_\varepsilon\left(-\mathrm{i}^\varepsilon \tau^\varepsilon \omega^\varepsilon\right)\right], \tag{3.23}$$

respectively.

This is a transfer pair referring to the nondifferentiable function, which is periodic, bounded, and locally fractional integrable on the interval $[-L, L]$.

By using (3.22) and (3.23), the local fractional Fourier formula for the nondifferentiable function becomes

$$\begin{aligned}\frac{\phi(\tau+0)+\phi(\tau-0)}{2} &= \frac{1}{(2\pi)^{\varepsilon}}\int_{-\infty}^{\infty}\left\{\left[\frac{1}{\Gamma(1+\varepsilon)}\int_{-\infty}^{\infty}\phi(\tau)E_{\varepsilon}\left(-\mathrm{i}^{\varepsilon}\tau^{\varepsilon}\omega^{\varepsilon}\right)(\mathrm{d}\tau)^{\varepsilon}\right]\right.\\ &\quad \left. E_{\varepsilon}\left(\mathrm{i}^{\varepsilon}\tau^{\varepsilon}\omega^{\varepsilon}\right)\right\}(\mathrm{d}\omega)^{\varepsilon}\\ &= \frac{\Gamma(1+\varepsilon)}{(2\pi)^{\varepsilon}}{}_{-\infty}I_{\infty}^{(\varepsilon)}\left\{\left[{}_{-\infty}I_{\infty}^{(\varepsilon)}\left(\phi(\tau)E_{\varepsilon}\left(-\mathrm{i}^{\varepsilon}\tau^{\varepsilon}\omega^{\varepsilon}\right)\right)\right]\right.\\ &\quad \left. E_{\varepsilon}\left(\mathrm{i}^{\varepsilon}\tau^{\varepsilon}\omega^{\varepsilon}\right)\right\}. \end{aligned} \tag{3.24}$$

Suppose that $\phi(\tau)$ is $2L$-periodic and the locally fractional continuous on the interval $[-L, L]$. Thus, the local fractional Fourier formula for the nondifferentiable function can be written in the form

$$\begin{aligned}\phi(\tau) &= \frac{1}{(2\pi)^{\varepsilon}}\int_{-\infty}^{\infty}\left\{\left[\frac{1}{\Gamma(1+\varepsilon)}\int_{-\infty}^{\infty}\phi(\tau)E_{\varepsilon}\left(-\mathrm{i}^{\varepsilon}\tau^{\varepsilon}\omega^{\varepsilon}\right)(\mathrm{d}\tau)^{\varepsilon}\right]E_{\varepsilon}\left(\mathrm{i}^{\varepsilon}\tau^{\varepsilon}\omega^{\varepsilon}\right)\right\}(\mathrm{d}\omega)^{\varepsilon}\\ &= \frac{\Gamma(1+\varepsilon)}{(2\pi)^{\varepsilon}}{}_{-\infty}I_{\infty}^{(\varepsilon)}\left\{\left[{}_{-\infty}I_{\infty}^{(\varepsilon)}\left(\phi(\tau)E_{\varepsilon}\left(-\mathrm{i}^{\varepsilon}\tau^{\varepsilon}\omega^{\varepsilon}\right)\right)\right]E_{\varepsilon}\left(\mathrm{i}^{\varepsilon}\tau^{\varepsilon}\omega^{\varepsilon}\right)\right\}. \end{aligned} \tag{3.25}$$

This is called the local fractional Fourier formula for the nondifferentiable function. Thus, we have completed the derivation of the local fractional Fourier transform operators.

3.2.2 Definitions of the local fractional Fourier transform operators

We now give the basic definitions of the local fractional Fourier transform operators via the local fractional integral operator (see [16, 21, 27, 76, 96, 97, 105–107]). In order to study the local fractional Fourier transforms, we define the generalized space $L_{\nu,\varepsilon}[\mathrm{R}]$ under the ν-norm given by

$$\|\theta\|_{\nu,\varepsilon} = \left(\frac{1}{\Gamma(1+\varepsilon)}\int_{-\infty}^{\infty}|\theta(\tau)|^{\nu}(\mathrm{d}\tau)^{\varepsilon}\right)^{1/\nu} \tag{3.26}$$

for $1 \leq \varepsilon < \infty$ and $1 \leq \nu < \infty$.

Definition 3.1. By setting

$$\theta \in L_{1,\varepsilon}[\mathrm{R}] \quad \text{and} \quad \|\theta\|_{1,\varepsilon} < \infty,$$

the local fractional Fourier transform operator, denoted by

$$\Im[\theta(\tau)] = \Theta(\omega)$$

is defined by

$$\Im\left[\theta\left(\tau\right)\right]=\Theta\left(\omega\right)=\frac{1}{\Gamma\left(1+\varepsilon\right)}\int_{-\infty}^{\infty}\theta\left(\tau\right)E_{\varepsilon}\left(-\mathrm{i}^{\varepsilon}\tau^{\varepsilon}\omega^{\varepsilon}\right)\left(\mathrm{d}\tau\right)^{\varepsilon},\tag{3.27}$$

where $\Im$ is called the local fractional Fourier transform operator.

Definition 3.2. The inverse local fractional Fourier transform operator, denoted by

$\Im^{-1}\left[\Theta\left(\omega\right)\right]=\theta\left(\tau\right)$

is defined by

$$\Im^{-1}\left[\Theta\left(\omega\right)\right]=\theta\left(\tau\right)=\frac{1}{\left(2\pi\right)^{\varepsilon}}\int_{-\infty}^{\infty}\Theta\left(\omega\right)E_{\varepsilon}\left(\mathrm{i}^{\varepsilon}\tau^{\varepsilon}\omega^{\varepsilon}\right)\left(\mathrm{d}\omega\right)^{\varepsilon},\tag{3.28}$$

where $\Im^{-1}$ is called the inverse local fractional Fourier transform operator.

Making use of (3.14) and (3.15), we present the local fractional Fourier formula in the following form:

$$\begin{aligned}\phi\left(\tau\right)&=\frac{1}{\Gamma\left(1+\varepsilon\right)}\int_{-\infty}^{\infty}\left(\int_{-\infty}^{\infty}\phi\left(\tau\right)E_{\varepsilon}\left[-\left(2\pi\ \mathrm{i}\right)^{\varepsilon}\tau^{\varepsilon}\varpi^{\varepsilon}\right]\left(\mathrm{d}\tau\right)^{\varepsilon}\right)\\&\quad E_{\varepsilon}\left[\left(2\pi\ \mathrm{i}\right)^{\varepsilon}\tau^{\varepsilon}\varpi^{\varepsilon}\right]\left(\mathrm{d}\varpi\right)^{\varepsilon}\\&=\Gamma\left(1+\varepsilon\right){}_{-\infty}I_{\infty}^{(\varepsilon)}\left\{{}_{-\infty}I_{\infty}^{(\varepsilon)}\phi\left(\tau\right)E_{\varepsilon}\left[-\left(2\pi\ \mathrm{i}\right)^{\varepsilon}\tau^{\varepsilon}\varpi^{\varepsilon}\right]\right\}E_{\varepsilon}\left[\left(2\pi\ \mathrm{i}\right)^{\varepsilon}\tau^{\varepsilon}\varpi^{\varepsilon}\right]\\&=\frac{1}{\sqrt{\Gamma\left(1+\varepsilon\right)}}\int_{-\infty}^{\infty}\left(\int_{-\infty}^{\infty}\frac{\phi\left(\tau\right)}{\sqrt{\Gamma\left(1+\varepsilon\right)}}E_{\varepsilon}\left[-\left(2\pi\ \mathrm{i}\right)^{\varepsilon}\tau^{\varepsilon}\varpi^{\varepsilon}\right]\left(\mathrm{d}\tau\right)^{\varepsilon}\right)\\&\quad E_{\varepsilon}\left[\left(2\pi\ \mathrm{i}\right)^{\varepsilon}\tau^{\varepsilon}\varpi^{\varepsilon}\right]\left(\mathrm{d}\varpi\right)^{\varepsilon}.\end{aligned}\tag{3.29}$$

In this case, we find the following quantity:

$$H\left(\varpi\right)=\int_{-\infty}^{\infty}\frac{\phi\left(\tau\right)}{\sqrt{\Gamma\left(1+\varepsilon\right)}}E_{\varepsilon}\left[-\left(2\pi\ \mathrm{i}\right)^{\varepsilon}\tau^{\varepsilon}\varpi^{\varepsilon}\right]\left(\mathrm{d}\tau\right)^{\varepsilon}.\tag{3.30}$$

We now revise (3.30) and we obtain the following alternative definition of local fractional Fourier transform operator.

Definition 3.3. Upon setting $g\in L_{1,\varepsilon}\left[\mathrm{R}\right]$ and $\|g\|_{1,\varepsilon}<\infty$, the generalized local fractional Fourier transform operator, denoted by

$\widetilde{\Im}\left[g\left(\tau\right)\right]=G\left(\varpi\right),$

is defined by

$$\begin{aligned}G\left(\varpi\right)&=\sqrt{\Gamma\left(1+\varepsilon\right)}{}_{-\infty}I_{\infty}^{(\varepsilon)}\left\{g\left(\tau\right)E_{\varepsilon}\left[-\left(2\pi\ \mathrm{i}\right)^{\varepsilon}\tau^{\varepsilon}\varpi^{\varepsilon}\right]\right\}\\&=\frac{1}{\sqrt{\Gamma\left(1+\varepsilon\right)}}\int_{-\infty}^{\infty}g\left(\tau\right)E_{\varepsilon}\left[-\left(2\pi\ \mathrm{i}\right)^{\varepsilon}\tau^{\varepsilon}\varpi^{\varepsilon}\right]\left(\mathrm{d}\tau\right)^{\varepsilon},\end{aligned}\tag{3.31}$$

where $\tau, \varpi \in R$, $\widetilde{\Im}$ is called the generalized local fractional Fourier transform operator.

Definition 3.4. The inverse generalized local fractional Fourier transform operator, denoted by

$$\widetilde{\Im}^{-1}\left[G(\varpi)\right] = g(\tau),$$

is defined by

$$\begin{aligned} g(\tau) &= \sqrt{\Gamma(1+\varepsilon)}{}_{-\infty}I_{\infty}^{(\varepsilon)}\left\{G(\varpi)\,E_{\varepsilon}\left[(2\pi\ \mathrm{i})^{\varepsilon}\,\tau^{\varepsilon}\varpi^{\varepsilon}\right]\right\} \\ &= \frac{1}{\sqrt{\Gamma(1+\varepsilon)}}\int_{-\infty}^{\infty} G(\varpi)\,E_{\varepsilon}\left[(2\pi\ \mathrm{i})^{\varepsilon}\,\tau^{\varepsilon}\varpi^{\varepsilon}\right](\mathrm{d}\varpi)^{\varepsilon}, \end{aligned} \tag{3.32}$$

where $\tau, \varpi \in R$ and $\widetilde{\Im}$ is called the inverse generalized local fractional Fourier transform operator.

In this case, we find that $\omega = 2\pi\varpi$.

For various definitions of the local fractional Fourier transform operators, we refer the reader to the earlier works [16, 21, 27, 76, 96, 97, 105–107]. Equation (3.17) provides the common definition of the local fractional Fourier transform operator to be applied to find the solutions of partial differential equations [27–105]. Equation (3.29) was utilized to handle the nondifferentiable problems in mathematical physics [100]. Therefore, we consider the expressions (3.27) and (3.31) throughout this chapter in order to present some examples.

3.2.3 Properties and theorems of local fractional Fourier transform operator

Theorem 3.1 (Fourier integral theorem for local fractional Fourier transform operator). *Suppose that $\theta(\tau)$ is local fractional continuous on the interval $[-\infty, \infty]$ (or $\theta(\tau), \Theta(\omega) \in L_{1,\varepsilon}[R]$). Thus, we have*

$$\theta(\tau) = \frac{\Gamma(1+\varepsilon)}{(2\pi)^{\varepsilon}}{}_{-\infty}I_{\infty}^{(\varepsilon)}\left\{\left[{}_{-\infty}I_{\infty}^{(\varepsilon)}\left(\theta(\tau)\,E_{\varepsilon}\left(-i^{\varepsilon}\tau^{\varepsilon}\omega^{\varepsilon}\right)\right)\right]E_{\varepsilon}\left(i^{\varepsilon}\tau^{\varepsilon}\omega^{\varepsilon}\right)\right\}. \tag{3.33}$$

This is the Fourier integral theorem for the local fractional Fourier transform operator.

Proof. Since $\theta(\tau)$ is locally fractional continuous on the interval $[-\infty, \infty]$, from (3.24), we conclude that

$$\phi(\tau+0) = \phi(\tau-0) = \phi(\tau), \tag{3.34}$$

so that

$$\theta(\tau) = \frac{1}{(2\pi)^{\varepsilon}} \int_{-\infty}^{\infty} \left\{ \left[\frac{1}{\Gamma(1+\varepsilon)} \int_{-\infty}^{\infty} \theta(\tau) E_{\varepsilon}\left(-i^{\varepsilon}\tau^{\varepsilon}\omega^{\varepsilon}\right)(d\tau)^{\varepsilon} \right] E_{\varepsilon}\left(i^{\varepsilon}\tau^{\varepsilon}\omega^{\varepsilon}\right) \right\} (d\omega)^{\varepsilon}$$
$$= \frac{\Gamma(1+\varepsilon)}{(2\pi)^{\varepsilon}} {}_{-\infty}I_{\infty}^{(\varepsilon)} \left\{ \left[{}_{-\infty}I_{\infty}^{(\varepsilon)} \left(\theta(\tau) E_{\varepsilon}\left(-i^{\varepsilon}\tau^{\varepsilon}\omega^{\varepsilon}\right)\right) \right] E_{\varepsilon}\left(i^{\varepsilon}\tau^{\varepsilon}\omega^{\varepsilon}\right) \right\}. \quad (3.35)$$

The asserted claim is thus proved. □

Property 7 (Linearity for the local fractional Fourier transform operator). *Suppose that*

$$\theta_1(\tau), \theta_2(\tau), \Theta_1(\omega), \Theta_2(\omega) \in L_{1,\varepsilon}[R],$$

$$\Im[\theta_1(\tau)] = \Theta_1(\omega)$$

and

$$\Im[\theta_2(\tau)] = \Theta_2(\omega).$$

Then,

$$\Im[a\theta_1(\tau) \pm b\theta_2(\tau)] = a\Theta_1(\omega) \pm b\Theta_2(\omega), \quad (3.36)$$

where a and b are constants.

Proof. By using the definition of the local fractional Fourier transform operator, we have

$$\Im[a\theta_1(\tau) \pm b\theta_2(\tau)] = \frac{1}{\Gamma(1+\varepsilon)} \int_{-\infty}^{\infty} [a\theta_1(\tau) \pm b\theta_2(\tau)] E_{\varepsilon}\left(-i^{\varepsilon}\tau^{\varepsilon}\omega^{\varepsilon}\right)(d\tau)^{\varepsilon}$$
$$= a\frac{1}{\Gamma(1+\varepsilon)} \int_{-\infty}^{\infty} \left[\theta_1(\tau) E_{\varepsilon}\left(-i^{\varepsilon}\tau^{\varepsilon}\omega^{\varepsilon}\right)\right](d\tau)^{\varepsilon}$$
$$\pm b\frac{1}{\Gamma(1+\varepsilon)} \int_{-\infty}^{\infty} \left[\theta_2(\tau) E_{\varepsilon}\left(-i^{\varepsilon}\tau^{\varepsilon}\omega^{\varepsilon}\right)\right](d\tau)^{\varepsilon}$$
$$= a\Theta_1(\omega) \pm b\Theta_2(\omega). \quad (3.37)$$

Thus, the asserted claim is proved. □

Property 8 (Shifting time for the local fractional Fourier transform operator). *Suppose that*

$$\theta(\tau), \Theta(\omega) \in L_{1,\varepsilon}[R],$$

$$\Im[\theta(\tau)] = \Theta(\omega),$$

and a is a constant. Then,

$$\Im[\theta(\tau - a)] = E_{\varepsilon}\left(-i^{\varepsilon}a^{\varepsilon}\omega^{\varepsilon}\right)\Im[\theta(\tau)]. \quad (3.38)$$

Proof. The definition of the local fractional Fourier transform operator leads us to

$$
\begin{aligned}
\Im\left[\theta\left(\tau-a\right)\right] &= \frac{1}{\Gamma\left(1+\varepsilon\right)}\int_{-\infty}^{\infty}\theta\left(\tau-a\right)E_{\varepsilon}\left(-\mathrm{i}^{\varepsilon}\tau^{\varepsilon}\omega^{\varepsilon}\right)\left(\mathrm{d}\tau\right)^{\varepsilon} \\
&= E_{\varepsilon}\left(-\mathrm{i}^{\varepsilon}a^{\varepsilon}\omega^{\varepsilon}\right)\frac{1}{\Gamma\left(1+\varepsilon\right)}\int_{-\infty}^{\infty}\theta\left(\tau-a\right)E_{\varepsilon}\left(-\mathrm{i}^{\varepsilon}\left(\tau-a\right)^{\varepsilon}\omega^{\varepsilon}\right)\left(\mathrm{d}\tau\right)^{\varepsilon} \\
&= E_{\varepsilon}\left(-\mathrm{i}^{\varepsilon}a^{\varepsilon}\omega^{\varepsilon}\right)\Theta\left(\omega\right) \\
&= E_{\varepsilon}\left(-\mathrm{i}^{\varepsilon}a^{\varepsilon}\omega^{\varepsilon}\right)\Im\left[\theta\left(\tau\right)\right],
\end{aligned}
\tag{3.39}
$$

which evidently completes the proof. □

Property 9 (Scaling time for local fractional Fourier transform operator)**.** *Suppose* $\theta\left(\tau\right),\Theta\left(\omega\right)\in L_{1,\varepsilon}\left[R\right]$, $\Im\left[\theta\left(\tau\right)\right]=\Theta\left(\omega\right)$, *and* a $(a>0)$ *is a constant, then, there is*

$$
\Im\left[\theta\left(a\tau\right)\right]=\frac{1}{a^{\varepsilon}}\Theta\left(\frac{\omega}{a}\right). \tag{3.40}
$$

Proof. Using the definition of the local fractional Fourier transform operator

$$
\begin{aligned}
\Im\left[\theta\left(a\tau\right)\right] &= \frac{1}{\Gamma\left(1+\varepsilon\right)}\int_{-\infty}^{\infty}\theta\left(a\tau\right)E_{\varepsilon}\left(-\mathrm{i}^{\varepsilon}\tau^{\varepsilon}\omega^{\varepsilon}\right)\left(\mathrm{d}\tau\right)^{\varepsilon} \\
&= \frac{1}{a^{\varepsilon}}\frac{1}{\Gamma\left(1+\varepsilon\right)}\int_{-\infty}^{\infty}\theta\left(a\tau\right)E_{\varepsilon}\left(-\mathrm{i}^{\varepsilon}\left(\frac{a\tau}{a}\right)^{\varepsilon}\omega^{\varepsilon}\right)\left(\mathrm{d}a\tau\right)^{\varepsilon} \\
&= \frac{1}{a^{\varepsilon}}\Theta\left(\frac{\omega}{a}\right).
\end{aligned}
\tag{3.41}
$$

Therefore, the proof is completed. □

Property 10 (Conjugate for local fractional Fourier transform operator)**.** *Suppose that*

$$
\theta\left(\tau\right),\Theta\left(\omega\right)\in L_{1,\varepsilon}\left[\mathrm{R}\right]
$$

and

$$
\Im\left[\theta\left(\tau\right)\right]=\Theta\left(\omega\right).
$$

Then,

$$
\Im\left[\overline{\theta\left(-\tau\right)}\right]=\overline{\Im\left[\theta\left(\tau\right)\right]}. \tag{3.42}
$$

Proof. By using the definition of the local fractional Fourier transform operator, we get

$$
\begin{aligned}
\Im\left[\overline{\theta\left(-\tau\right)}\right] &= \frac{1}{\Gamma\left(1+\varepsilon\right)}\int_{-\infty}^{\infty}\overline{\theta\left(-\tau\right)}E_{\varepsilon}\left(-\mathrm{i}^{\varepsilon}\tau^{\varepsilon}\omega^{\varepsilon}\right)\left(\mathrm{d}\tau\right)^{\varepsilon} \\
&= \frac{1}{\Gamma\left(1+\varepsilon\right)}\int_{-\infty}^{\infty}\overline{\theta\left(-\tau\right)E_{\varepsilon}\left(\mathrm{i}^{\varepsilon}\tau^{\varepsilon}\omega^{\varepsilon}\right)\left(\mathrm{d}\tau\right)^{\varepsilon}}
\end{aligned}
$$

$$
\begin{aligned}
&= \overline{\frac{1}{\Gamma(1+\varepsilon)} \int_{-\infty}^{\infty} \theta(-\tau) E_{\varepsilon}\left(-\mathrm{i}^{\varepsilon}(-\tau)^{\varepsilon} \omega^{\varepsilon}\right)(\mathrm{d}\tau)^{\varepsilon}} \\
&= \overline{\frac{1}{\Gamma(1+\varepsilon)} \int_{-\infty}^{\infty} \theta(\tau) E_{\varepsilon}\left(-\mathrm{i}^{\varepsilon} \tau^{\varepsilon} \omega^{\varepsilon}\right)(\mathrm{d}\tau)^{\varepsilon}} \\
&= \overline{\Im[\theta(\tau)]}.
\end{aligned} \tag{3.43}
$$

Thus, we have completed the proof. □

Property 11 (Translation for local fractional Fourier transform operator). *Suppose that*

$$
\begin{aligned}
&\theta(\tau), \Theta(\omega) \in L_{1,\varepsilon}[R], \\
&\quad \Im[\theta(\tau)] = \Theta(\omega),
\end{aligned}
$$

and a is a constant. Then,

$$
\Im\left[E_{\varepsilon}\left(\mathrm{i}^{\varepsilon} a^{\varepsilon} \tau^{\varepsilon}\right) \theta(\tau)\right] = \Theta(\omega - a). \tag{3.44}
$$

Proof. Once again, by using the definition of the local fractional Fourier transform operator, we conclude that

$$
\begin{aligned}
\Im\left[E_{\varepsilon}\left(-\mathrm{i}^{\varepsilon} a^{\varepsilon} \tau^{\varepsilon}\right) \theta(\tau)\right] &= \frac{1}{\Gamma(1+\varepsilon)} \int_{-\infty}^{\infty}\left[\theta(\tau) E_{\varepsilon}\left(\mathrm{i}^{\varepsilon} a^{\varepsilon} \tau^{\varepsilon}\right)\right] E_{\varepsilon}\left(-\mathrm{i}^{\varepsilon} \tau^{\varepsilon} \omega^{\varepsilon}\right)(\mathrm{d}\tau)^{\varepsilon} \\
&= \frac{1}{\Gamma(1+\varepsilon)} \int_{-\infty}^{\infty} \theta(\tau) E_{\varepsilon}\left(-\mathrm{i}^{\varepsilon}(\omega-a)^{\varepsilon} \tau^{\varepsilon}\right)(\mathrm{d}\tau)^{\varepsilon} \\
&= \Theta(\omega-a).
\end{aligned} \tag{3.45}
$$

Therefore, we have proved the claim. □

Property 12 (Duality for local fractional Fourier transform operator). *Suppose that*

$$
\theta(\tau), \Theta(\omega) \in L_{1,\varepsilon}[R]
$$

and

$$
\Im[\theta(\tau)] = \Theta(\omega).
$$

Then,

$$
\Im[\theta(-\tau)] = \frac{\Gamma(1+\varepsilon)}{(2\pi)^{\varepsilon}} \Im[\Theta(\tau)]. \tag{3.46}
$$

Proof. It follows directly from the definition of the local fractional Fourier transform operator that

$$
\theta(\tau) = \frac{1}{(2\pi)^{\varepsilon}} \int_{-\infty}^{\infty} \Theta(\omega) E_{\varepsilon}\left(\mathrm{i}^{\varepsilon} \tau^{\varepsilon} \omega^{\varepsilon}\right)(\mathrm{d}\omega)^{\varepsilon} = \Im^{-1}[\Theta(\omega)]. \tag{3.47}
$$

Upon interchanging τ and ω, from (3.47), we have

$$\theta(\omega) = \frac{1}{(2\pi)^{\varepsilon}} \int_{-\infty}^{\infty} \Theta(\tau) E_{\varepsilon}\left(\mathrm{i}^{\varepsilon}\omega^{\varepsilon}\tau^{\varepsilon}\right)(\mathrm{d}\tau)^{\varepsilon}. \tag{3.48}$$

By replacing ω by $-\omega$, we obtain

$$\begin{aligned}\theta(-\omega) &= \frac{\Gamma(1+\varepsilon)}{(2\pi)^{\varepsilon}}\left[\frac{1}{\Gamma(1+\varepsilon)}\int_{-\infty}^{\infty}\Theta(\tau)E_{\varepsilon}\left(-\mathrm{i}^{\varepsilon}\omega^{\varepsilon}\tau^{\varepsilon}\right)(\mathrm{d}\tau)^{\varepsilon}\right]\\ &= \frac{\Gamma(1+\varepsilon)}{(2\pi)^{\varepsilon}}\Im[\Theta(\tau)].\end{aligned} \tag{3.49}$$

This proves the result (3.46). □

Property 13 (Composition for the local fractional Fourier transform operator). *Suppose that*

$$\theta_1(\tau), \theta_2(\tau), \Theta_1(\omega), \Theta_2(\omega) \in L_{1,\varepsilon}[R],$$

$$\Im[\theta_1(\tau)] = \Theta_1(\omega),$$

and

$$\Im[\theta_2(\tau)] = \Theta_2(\omega).$$

Then,

$$\int_{-\infty}^{\infty}\theta_1(\omega)\Theta_2(\omega)E_{\varepsilon}\left(\mathrm{i}^{\varepsilon}\tau^{\varepsilon}\omega^{\varepsilon}\right)(\mathrm{d}\omega)^{\varepsilon} = \int_{-\infty}^{\infty}\theta_2(\eta)\Theta_1(\eta-\tau)(\mathrm{d}\eta)^{\varepsilon}. \tag{3.50}$$

Proof. We write the left-hand side of (3.50) in the following form:

$$\begin{aligned}&\int_{-\infty}^{\infty}\theta_1(\omega)\Theta_2(\omega)E_{\varepsilon}\left(\mathrm{i}^{\varepsilon}\tau^{\varepsilon}\omega^{\varepsilon}\right)(\mathrm{d}\omega)^{\varepsilon}\\ &= \int_{-\infty}^{\infty}\theta_1(\omega)E_{\varepsilon}\left(\mathrm{i}^{\varepsilon}\tau^{\varepsilon}\omega^{\varepsilon}\right)\left[\frac{1}{\Gamma(1+\varepsilon)}\int_{-\infty}^{\infty}\theta_2(\eta)E_{\varepsilon}\left(-\mathrm{i}^{\varepsilon}\eta^{\varepsilon}\omega^{\varepsilon}\right)(\mathrm{d}\eta)^{\varepsilon}\right](\mathrm{d}\omega)^{\varepsilon}\\ &= \int_{-\infty}^{\infty}\theta_2(\eta)(\mathrm{d}\eta)^{\varepsilon}\left[\frac{1}{\Gamma(1+\varepsilon)}\int_{-\infty}^{\infty}\theta_1(\omega)E_{\varepsilon}\left(-\mathrm{i}^{\varepsilon}(\eta-\tau)^{\varepsilon}\omega^{\varepsilon}\right)(\mathrm{d}\omega)^{\varepsilon}\right]\\ &= \int_{-\infty}^{\infty}\theta_2(\eta)\Theta_1(\eta-\tau)(\mathrm{d}\eta)^{\varepsilon}.\end{aligned} \tag{3.51}$$

Therefore, we have given the proof of (3.50). □

It easily follows from (3.51) that

$$\int_{-\infty}^{\infty}\theta_1(\omega)\Theta_2(\omega)(\mathrm{d}\omega)^{\varepsilon} = \int_{-\infty}^{\infty}\Theta_1(\eta)\theta_2(\eta)(\mathrm{d}\eta)^{\varepsilon}, \tag{3.52}$$

where $\tau = 0$.

Theorem 3.2. *Suppose that*

$$\theta(\tau), \Theta(\omega) \in L_{1,\varepsilon}[R],$$

$$\Im[\theta(\tau)] = \Theta(\omega),$$

and

$$\lim_{|\tau|\to\infty} \theta(\tau) = 0.$$

Then,

$$\Im\left[\theta^{(\varepsilon)}(\tau)\right] = (i\omega)^{\varepsilon}\,\Im[\theta(\tau)] = i^{\varepsilon}\omega^{\varepsilon}\Im[\theta(\tau)]. \tag{3.53}$$

Proof. The local fractional Fourier transform definition leads us to the operator given by

$$\Im\left[\theta^{(\varepsilon)}(\tau)\right] = \frac{1}{\Gamma(1+\varepsilon)}\int_{-\infty}^{\infty}\theta^{(\varepsilon)}(\tau)\,E_{\varepsilon}\left(-\mathrm{i}^{\varepsilon}\tau^{\varepsilon}\omega^{\varepsilon}\right)(\mathrm{d}\tau)^{\varepsilon}. \tag{3.54}$$

Now, upon integration by part for the local fractional integral operator in (3.54), we conclude that

$$\begin{aligned}\Im\left[\theta^{(\varepsilon)}(\tau)\right] &= \frac{1}{\Gamma(1+\varepsilon)}\int_{-\infty}^{\infty}\theta^{(\varepsilon)}(\tau)\,E_{\varepsilon}\left(-\mathrm{i}^{\varepsilon}\tau^{\varepsilon}\omega^{\varepsilon}\right)(\mathrm{d}\tau)^{\varepsilon}\\ &=\left[\theta(\tau)\,E_{\varepsilon}\left(-\mathrm{i}^{\varepsilon}\tau^{\varepsilon}\omega^{\varepsilon}\right)\right]_{-\infty}^{\infty} - \frac{\mathrm{i}^{\varepsilon}\omega^{\varepsilon}}{\Gamma(1+\varepsilon)}\int_{-\infty}^{\infty}\theta(\tau)\,E_{\varepsilon}\left(-\mathrm{i}^{\varepsilon}\tau^{\varepsilon}\omega^{\varepsilon}\right)(\mathrm{d}\tau)^{\varepsilon},\end{aligned} \tag{3.55}$$

which, by using

$$\lim_{|\tau|\to\infty} \theta(\tau) = 0,$$

yields

$$\begin{aligned}\Im\left[\theta^{(\varepsilon)}(\tau)\right] &= \mathrm{i}^{\varepsilon}\omega^{\varepsilon}\left[\frac{1}{\Gamma(1+\varepsilon)}\int_{-\infty}^{\infty}\theta(\tau)\,E_{\varepsilon}\left(-\mathrm{i}^{\varepsilon}\tau^{\varepsilon}\omega^{\varepsilon}\right)(\mathrm{d}\tau)^{\varepsilon}\right]\\ &= \mathrm{i}^{\varepsilon}\omega^{\varepsilon}\Im[\theta(\tau)].\end{aligned} \tag{3.56}$$

Therefore, the result (3.53) follows. □

In a similar manner, by repeating this process, we have

$$\Im\left[\theta^{(k\varepsilon)}(\tau)\right] = (\mathrm{i}\omega)^{k\varepsilon}\,\Im[\theta(\tau)], \quad k \in \mathrm{N}, \tag{3.57}$$

where

$$\theta^{((k-1)\varepsilon)}(0) = \cdots = \theta^{(\varepsilon)}(0) = \theta(0) = 0. \tag{3.58}$$

Theorem 3.3. *Suppose that*

$$\theta(\tau), \Theta(\omega) \in L_{1,\varepsilon}[R],$$

$$\Im[\theta(\tau)] = \Theta(\omega),$$

and

$$\lim_{\tau\to\infty} {}_{-\infty}I_{\tau}^{(\varepsilon)}\theta(\tau) = 0.$$

Then,

$$\Im\left[{}_{-\infty}I_{\tau}^{(\varepsilon)}\theta(\tau)\right] = \frac{1}{(i\omega)^{\varepsilon}}\Im[\theta(\tau)]. \tag{3.59}$$

Proof. In accordance with the definition of the local fractional Fourier transform operator, we have

$$\Im\left[{}_{-\infty}I_{\tau}^{(\varepsilon)}\theta(\tau)\right] = \frac{1}{\Gamma(1+\varepsilon)}\int_{-\infty}^{\infty}\left({}_{-\infty}I_{\tau}^{(\varepsilon)}\theta(\tau)\right)E_{\varepsilon}\left(-\mathrm{i}^{\varepsilon}\tau^{\varepsilon}\omega^{\varepsilon}\right)(\mathrm{d}\tau)^{\varepsilon}, \tag{3.60}$$

which, by using integration by part for the local fractional integral operator, yields

$$\begin{aligned}\Im\left[{}_{-\infty}I_{\tau}^{(\varepsilon)}\theta(\tau)\right] &= \left[{}_{-\infty}I_{\tau}^{(\varepsilon)}\theta(\tau)E_{\varepsilon}\left(-\mathrm{i}^{\varepsilon}\tau^{\varepsilon}\omega^{\varepsilon}\right)\right]\Big|_{-\infty}^{\infty} \\ &\quad + \left[\frac{(\mathrm{i}\omega)^{\varepsilon}}{\Gamma(1+\varepsilon)}\int_{-\infty}^{\infty}\theta(\tau)E_{\varepsilon}\left(-\mathrm{i}^{\varepsilon}\tau^{\varepsilon}\omega^{\varepsilon}\right)(\mathrm{d}\tau)^{\varepsilon}\right].\end{aligned} \tag{3.61}$$

Now, by taking

$$\lim_{\tau\to\infty} {}_{-\infty}I_{\tau}^{(\varepsilon)}\theta(\tau) = 0$$

into account in (3.61), we conclude that

$$\begin{aligned}\Im\left[{}_{-\infty}I_{\tau}^{(\varepsilon)}\theta(\tau)\right] &= \left[\frac{(\mathrm{i}\omega)^{\varepsilon}}{\Gamma(1+\varepsilon)}\int_{-\infty}^{\infty}\theta(\tau)E_{\varepsilon}\left(-\mathrm{i}^{\varepsilon}\tau^{\varepsilon}\omega^{\varepsilon}\right)(\mathrm{d}\tau)^{\varepsilon}\right] \\ &= (\mathrm{i}\omega)^{\varepsilon}\,\Im[\theta(\tau)].\end{aligned} \tag{3.62}$$

Thus, the claim has been proved. □

Continuing in the same manner, repeating the above methodology, we obtain

$$\Im\left[{}_{-\infty}I_{\tau}^{(k\varepsilon)}\theta(\tau)\right] = \frac{1}{(\mathrm{i}\omega)^{k\varepsilon}}\Im[\theta(\tau)], \tag{3.63}$$

where

$$\lim_{\tau\to\infty} {}_{-\infty}I_{\tau}^{(k\varepsilon)}\theta(\tau) = 0. \tag{3.64}$$

Theorem 3.4. *Suppose that*

$$\theta(\tau), \Theta(\omega) \in L_{1,\varepsilon}[R]$$

and

$$\Im\left[\theta\left(\tau\right)\right]=\Theta\left(\omega\right).$$

Then,

$$\Im\left[\tau^{\varepsilon}\theta\left(\tau\right)\right]=i^{\varepsilon}\Theta^{(\varepsilon)}\left(\omega\right). \tag{3.65}$$

Proof. We consider

$$\Im^{-1}\left[\mathrm{i}^{\varepsilon}\Theta^{(\varepsilon)}\left(\omega\right)\right]=\frac{1}{(2\pi)^{\varepsilon}}\int_{-\infty}^{\infty}\left[\mathrm{i}^{\varepsilon}\Theta^{(\varepsilon)}\left(\omega\right)\right]E_{\varepsilon}\left(\mathrm{i}^{\varepsilon}\tau^{\varepsilon}\omega^{\varepsilon}\right)(\mathrm{d}\omega)^{\varepsilon}, \tag{3.66}$$

which, by integrating by part for local fractional integral operator, leads us to

$$\begin{aligned}\Im^{-1}\left[\mathrm{i}^{\varepsilon}\Theta^{(\varepsilon)}\left(\omega\right)\right]&=\frac{\Gamma\left(1+\varepsilon\right)}{(2\pi)^{\varepsilon}}\left[\mathrm{i}^{\varepsilon}\Theta\left(\omega\right)E_{\varepsilon}\left(\mathrm{i}^{\varepsilon}\tau^{\varepsilon}\omega^{\varepsilon}\right)\right]_{-\infty}^{\infty}\\&\quad+\tau^{\varepsilon}\left[\frac{1}{(2\pi)^{\varepsilon}}\int_{-\infty}^{\infty}\Theta\left(\omega\right)E_{\varepsilon}\left(\mathrm{i}^{\varepsilon}\tau^{\varepsilon}\omega^{\varepsilon}\right)(\mathrm{d}\omega)^{\varepsilon}\right].\end{aligned} \tag{3.67}$$

Taking

$$\lim_{|\omega|\to\infty}\Theta\left(\omega\right)=0$$

into account in (3.65), we get

$$\begin{aligned}\Im^{-1}\left[\mathrm{i}^{\varepsilon}\Theta^{(\varepsilon)}\left(\omega\right)\right]&=\tau^{\varepsilon}\left[\frac{1}{(2\pi)^{\varepsilon}}\int_{-\infty}^{\infty}\Theta\left(\omega\right)E_{\varepsilon}\left(\mathrm{i}^{\varepsilon}\tau^{\varepsilon}\omega^{\varepsilon}\right)(\mathrm{d}\omega)^{\varepsilon}\right]\\&=\tau^{\varepsilon}\theta\left(\tau\right),\end{aligned} \tag{3.68}$$

which evidently completes the proof. □

In a similar manner, by repeating this process, we get

$$\Im\left[\tau^{k\varepsilon}\theta\left(\tau\right)\right]=\mathrm{i}^{k\varepsilon}\Theta^{(k\varepsilon)}\left(\omega\right), \tag{3.69}$$

where

$$\lim_{|\omega|\to\infty}\Theta^{(k\varepsilon)}\left(\omega\right)=0,\quad k\in\mathrm{N}. \tag{3.70}$$

Definition 3.5. The local fractional convolution of two functions $\theta_1(\tau)$ and $\theta_2(\tau)$ via the local fractional integral operator, denoted by

$$\left(\theta_1*\theta_2\right)(\tau)=\theta_1(\tau)*\theta_2(\tau),$$

is defined by

$$\begin{aligned}\left(\theta_1*\theta_2\right)(\tau)&=\theta_1(\tau)*\theta_2(\tau)\\&={}_{-\infty}I_{\infty}^{(\varepsilon)}\left[\theta_1(t)\,\theta_2(\tau-t)\right]\\&=\frac{1}{\Gamma\left(1+\varepsilon\right)}\int_{-\infty}^{\infty}\theta_1(t)\,\theta_2(\tau-t)\,(\mathrm{d}t)^{\varepsilon}.\end{aligned} \tag{3.71}$$

Definition 3.6. The local fractional convolution of the local fractional Fourier transform operators $\Theta_1(\omega)$ and $\Theta_2(\omega)$, denoted by

$$(\Theta_1 * \Theta_2)(\omega) = \Theta_1(\omega) * \Theta_2(\omega),$$

is defined by

$$\begin{aligned}(\Theta_1 * \Theta_2)(\omega) &= \Theta_1(\omega) * \Theta_2(\omega) \\ &= \frac{\Gamma(1+\varepsilon)}{(2\pi)^{\varepsilon}} {}_{-\infty}I_{\infty}^{(\varepsilon)} \left[\Theta_1(\varpi)\Theta_2(\omega-\varpi)\right] \\ &= \frac{1}{(2\pi)^{\varepsilon}} \int_{-\infty}^{\infty} \Theta_1(\varpi)\Theta_2(\omega-\varpi)(\mathrm{d}\varpi)^{\varepsilon}. \end{aligned} \tag{3.72}$$

From the definition of the local fractional convolution of two functions, we have the following properties:

(a) $\theta_1(\tau) * \theta_2(\tau) = \theta_2(\tau) * \theta_1(\tau)$ and
(b) $\theta_1(\tau) * (\theta_2(\tau) + \theta_3(\tau)) = \theta_1(\tau) * \theta_2(\tau) + \theta_1(\tau) * \theta_3(\tau)$.

Theorem 3.5. *Suppose that*

$$\theta_1(\tau), \theta_2(\tau), \Theta_1(\omega), \Theta_2(\omega) \in L_{1,\varepsilon}[R],$$

$$\Im[\theta_1(\tau)] = \Theta_1(\omega),$$

and

$$\Im[\theta_2(\tau)] = \Theta_2(\omega).$$

Then,

$$\Im[\theta_1(\tau) * \theta_2(\tau)] = \Theta_1(\omega)\Theta_2(\omega) \tag{3.73}$$

or

$$\theta_1(\tau) * \theta_2(\tau) = \Im^{-1}\left[\Theta_1(\omega)\Theta_2(\omega)\right] \tag{3.74}$$

or, equivalently,

$$\frac{1}{\Gamma(1+\varepsilon)} \int_{-\infty}^{\infty} \theta_1(t)\theta_2(\tau-t)(dt)^{\varepsilon} = \frac{1}{(2\pi)^{\varepsilon}} \int_{-\infty}^{\infty} \Theta_1(\omega)\Theta_2(\omega) E_{\varepsilon}\left(i^{\varepsilon}\tau^{\varepsilon}\omega^{\varepsilon}\right)(d\omega)^{\varepsilon}. \tag{3.75}$$

Proof. With the help of the definition of the local fractional Fourier transform operator, we have

$$\begin{aligned}\Im[\theta_1(\tau) * \theta_2(\tau)] &= \frac{1}{\Gamma(1+\varepsilon)} \int_{-\infty}^{\infty} E_{\varepsilon}\left(-\mathrm{i}^{\varepsilon}\tau^{\varepsilon}\omega^{\varepsilon}\right)(\mathrm{d}\tau)^{\varepsilon} \\ &\quad \times \left[\frac{1}{\Gamma(1+\varepsilon)} \int_{-\infty}^{\infty} \theta_1(\eta)\theta_2(\tau-\eta)(\mathrm{d}\eta)^{\varepsilon}\right] \\ &= \frac{1}{\Gamma(1+\varepsilon)} \int_{-\infty}^{\infty} \theta_1(\eta) E_{\varepsilon}\left(-\mathrm{i}^{\varepsilon}\eta^{\varepsilon}\omega^{\varepsilon}\right) \\ &\quad \times \left[\frac{1}{\Gamma(1+\varepsilon)} \int_{-\infty}^{\infty} E_{\varepsilon}\left(-\mathrm{i}^{\varepsilon}(\tau-\eta)^{\varepsilon}\omega^{\varepsilon}\right)\theta_2(\tau-\eta)(\mathrm{d}\tau)^{\varepsilon}\right](\mathrm{d}\eta)^{\varepsilon}, \end{aligned} \tag{3.76}$$

where

$$\Theta_2(\omega) = \frac{1}{\Gamma(1+\varepsilon)} \int_{-\infty}^{\infty} E_\varepsilon\left(-\mathrm{i}^\varepsilon (\tau-\eta)^\varepsilon \omega^\varepsilon\right) \theta_2(\tau-\eta) (\mathrm{d}\tau)^\varepsilon \tag{3.77}$$

and

$$\begin{aligned} \Im[\theta_1(\tau) * \theta_2(\tau)] &= \frac{1}{\Gamma(1+\varepsilon)} \int_{-\infty}^{\infty} \theta_1(\eta) E_\varepsilon\left(-\mathrm{i}^\varepsilon \eta^\varepsilon \omega^\varepsilon\right) \Theta_2(\omega) (\mathrm{d}\eta)^\varepsilon \\ &= \Theta_2(\omega) \frac{1}{\Gamma(1+\varepsilon)} \int_{-\infty}^{\infty} \theta_1(\eta) E_\varepsilon\left(-\mathrm{i}^\varepsilon \eta^\varepsilon \omega^\varepsilon\right) (\mathrm{d}\eta)^\varepsilon \\ &= \Theta_1(\omega) \Theta_2(\omega). \end{aligned} \tag{3.78}$$

Thus, we have proved the result. □

Theorem 3.6. *Suppose that*

$$\theta_1(\tau), \theta_2(\tau), \Theta_1(\omega), \Theta_2(\omega) \in L_{1,\varepsilon}[R],$$

$$\Im[\theta_1(\tau)] = \Theta_1(\omega),$$

and

$$\Im[\theta_2(\tau)] = \Theta_2(\omega).$$

Then,

$$\Im[\theta_1(\tau)\theta_2(\tau)] = \Theta_1(\omega) * \Theta_2(\omega) \tag{3.79}$$

or

$$\theta_1(\tau)\theta_2(\tau) = \Im^{-1}[\Theta_1(\omega) * \Theta_2(\omega)] \tag{3.80}$$

or, equivalently,

$$\frac{1}{\Gamma(1+\varepsilon)} \int_{-\infty}^{\infty} \theta_1(\tau)\theta_2(\tau) E_\varepsilon\left(-i^\varepsilon \tau^\varepsilon \omega^\varepsilon\right) (d\tau)^\varepsilon = \frac{1}{(2\pi)^\varepsilon} \int_{-\infty}^{\infty} \Theta_1(t)\Theta_2(\omega-t) (dt)^\varepsilon. \tag{3.81}$$

Proof. By directly using the definition of the local fractional Fourier transform operator, we conclude that

$$\begin{aligned} \Im^{-1}[\Theta_1(\omega) * \Theta_2(\omega)] &= \frac{1}{(2\pi)^\varepsilon} \int_{-\infty}^{\infty} \left[\frac{1}{(2\pi)^\varepsilon} \int_{-\infty}^{\infty} \Theta_1(\varpi)\Theta_2(\omega-\varpi) (\mathrm{d}\varpi)^\varepsilon\right] \\ &\quad \times E_\varepsilon\left(\mathrm{i}^\varepsilon \tau^\varepsilon \omega^\varepsilon\right) (\mathrm{d}\omega)^\varepsilon \\ &= \frac{1}{(2\pi)^\varepsilon} \int_{-\infty}^{\infty} E_\varepsilon\left(\mathrm{i}^\varepsilon \tau^\varepsilon \varpi^\varepsilon\right) \Theta_1(\varpi) \\ &\quad \times \left[\frac{1}{(2\pi)^\varepsilon} \int_{-\infty}^{\infty} \Theta_2(\omega-\varpi) E_\varepsilon\left(\mathrm{i}^\varepsilon \tau^\varepsilon (\omega-\varpi)^\varepsilon\right) (\mathrm{d}\omega)^\varepsilon\right] (\mathrm{d}\varpi)^\varepsilon, \end{aligned} \tag{3.82}$$

which yields

$$\theta_2(\tau) = \frac{1}{(2\pi)^\varepsilon}\int_{-\infty}^{\infty}\Theta_2(\omega-\varpi)\,E_\varepsilon\left(\mathrm{i}^\varepsilon\tau^\varepsilon(\omega-\varpi)^\varepsilon\right)(\mathrm{d}\omega)^\varepsilon \tag{3.83}$$

and

$$\begin{aligned}\Im^{-1}\left[\Theta_1(\omega)*\Theta_2(\omega)\right] &= \frac{1}{(2\pi)^\varepsilon}\int_{-\infty}^{\infty}E_\varepsilon\left(\mathrm{i}^\varepsilon\tau^\varepsilon\varpi^\varepsilon\right)\Theta_1(\varpi)\,\theta_2(\tau)\,(\mathrm{d}\varpi)^\varepsilon\\ &= \left[\frac{1}{(2\pi)^\varepsilon}\int_{-\infty}^{\infty}E_\varepsilon\left(\mathrm{i}^\varepsilon\tau^\varepsilon\varpi^\varepsilon\right)\Theta_1(\varpi)\,(\mathrm{d}\varpi)^\varepsilon\right]\theta_2(\tau)\\ &= \theta_1(\tau)\,\theta_2(\tau).\end{aligned} \tag{3.84}$$

Therefore, we have completed the proof. □

Theorem 3.7 (Convolution theorem for the local fractional Fourier transform operator). *Suppose that*

$$\theta_1(\tau),\theta_2(\tau),\Theta_1(\omega),\Theta_2(\omega)\in L_{1,\varepsilon}[R],$$

$$\Im\left[\theta_1(\tau)\right] = \Theta_1(\omega),$$

and

$$\Im\left[\theta_2(\tau)\right] = \Theta_2(\omega).$$

Then,

$$\frac{1}{\Gamma(1+\varepsilon)}\int_{-\infty}^{\infty}\theta_1(\tau)\,\overline{\theta_2(\tau)}\,(d\tau)^\varepsilon = \frac{1}{(2\pi)^\varepsilon}\int_{-\infty}^{\infty}\Theta_1(\omega)\,\overline{\Theta_2(\omega)}\,(d\omega)^\varepsilon. \tag{3.85}$$

Proof. We consider

$$\begin{aligned}\overline{\theta_2(\tau)} &= \overline{\frac{1}{(2\pi)^\varepsilon}\int_{-\infty}^{\infty}\Theta_2(\omega)\,E_\varepsilon\left(\mathrm{i}^\varepsilon\tau^\varepsilon\omega^\varepsilon\right)(\mathrm{d}\omega)^\varepsilon}\\ &= \frac{1}{(2\pi)^\varepsilon}\int_{-\infty}^{\infty}\overline{\Theta_2(\omega)}E_\varepsilon\left(-\mathrm{i}^\varepsilon\tau^\varepsilon\omega^\varepsilon\right)(\mathrm{d}\omega)^\varepsilon.\end{aligned} \tag{3.86}$$

In this case, from (3.86), we observe that

$$\begin{aligned}\frac{1}{\Gamma(1+\varepsilon)}\int_{-\infty}^{\infty}\theta_1(\tau)\,\overline{\theta_2(\tau)}\,(\mathrm{d}\tau)^\varepsilon &= \frac{1}{\Gamma(1+\varepsilon)}\int_{-\infty}^{\infty}\theta_1(\tau)\\ &\quad\times\left[\frac{1}{(2\pi)^\varepsilon}\int_{-\infty}^{\infty}\overline{\Theta_2(\omega)}E_\varepsilon\left(-\mathrm{i}^\varepsilon\tau^\varepsilon\omega^\varepsilon\right)(\mathrm{d}\omega)^\varepsilon\right](\mathrm{d}\tau)^\varepsilon\\ &= \left[\frac{1}{(2\pi)^\varepsilon}\int_{-\infty}^{\infty}\overline{\Theta_2(\omega)}\right.\\ &\quad\left.\times\left[\frac{1}{\Gamma(1+\varepsilon)}\int_{-\infty}^{\infty}\theta_1(\tau)\,E_\varepsilon\left(-\mathrm{i}^\varepsilon\tau^\varepsilon\omega^\varepsilon\right)(\mathrm{d}\tau)^\varepsilon\right](\mathrm{d}\omega)^\varepsilon\right],\end{aligned} \tag{3.87}$$

which yields

$$\Theta_1(\omega) = \frac{1}{\Gamma(1+\varepsilon)} \int_{-\infty}^{\infty} \theta_1(\tau) E_\varepsilon\left(-\mathrm{i}^\varepsilon \tau^\varepsilon \omega^\varepsilon\right) (\mathrm{d}\tau)^\varepsilon \tag{3.88}$$

and

$$\frac{1}{\Gamma(1+\varepsilon)} \int_{-\infty}^{\infty} \theta_1(\tau) \overline{\theta_2(\tau)} (\mathrm{d}\tau)^\varepsilon = \frac{1}{(2\pi)^\varepsilon} \int_{-\infty}^{\infty} \Theta_1(\omega) \overline{\Theta_2(\omega)} (\mathrm{d}\omega)^\varepsilon. \tag{3.89}$$

Therefore, we have completed the proof. □

Theorem 3.8 (Parseval's theorem for local fractional Fourier transform operator). *Suppose that*

$$\theta(\tau) \in L_{1,\varepsilon}[R]$$

and

$$\Im[\theta(\tau)] = \Theta(\omega).$$

Then,

$$\frac{1}{\Gamma(1+\varepsilon)} \int_{-\infty}^{\infty} |\theta(\tau)|^2 (d\tau)^\varepsilon = \frac{1}{(2\pi)^\varepsilon} \int_{-\infty}^{\infty} |\Theta(\omega)|^2 (d\omega)^\varepsilon. \tag{3.90}$$

Proof. Considering

$$\theta_1(\tau) = \theta_2(\tau) = \theta(\tau)$$

in (3.85), we have

$$\theta_1(\tau) \overline{\theta_2(\tau)} = \theta(\tau) \overline{\theta(\tau)} = |\theta(\tau)|^2 \tag{3.91}$$

and

$$\Theta_1(\omega) \overline{\Theta_2(\omega)} = \Theta(\omega) \overline{\Theta(\omega)} = |\Theta(\omega)|^2, \tag{3.92}$$

which lead us to the result (3.90). □

3.2.4 Properties and theorems of the generalized local fractional Fourier transform operator

Theorem 3.9 (Fourier integral theorem for generalized local fractional Fourier transform operator). *Suppose that $g(\tau)$ is local fractional continuous on the interval $[-\infty, \infty]$ (or $g(\tau), G(\varpi) \in L_{1,\varepsilon}[R]$). Then,*

$$\theta(\tau) = \Gamma(1+\varepsilon) \, {}_{-\infty}I_{\infty}^{(\varepsilon)} \left\{ \left[{}_{-\infty}I_{\infty}^{(\varepsilon)} \left(\theta(\tau) E_\varepsilon\left(-(2\pi\, i)^\varepsilon \tau^\varepsilon \omega^\varepsilon\right)\right)\right] E_\varepsilon\left((2\pi\, i)^\varepsilon \tau^\varepsilon \omega^\varepsilon\right)\right\}. \tag{3.93}$$

This is the Fourier integral theorem for the local fractional Fourier transform operator.

Proof. When $\omega = 2\pi\varpi$, we can transform (3.33) into

$$\theta(\tau_0) = \frac{\Gamma(1+\varepsilon)}{(2\pi)^{\varepsilon}} {}_{-\infty}I_{\infty}^{(\varepsilon)} \left\{ \left[{}_{-\infty}I_{\infty}^{(\varepsilon)} \left(\theta(\tau_0) E_{\varepsilon}\left(-\mathrm{i}^{\varepsilon}\tau_0^{\varepsilon} (2\pi\varpi)^{\varepsilon} \right) \right) \right] E_{\varepsilon}\left(\mathrm{i}^{\varepsilon}\tau_0^{\varepsilon} (2\pi\varpi)^{\varepsilon} \right) \right\}, \tag{3.94}$$

which leads us to

$$g(\tau) = \Gamma(1+\varepsilon) {}_{-\infty}I_{\infty}^{(\varepsilon)} \left\{ \left[{}_{-\infty}I_{\infty}^{(\varepsilon)} \left(g(\tau) E_{\varepsilon}\left(-(2\pi\,\mathrm{i})^{\varepsilon} \tau^{\varepsilon}\varpi^{\varepsilon} \right) \right) \right] E_{\varepsilon}\left((2\pi\,\mathrm{i})^{\varepsilon} \tau^{\varepsilon}\varpi^{\varepsilon} \right) \right\}, \tag{3.95}$$

where

$$\tau_0 = \frac{\tau}{2\pi},$$

$$G(\varpi) = \frac{1}{\sqrt{\Gamma(1+\varepsilon)}} \int_{-\infty}^{\infty} g(\tau) E_{\varepsilon}\left[-(2\pi\,\mathrm{i})^{\varepsilon} \tau^{\varepsilon}\varpi^{\varepsilon} \right] (\mathrm{d}\tau)^{\varepsilon} = \sqrt{\Gamma(1+\varepsilon)} {}_{-\infty}I_{\infty}^{(\varepsilon)} \left\{ g(\tau) E_{\varepsilon}\left[-(2\pi\,\mathrm{i})^{\varepsilon} \tau^{\varepsilon}\varpi^{\varepsilon} \right] \right\}, \tag{3.96}$$

and

$$g(\tau) = \frac{1}{\sqrt{\Gamma(1+\varepsilon)}} \int_{-\infty}^{\infty} G(\varpi) E_{\varepsilon}\left[(2\pi\,\mathrm{i})^{\varepsilon} \tau^{\varepsilon}\varpi^{\varepsilon} \right] (\mathrm{d}\varpi)^{\varepsilon} = \sqrt{\Gamma(1+\varepsilon)} {}_{-\infty}I_{\infty}^{(\varepsilon)} \left\{ G(\varpi) E_{\varepsilon}\left[(2\pi\,\mathrm{i})^{\varepsilon} \tau^{\varepsilon}\varpi^{\varepsilon} \right] \right\}. \tag{3.97}$$

Hence, we have completed the proof. □

Property 14 (Linearity for generalized local fractional Fourier transform operator**).** *Suppose that*

$$g_1(\tau), g_2(\tau), G_1(\varpi), G_2(\varpi) \in L_{1,\varepsilon}[R],$$

$$\overline{\Im}[g_1(\tau)] = G_1(\varpi),$$

and

$$\overline{\Im}[g_2(\tau)] = G_2(\varpi).$$

Then,

$$\overline{\Im}[a g_1(\tau) \pm b g_2(\tau)] = a G_1(\varpi) \pm b G_2(\varpi), \tag{3.98}$$

where a and b are constants.

Proof. We observe that

$$\begin{aligned}
\Im\left[ag_1(\tau) \pm bg_2(\tau)\right] &= \frac{1}{\sqrt{\Gamma(1+\varepsilon)}} \int_{-\infty}^{\infty} \left[ag_1(\tau) \pm bg_2(\tau)\right] \\
&\quad E_\varepsilon\left(-(2\pi\,\mathrm{i})^\varepsilon \tau^\varepsilon \varpi^\varepsilon\right)(\mathrm{d}\tau)^\varepsilon \\
&= a\frac{1}{\sqrt{\Gamma(1+\varepsilon)}} \int_{-\infty}^{\infty} \left[g_1(\tau) E_\varepsilon\left(-(2\pi\,\mathrm{i})^\varepsilon \tau^\varepsilon \varpi^\varepsilon\right)\right](\mathrm{d}\tau)^\varepsilon \\
&\quad \pm b\frac{1}{\sqrt{\Gamma(1+\varepsilon)}} \int_{-\infty}^{\infty} \left[g_2(\tau) E_\varepsilon\left(-(2\pi\,\mathrm{i})^\varepsilon \tau^\varepsilon \varpi^\varepsilon\right)\right](\mathrm{d}\tau)^\varepsilon \\
&= aG_1(\varpi) \pm bG_2(\varpi).
\end{aligned} \tag{3.99}$$

Thus, we have completed the proof. □

Property 15 (Shifting time for the generalized local fractional Fourier transform operator)**.** *Suppose that*

$$g(\tau), G(\varpi) \in L_{1,\varepsilon}[R],$$

$$\Im\left[g(\tau)\right] = G(\varpi),$$

and a is a constant. Then,

$$\Im\left[g(\tau-a)\right] = E_\varepsilon\left(-(2\pi\,\mathrm{i})^\varepsilon a^\varepsilon \varpi^\varepsilon\right) \Im\left[g(\tau)\right]. \tag{3.100}$$

Proof. By making use of the definition of the generalized local fractional Fourier transform operator, we have

$$\begin{aligned}
\Im\left[g(\tau-a)\right] &= \frac{1}{\sqrt{\Gamma(1+\varepsilon)}} \int_{-\infty}^{\infty} g(\tau-a) E_\varepsilon\left(-(2\pi\,\mathrm{i})^\varepsilon \tau^\varepsilon \varpi^\varepsilon\right)(\mathrm{d}\tau)^\varepsilon \\
&= E_\varepsilon\left(-(2\pi\,\mathrm{i})^\varepsilon a^\varepsilon \varpi^\varepsilon\right) \frac{1}{\sqrt{\Gamma(1+\varepsilon)}} \int_{-\infty}^{\infty} g(\tau-a) \\
&\quad E_\varepsilon\left(-(2\pi\,\mathrm{i})^\varepsilon (\tau-a)^\varepsilon \varpi^\varepsilon\right)(\mathrm{d}\tau)^\varepsilon \\
&= E_\varepsilon\left(-(2\pi\,\mathrm{i})^\varepsilon a^\varepsilon \varpi^\varepsilon\right) G(\varpi) \\
&= E_\varepsilon\left(-(2\pi\,\mathrm{i})^\varepsilon a^\varepsilon \varpi^\varepsilon\right) \Im\left[g(\tau)\right].
\end{aligned} \tag{3.101}$$

Thus, the claim has been proved. □

Property 16 (Scaling time for the local fractional Fourier transform operator)**.** *Suppose that*

$$g(\tau), G(\varpi) \in L_{1,\varepsilon}[R],$$

$$\Im\left[g(\tau)\right] = G(\varpi),$$

and a is a positive constant. Then

$$\overline{\Im}\left[g\left(a\tau\right)\right]=\frac{1}{a^{\varepsilon}}G\left(\frac{\varpi}{a}\right). \tag{3.102}$$

Proof. By the definition of the local fractional Fourier transform operator, we obtain

$$\begin{aligned}\overline{\Im}\left[g\left(a\tau\right)\right]&=\frac{1}{\sqrt{\Gamma\left(1+\varepsilon\right)}}\int_{-\infty}^{\infty}g\left(a\tau\right)E_{\varepsilon}\left(-\left(2\pi\ \mathrm{i}\right)^{\varepsilon}\tau^{\varepsilon}\varpi^{\varepsilon}\right)\left(\mathrm{d}\tau\right)^{\varepsilon}\\&=\frac{1}{a^{\varepsilon}}\frac{1}{\sqrt{\Gamma\left(1+\varepsilon\right)}}\int_{-\infty}^{\infty}\theta\left(a\tau\right)E_{\varepsilon}\left(-\left(2\pi\ \mathrm{i}\right)^{\varepsilon}\left(\frac{a\tau}{a}\right)^{\varepsilon}\varpi^{\varepsilon}\right)\left(\mathrm{d}a\tau\right)^{\varepsilon}\\&=\frac{1}{a^{\varepsilon}}\Theta\left(\frac{\varpi}{a}\right),\end{aligned} \tag{3.103}$$

which evidently completes the proof. □

Property 17 (Conjugate for the local fractional Fourier transform operator)**.** *Suppose that*

$$g\left(\tau\right),G\left(\varpi\right)\in L_{1,\varepsilon}\left[R\right]$$

and

$$\overline{\Im}\left[g\left(\tau\right)\right]=G\left(\varpi\right).$$

Then

$$\overline{\Im}\left[\overline{g\left(-\tau\right)}\right]=\overline{\overline{\Im}\left[g\left(\tau\right)\right]}. \tag{3.104}$$

Proof. The definition of the generalized local fractional Fourier transform operator implies that

$$\begin{aligned}\Im\left[\overline{g\left(-\tau\right)}\right]&=\frac{1}{\sqrt{\Gamma\left(1+\varepsilon\right)}}\int_{-\infty}^{\infty}\overline{g\left(-\tau\right)}E_{\varepsilon}\left(-\left(2\pi\ \mathrm{i}\right)^{\varepsilon}\tau^{\varepsilon}\varpi^{\varepsilon}\right)\left(\mathrm{d}\tau\right)^{\varepsilon}\\&=\frac{1}{\sqrt{\Gamma\left(1+\varepsilon\right)}}\int_{-\infty}^{\infty}\overline{\theta\left(-\tau\right)E_{\varepsilon}\left(\left(2\pi\ \mathrm{i}\right)^{\varepsilon}\tau^{\varepsilon}\varpi^{\varepsilon}\right)}\left(\mathrm{d}\tau\right)^{\varepsilon}\\&=\overline{\frac{1}{\sqrt{\Gamma\left(1+\varepsilon\right)}}\int_{-\infty}^{\infty}g\left(-\tau\right)E_{\varepsilon}\left(-\left(2\pi\ \mathrm{i}\right)^{\varepsilon}\left(-\tau\right)^{\varepsilon}\varpi^{\varepsilon}\right)\left(\mathrm{d}\tau\right)^{\varepsilon}}\\&=\overline{\frac{1}{\sqrt{\Gamma\left(1+\varepsilon\right)}}\int_{-\infty}^{\infty}g\left(\tau\right)E_{\varepsilon}\left(-\left(2\pi\ \mathrm{i}\right)^{\varepsilon}\tau^{\varepsilon}\varpi^{\varepsilon}\right)\left(\mathrm{d}\tau\right)^{\varepsilon}}\\&=\overline{\overline{\Im}\left[g\left(\tau\right)\right]},\end{aligned} \tag{3.105}$$

which proves the asserted result. □

Property 18 (Translation for the local fractional Fourier transform operator). *Suppose that*

$$g(\tau), G(\varpi) \in L_{1,\varepsilon}[R],$$

$$\Im[g(\tau)] = G(\varpi),$$

and a is a constant. Then,

$$\Im\left[E_\varepsilon\left((2\pi\,\mathrm{i})^\varepsilon a^\varepsilon \tau^\varepsilon\right) g(\tau)\right] = G(\varpi - a). \tag{3.106}$$

Proof. From the definition of the generalized local fractional Fourier transform operator, we get

$$\begin{aligned}\Im\left[E_\varepsilon\left(-(2\pi\,\mathrm{i})^\varepsilon a^\varepsilon \tau^\varepsilon\right) g(\tau)\right] &= \frac{1}{\sqrt{\Gamma(1+\varepsilon)}}\int_{-\infty}^{\infty}\left[g(\tau) E_\varepsilon\left((2\pi\,\mathrm{i})^\varepsilon a^\varepsilon \tau^\varepsilon\right)\right] \\ &\quad E_\varepsilon\left(-(2\pi\,\mathrm{i})^\varepsilon \tau^\varepsilon \varpi^\varepsilon\right)(d\tau)^\varepsilon \\ &= \frac{1}{\sqrt{\Gamma(1+\varepsilon)}}\int_{-\infty}^{\infty} g(\tau) E_\varepsilon\left(-(2\pi\,\mathrm{i})^\varepsilon\right. \\ &\quad \left.(\varpi - a)^\varepsilon \tau^\varepsilon\right)(d\tau)^\varepsilon \\ &= G(\varpi - a),\end{aligned} \tag{3.107}$$

which completes the proof. □

Property 19 (Duality for local fractional Fourier transform operator). *Suppose that*

$$g(\tau), G(\varpi) \in L_{1,\varepsilon}[R]$$

and

$$\Im[g(\tau)] = G(\varpi).$$

Then,

$$\Im[g(-\tau)] = \Im[G(\tau)]. \tag{3.108}$$

Proof. From the definition of the generalized local fractional Fourier transform operator, we get

$$g(\tau) = \frac{1}{\sqrt{\Gamma(1+\varepsilon)}}\int_{-\infty}^{\infty} G(\varpi) E_\varepsilon\left((2\pi\,\mathrm{i})^\varepsilon \tau^\varepsilon \varpi^\varepsilon\right)(d\varpi)^\varepsilon = \Im^{-1}[G(\varpi)]. \tag{3.109}$$

Now, upon interchanging τ and ϖ, from (3.109), we arrive at

$$g(\varpi) = \frac{1}{\sqrt{\Gamma(1+\varepsilon)}}\int_{-\infty}^{\infty} G(\tau) E_\varepsilon\left((2\pi\,\mathrm{i})^\varepsilon \tau^\varepsilon \varpi^\varepsilon\right)(d\tau)^\varepsilon. \tag{3.110}$$

Replacing ϖ by $-\varpi$, we obtain

$$g(-\varpi) = \frac{1}{\sqrt{\Gamma(1+\varepsilon)}} \int_{-\infty}^{\infty} G(\tau) E_{\varepsilon}\left(-(2\pi \, i)^{\varepsilon} \tau^{\varepsilon} \varpi^{\varepsilon}\right) (d\tau)^{\varepsilon}, \tag{3.111}$$

which evidently proves the result (3.108). □

Property 20 (Composition for the local fractional Fourier transform operator). *Suppose that*

$$g_1(\tau), g_2(\tau), G_1(\varpi), G_2(\varpi) \in L_{1,\varepsilon}[R],$$

$$\Im[g_1(\tau)] = G_1(\varpi),$$

and

$$\Im[g_2(\tau)] = G_2(\varpi).$$

Then,

$$\int_{-\infty}^{\infty} g_1(\varpi) G_2(\varpi) E_{\varepsilon}\left((2\pi \, i)^{\varepsilon} \tau^{\varepsilon} \varpi^{\varepsilon}\right) (d\varpi)^{\varepsilon} = \int_{-\infty}^{\infty} g_2(\eta) G_1(\eta - \tau) (d\eta)^{\varepsilon}. \tag{3.112}$$

Proof. Let us write the left-hand side of (3.112) as follows:

$$\begin{aligned}
&\int_{-\infty}^{\infty} g_1(\varpi) G_2(\varpi) E_{\varepsilon}\left((2\pi \, i)^{\varepsilon} \tau^{\varepsilon} \varpi^{\varepsilon}\right) (d\varpi)^{\varepsilon} \\
&= \int_{-\infty}^{\infty} g_1(\varpi) \left[\frac{1}{\sqrt{\Gamma(1+\varepsilon)}} \int_{-\infty}^{\infty} g_2(\eta) E_{\varepsilon}\left(-(2\pi \, i)^{\varepsilon} \eta^{\varepsilon} \varpi^{\varepsilon}\right) (d\eta)^{\varepsilon}\right] \\
&\quad E_{\varepsilon}\left((2\pi \, i)^{\varepsilon} \tau^{\varepsilon} \varpi^{\varepsilon}\right) (d\varpi)^{\varepsilon} \\
&= \int_{-\infty}^{\infty} \left[\frac{1}{\sqrt{\Gamma(1+\varepsilon)}} \int_{-\infty}^{\infty} g_1(\varpi) E_{\varepsilon}\left(-(2\pi \, i)^{\varepsilon} (\eta-\tau)^{\varepsilon} \varpi^{\varepsilon}\right) (d\varpi)^{\varepsilon}\right] g_2(\eta) (d\eta)^{\varepsilon} \\
&= \int_{-\infty}^{\infty} G_1(\eta - \tau) g_2(\eta) (d\eta)^{\varepsilon},
\end{aligned} \tag{3.113}$$

which obviously establishes the result (3.112). □

Theorem 3.10. *Suppose that*

$$g(\tau), G(\varpi) \in L_{1,\varepsilon}[R],$$

$$\Im[g(\tau)] = G(\varpi),$$

and

$$\lim_{|\iota| \to \infty} g(\tau) = 0.$$

Then,

$$\overline{\Im}\left[g^{(\varepsilon)}(\tau)\right] = (2\pi\, i\varpi)^{\varepsilon}\, \overline{\Im}\left[g(\tau)\right]. \tag{3.114}$$

Proof. From the definition of the generalized local fractional Fourier transform operator, we find that

$$\overline{\Im}\left[g^{(\varepsilon)}(\tau)\right] = \frac{1}{\sqrt{\Gamma(1+\varepsilon)}} \int_{-\infty}^{\infty} g^{(\varepsilon)}(\tau)\, E_{\varepsilon}\left(-(2\pi\,\mathrm{i})^{\varepsilon}\,\tau^{\varepsilon}\varpi^{\varepsilon}\right)(\mathrm{d}\tau)^{\varepsilon}. \tag{3.115}$$

Now, by integration by part for the local fractional integral operator in (3.115), we obtain

$$\begin{aligned}\overline{\Im}\left[g^{(\varepsilon)}(\tau)\right] &= \frac{1}{\sqrt{\Gamma(1+\varepsilon)}} \int_{-\infty}^{\infty} g^{(\varepsilon)}(\tau)\, E_{\varepsilon}\left(-(2\pi\,\mathrm{i})^{\varepsilon}\,\tau^{\varepsilon}\varpi^{\varepsilon}\right)(\mathrm{d}\tau)^{\varepsilon} \\ &= \left[\sqrt{\Gamma(1+\varepsilon)} g(\tau)\, E_{\varepsilon}\left(-(2\pi\,\mathrm{i})^{\varepsilon}\,\tau^{\varepsilon}\varpi^{\varepsilon}\right)\right]_{-\infty}^{\infty} \\ &\quad + \frac{(2\pi\,\mathrm{i})^{\varepsilon}\,\varpi^{\varepsilon}}{\sqrt{\Gamma(1+\varepsilon)}} \int_{-\infty}^{\infty} g(\tau)\, E_{\varepsilon}\left(-(2\pi\,\mathrm{i})^{\varepsilon}\,\tau^{\varepsilon}\varpi^{\varepsilon}\right)(\mathrm{d}\tau)^{\varepsilon},\end{aligned} \tag{3.116}$$

which, by using

$$\lim_{|\tau|\to\infty} g(\tau) = 0,$$

leads us to

$$\begin{aligned}\overline{\Im}\left[g^{(\varepsilon)}(\tau)\right] &= (2\pi\,\mathrm{i})^{\varepsilon}\,\varpi^{\varepsilon}\left[\frac{1}{\sqrt{\Gamma(1+\varepsilon)}} \int_{-\infty}^{\infty} g(\tau)\, E_{\varepsilon}\left(-(2\pi\,\mathrm{i})^{\varepsilon}\,\tau^{\varepsilon}\varpi^{\varepsilon}\right)(\mathrm{d}\tau)^{\varepsilon}\right] \\ &= (2\pi\,\mathrm{i}\varpi)^{\varepsilon}\,\overline{\Im}\left[g(\tau)\right].\end{aligned} \tag{3.117}$$

Thus, we get the asserted result (3.114). □

In similar manner, by repeating this process, we get

$$\overline{\Im}\left[g^{(k\varepsilon)}(\tau)\right] = (2\pi\,\mathrm{i}\varpi)^{k\varepsilon}\,\overline{\Im}\left[g(\tau)\right] \quad (k \in \mathrm{N}), \tag{3.118}$$

where

$$g^{((k-1)\varepsilon)}(0) = \cdots = g^{(\varepsilon)}(0) = g(0) = 0. \tag{3.119}$$

Theorem 3.11. *Suppose that*

$$g(\tau), G(\varpi) \in L_{1,\varepsilon}[R],$$

$$\overline{\Im}\left[g(\tau)\right] = G(\varpi),$$

and

$$\lim_{\tau\to\infty} -\infty I_{\tau}^{(\varepsilon)} g(\tau) = 0.$$

Then,

$$\overline{\Im}\left[{}_{-\infty}I_{\tau}^{(\varepsilon)}g(\tau)\right]=\frac{1}{(2\pi\ i\varpi)^{\varepsilon}}\overline{\Im}\left[g(\tau)\right]. \tag{3.120}$$

Proof. In view of the definition of the generalized local fractional Fourier transform operator, we conclude that

$$\overline{\Im}\left[{}_{-\infty}I_{\tau}^{(\varepsilon)}g(\tau)\right]=\frac{1}{\sqrt{\Gamma(1+\varepsilon)}}\int_{-\infty}^{\infty}\left({}_{-\infty}I_{\tau}^{(\varepsilon)}g(\tau)\right)E_{\varepsilon}\left(-(2\pi\ \mathrm{i})^{\varepsilon}\tau^{\varepsilon}\varpi^{\varepsilon}\right)(\mathrm{d}\tau)^{\varepsilon}, \tag{3.121}$$

which, by integrating by part for the local fractional integral operator, yields

$$\begin{aligned}\overline{\Im}\left[{}_{-\infty}I_{\tau}^{(\varepsilon)}g(\tau)\right]&=\left[\sqrt{\Gamma(1+\varepsilon)}{}_{-\infty}I_{\tau}^{(\varepsilon)}g(\tau)E_{\varepsilon}\left(-(2\pi\ \mathrm{i})^{\varepsilon}\tau^{\varepsilon}\varpi^{\varepsilon}\right)\right]\Big|_{-\infty}^{\infty}\\&\quad+\left[\frac{1}{(2\pi\varpi\mathrm{i})^{\varepsilon}\sqrt{\Gamma(1+\varepsilon)}}\int_{-\infty}^{\infty}g(\tau)E_{\varepsilon}\left(-(2\pi\ \mathrm{i})^{\varepsilon}\tau^{\varepsilon}\varpi^{\varepsilon}\right)(\mathrm{d}\tau)^{\varepsilon}\right].\end{aligned} \tag{3.122}$$

Now, by taking

$$\lim_{\tau\to\infty}{}_{-\infty}I_{\tau}^{(\varepsilon)}g(\tau)=0$$

into account in (3.122), we have

$$\begin{aligned}\overline{\Im}\left[{}_{-\infty}I_{\tau}^{(\varepsilon)}g(\tau)\right]&=\frac{1}{(2\pi\varpi\mathrm{i})^{\varepsilon}\sqrt{\Gamma(1+\varepsilon)}}\int_{-\infty}^{\infty}g(\tau)E_{\varepsilon}\left(-(2\pi\ \mathrm{i})^{\varepsilon}\tau^{\varepsilon}\varpi^{\varepsilon}\right)(\mathrm{d}\tau)^{\varepsilon}\\&=\frac{1}{(2\pi\varpi\mathrm{i})^{\varepsilon}}\overline{\Im}\left[g(\tau)\right],\end{aligned} \tag{3.123}$$

which completes the proof. □

In a similar manner, by repeating this process, we have

$$\overline{\Im}\left[{}_{-\infty}I_{\tau}^{(k\varepsilon)}g(\tau)\right]=\frac{1}{(2\pi\varpi\mathrm{i})^{k\varepsilon}}\overline{\Im}\left[g(\tau)\right], \tag{3.124}$$

where

$$\lim_{\tau\to\infty}{}_{-\infty}I_{\tau}^{(k\varepsilon)}g(\tau)=0. \tag{3.125}$$

Theorem 3.12. *Suppose that*

$$g(\tau),G(\varpi)\in L_{1,\varepsilon}[R]$$

and

$$\overline{\Im}\left[g(\tau)\right]=G(\varpi).$$

Then,

$$\mathfrak{F}\left[\tau^{\varepsilon} g(\tau)\right]=\left(\frac{i}{2\pi}\right)^{\varepsilon} G^{(\varepsilon)}(\varpi). \tag{3.126}$$

Proof. From the definition of the inverse generalized local fractional Fourier transform operator, we observe that

$$\mathfrak{F}^{-1}\left[\left(\frac{\mathrm{i}}{2\pi}\right)^{\varepsilon} G^{(\varepsilon)}(\varpi)\right]=\frac{1}{\sqrt{\Gamma(1+\varepsilon)}}\int_{-\infty}^{\infty}\left[\left(\frac{\mathrm{i}}{2\pi}\right)^{\varepsilon} G^{(\varepsilon)}(\varpi)\right] E_{\varepsilon}\left((2\pi\,\mathrm{i})^{\varepsilon}\tau^{\varepsilon}\varpi^{\varepsilon}\right)(\mathrm{d}\varpi)^{\varepsilon}, \tag{3.127}$$

which, after integrating by part for local fractional integral operator, yields

$$\mathfrak{F}^{-1}\left[\left(\frac{\mathrm{i}}{2\pi}\right)^{\varepsilon} G^{(\varepsilon)}(\varpi)\right]=\left[\left(\frac{\mathrm{i}}{2\pi}\right)^{\varepsilon} G(\varpi) E_{\varepsilon}\left((2\pi\,\mathrm{i})^{\varepsilon}\tau^{\varepsilon}\varpi^{\varepsilon}\right)\right]_{-\infty}^{\infty} + \tau^{\varepsilon}\left[\frac{1}{\sqrt{\Gamma(1+\varepsilon)}}\int_{-\infty}^{\infty} G(\varpi) E_{\varepsilon}\left((2\pi\,\mathrm{i})^{\varepsilon}\tau^{\varepsilon}\varpi^{\varepsilon}\right)(\mathrm{d}\varpi)^{\varepsilon}\right]. \tag{3.128}$$

Now, by taking

$$\lim_{|\omega|\to\infty}\Theta(\omega)=0$$

into account in (3.128), we have

$$\begin{aligned}\mathfrak{F}^{-1}\left[\left(\frac{\mathrm{i}}{2\pi}\right)^{\varepsilon} G^{(\varepsilon)}(\varpi)\right]&=\tau^{\varepsilon}\left[\frac{1}{\sqrt{\Gamma(1+\varepsilon)}}\int_{-\infty}^{\infty} G(\varpi) E_{\varepsilon}\left((2\pi\,\mathrm{i})^{\varepsilon}\tau^{\varepsilon}\varpi^{\varepsilon}\right)(\mathrm{d}\varpi)^{\varepsilon}\right]\\&=\tau^{\varepsilon} g(\tau),\end{aligned} \tag{3.129}$$

which proves the claimed result. □

In a similar way, we obtain

$$\mathfrak{F}\left[\tau^{k\varepsilon} g(\tau)\right]=\left(\frac{\mathrm{i}}{2\pi}\right)^{k\varepsilon} G^{(k\varepsilon)}(\varpi). \tag{3.130}$$

Definition 3.7. The local fractional convolution of two functions $g_1(\tau)$ and $g_2(\tau)$ via local fractional integral operator, denoted by

$$(g_1 * g_2)(\tau)=g_1(\tau) * g_2(\tau)$$

is defined as follows:

$$\begin{aligned}(g_1 * g_2)(\tau)&=g_1(\tau) * g_2(\tau)\\&=\sqrt{\Gamma(1+\varepsilon)}{}_{-\infty}I_{\infty}^{(\varepsilon)}\left[g_1(t)\, g_2(\tau-t)\right]\\&=\frac{1}{\sqrt{\Gamma(1+\varepsilon)}}\int_{-\infty}^{\infty} g_1(t)\, g_2(\tau-t)(\mathrm{d}t)^{\varepsilon}.\end{aligned} \tag{3.131}$$

By using the above definition of the local fractional convolution of two functions, we have the following properties:

(a) $g_1(\tau) * g_2(\tau) = g_2(\tau) * g_1(\tau)$ and
(b) $g_1(\tau) * (g_2(\tau) + g_3(\tau)) = g_1(\tau) * g_2(\tau) + g_1(\tau) * g_3(\tau)$.

Theorem 3.13. *Suppose that*

$$g_1(\tau), g_2(\tau), G_1(\varpi), G_2(\varpi) \in L_{1,\varepsilon}[R],$$

$$\overline{\Im}[g_1(\tau)] = G_1(\varpi),$$

and

$$\overline{\Im}[g_2(\tau)] = G_2(\varpi).$$

Then,

$$\overline{\Im}[g_1(\tau) * g_2(\tau)] = G_1(\varpi)G_2(\varpi) \tag{3.132}$$

or

$$g_1(\tau) * g_2(\tau) = \overline{\Im}^{-1}[G_1(\varpi)G_2(\varpi)] \tag{3.133}$$

or, equivalently,

$$\frac{1}{\sqrt{\Gamma(1+\varepsilon)}}\int_{-\infty}^{\infty} g_1(t)\, g_2(\tau - t)\,(dt)^{\varepsilon} = \frac{1}{\sqrt{\Gamma(1+\varepsilon)}}\int_{-\infty}^{\infty} G_1(\varpi)\, G_2(\varpi)\,(d\varpi)^{\varepsilon}. \tag{3.134}$$

Proof. From the definition of the generalized local fractional Fourier transform operator, we get

$$\begin{aligned}\overline{\Im}[g_1(\tau) * g_2(\tau)] &= \frac{1}{\sqrt{\Gamma(1+\varepsilon)}}\int_{-\infty}^{\infty} E_\varepsilon\left(-(2\pi\,\mathrm{i})^{\varepsilon}\tau^{\varepsilon}\varpi^{\varepsilon}\right)(\mathrm{d}\tau)^{\varepsilon} \\ &\quad\times\left[\frac{1}{\sqrt{\Gamma(1+\varepsilon)}}\int_{-\infty}^{\infty} g_1(\eta)\, g_2(\tau-\eta)\,(\mathrm{d}\eta)^{\varepsilon}\right] \\ &= \frac{1}{\sqrt{\Gamma(1+\varepsilon)}}\int_{-\infty}^{\infty} g_1(\eta)\, E_\varepsilon\left(-(2\pi\,\mathrm{i})^{\varepsilon}\eta^{\varepsilon}\varpi^{\varepsilon}\right) \\ &\quad\times\left[\frac{\int_{-\infty}^{\infty} E_\varepsilon\left(-(2\pi\,\mathrm{i})^{\varepsilon}(\tau-\eta)^{\varepsilon}\varpi^{\varepsilon}\right)\theta_2(\tau-\eta)\,(\mathrm{d}\tau)^{\varepsilon}}{\sqrt{\Gamma(1+\varepsilon)}}\right](\mathrm{d}\eta)^{\varepsilon},\end{aligned} \tag{3.135}$$

where

$$G_2(\varpi) = \frac{1}{\sqrt{\Gamma(1+\varepsilon)}}\int_{-\infty}^{\infty} E_\varepsilon\left(-(2\pi\,\mathrm{i})^{\varepsilon}(\tau-\eta)^{\varepsilon}\varpi^{\varepsilon}\right)\theta_2(\tau-\eta)\,(\mathrm{d}\tau)^{\varepsilon} \tag{3.136}$$

and

$$\begin{aligned}\Im[g_1(\tau) * g_2(\tau)] &= \frac{1}{\sqrt{\Gamma(1+\varepsilon)}}\int_{-\infty}^{\infty} g_1(\eta) E_\varepsilon\left(-(2\pi \text{ i})^\varepsilon \eta^\varepsilon \varpi^\varepsilon\right) G_2(\varpi)(\mathrm{d}\eta)^\varepsilon \\ &= G_2(\varpi)\frac{1}{\sqrt{\Gamma(1+\varepsilon)}}\int_{-\infty}^{\infty} g_1(\eta) E_\varepsilon\left(-(2\pi \text{ i})^\varepsilon \eta^\varepsilon \varpi^\varepsilon\right)(\mathrm{d}\eta)^\varepsilon \\ &= \Theta_1(\varpi)\Theta_2(\varpi).\end{aligned} \tag{3.137}$$

This completes the proof. □

Theorem 3.14. *Suppose that*

$$g_1(\tau), g_2(\tau), G_1(\varpi), G_2(\varpi) \in L_{1,\varepsilon}[R],$$

$$\Im[g_1(\tau)] = G_1(\varpi),$$

and

$$\Im[g_2(\tau)] = G_2(\varpi).$$

$$\Im[g_1(\tau) g_2(\tau)] = G_1(\varpi) * G_2(\varpi) \tag{3.138}$$

or

$$g_1(\tau) g_2(\tau) = \Im^{-1}[G_1(\varpi) * G_2(\varpi)] \tag{3.139}$$

or, equivalently,

$$\frac{1}{\sqrt{\Gamma(1+\varepsilon)}}\int_{-\infty}^{\infty} g_1(\tau) g_2(\tau)(d\tau)^\varepsilon = \frac{1}{\sqrt{\Gamma(1+\varepsilon)}}\int_{-\infty}^{\infty} G_1(t) G_2(\varpi - t)(dt)^\varepsilon. \tag{3.140}$$

Proof. From the definition of the generalized local fractional Fourier transform operator, we have

$$\begin{aligned}\Im^{-1}[G_1(\varpi) * G_2(\varpi)] &= \frac{1}{\sqrt{\Gamma(1+\varepsilon)}}\int_{-\infty}^{\infty} \\ &\quad \times \left[\frac{1}{\sqrt{\Gamma(1+\varepsilon)}}\int_{-\infty}^{\infty} G_1(\varpi) G_2(\omega-\varpi)(\mathrm{d}\varpi)^\varepsilon\right] \\ &\quad E_\varepsilon\left((2\pi \text{ i})^\varepsilon \tau^\varepsilon \omega^\varepsilon\right)(\mathrm{d}\omega)^\varepsilon \\ &= \frac{1}{\sqrt{\Gamma(1+\varepsilon)}}\int_{-\infty}^{\infty} E_\varepsilon\left((2\pi \text{ i})^\varepsilon \tau^\varepsilon \varpi^\varepsilon\right) G_1(\varpi) \\ &\quad \times \left[\frac{\int_{-\infty}^{\infty} G_2(\omega-\varpi) E_\varepsilon\left((2\pi \text{ i})^\varepsilon \tau^\varepsilon (\omega-\varpi)^\varepsilon\right)(\mathrm{d}\omega)^\varepsilon}{\sqrt{\Gamma(1+\varepsilon)}}\right](\mathrm{d}\varpi)^\varepsilon,\end{aligned} \tag{3.141}$$

where

$$g_2(\tau) = \frac{1}{\sqrt{\Gamma(1+\varepsilon)}} \int_{-\infty}^{\infty} G_2(\omega - \varpi) E_\varepsilon\left((2\pi\, \mathrm{i})^\varepsilon \tau^\varepsilon (\omega - \varpi)^\varepsilon\right) (\mathrm{d}\omega)^\varepsilon \tag{3.142}$$

and

$$\begin{aligned} \overline{\Im}^{-1}\left[G_1(\varpi) * G_2(\varpi)\right] &= \frac{1}{\sqrt{\Gamma(1+\varepsilon)}} \int_{-\infty}^{\infty} E_\varepsilon\left((2\pi\, \mathrm{i})^\varepsilon \tau^\varepsilon \varpi^\varepsilon\right) G_1(\varpi)\, g_2(\tau) (\mathrm{d}\varpi)^\varepsilon \\ &= \left[\frac{1}{\sqrt{\Gamma(1+\varepsilon)}} \int_{-\infty}^{\infty} E_\varepsilon\left((2\pi\, \mathrm{i})^\varepsilon \tau^\varepsilon \varpi^\varepsilon\right) \Theta_1(\varpi) (\mathrm{d}\varpi)^\varepsilon\right] g_2(\tau) \\ &= g_1(\tau)\, g_2(\tau). \end{aligned} \tag{3.143}$$

Thus, we have completed the proof. □

Theorem 3.15 (Convolution theorem for generalized local fractional Fourier transform operator)**.** *Suppose that*

$$g_1(\tau), g_2(\tau), G_1(\varpi), G_2(\varpi) \in L_{1,\varepsilon}[R],$$

$$\overline{\Im}\left[g_1(\tau)\right] = G_1(\varpi),$$

and

$$\overline{\Im}\left[g_2(\tau)\right] = G_2(\varpi).$$

Then,

$$\frac{1}{\sqrt{\Gamma(1+\varepsilon)}} \int_{-\infty}^{\infty} g_1(\tau)\, \overline{g_2(\tau)}\, (d\tau)^\varepsilon = \frac{1}{\sqrt{\Gamma(1+\varepsilon)}} \int_{-\infty}^{\infty} G_1(\varpi)\, \overline{G_2(\varpi)}\, (d\varpi)^\varepsilon. \tag{3.144}$$

Proof. We consider

$$\begin{aligned} \overline{g_2(\tau)} &= \overline{\frac{1}{\sqrt{\Gamma(1+\varepsilon)}} \int_{-\infty}^{\infty} G_2(\varpi) E_\varepsilon\left((2\pi\, \mathrm{i})^\varepsilon \tau^\varepsilon \varpi^\varepsilon\right) (\mathrm{d}\varpi)^\varepsilon} \\ &= \frac{1}{\sqrt{\Gamma(1+\varepsilon)}} \int_{-\infty}^{\infty} \overline{G_2(\varpi)} E_\varepsilon\left(-(2\pi\, \mathrm{i})^\varepsilon \tau^\varepsilon \varpi^\varepsilon\right) (\mathrm{d}\varpi)^\varepsilon, \end{aligned} \tag{3.145}$$

which, in view of (3.145), yields

$$\begin{aligned} &\frac{1}{\sqrt{\Gamma(1+\varepsilon)}} \int_{-\infty}^{\infty} g_1(\tau)\, \overline{g_2(\tau)}\, (\mathrm{d}\tau)^\varepsilon \\ &= \frac{1}{\sqrt{\Gamma(1+\varepsilon)}} \int_{-\infty}^{\infty} g_1(\tau) \left(\frac{1}{\sqrt{\Gamma(1+\varepsilon)}} \int_{-\infty}^{\infty} \overline{G_2(\varpi)} E_\varepsilon\left(-(2\pi\, \mathrm{i})^\varepsilon \tau^\varepsilon \varpi^\varepsilon\right) (\mathrm{d}\varpi)^\varepsilon\right) (\mathrm{d}\tau)^\varepsilon \\ &= \frac{1}{\sqrt{\Gamma(1+\varepsilon)}} \int_{-\infty}^{\infty} \overline{G_2(\varpi)} \left(\frac{1}{\sqrt{\Gamma(1+\varepsilon)}} \int_{\infty}^{\infty} g_1(\tau) E_\varepsilon\left(-(2\pi\, \mathrm{i})^\varepsilon \tau^\varepsilon \varpi^\varepsilon\right) (\mathrm{d}\tau)^\varepsilon\right) (\mathrm{d}\varpi)^\varepsilon, \end{aligned} \tag{3.146}$$

where

$$G_1(\varpi) = \frac{1}{\sqrt{\Gamma(1+\varepsilon)}} \int_{-\infty}^{\infty} g_1(\tau) E_\varepsilon\left(-(2\pi\, i)^\varepsilon \tau^\varepsilon \varpi^\varepsilon\right)(d\tau)^\varepsilon \tag{3.147}$$

and

$$\frac{1}{\sqrt{\Gamma(1+\varepsilon)}} \int_{-\infty}^{\infty} g_1(\tau)\overline{g_2(\tau)}(d\tau)^\varepsilon = \frac{1}{\sqrt{\Gamma(1+\varepsilon)}} \int_{-\infty}^{\infty} G_1(\varpi)\overline{G_2(\varpi)}(d\varpi)^\varepsilon. \tag{3.148}$$

Therefore, we have completed the proof. □

Theorem 3.16 (Parseval's theorem for generalized local fractional Fourier transform operator)**.** *Suppose that*

$$g(\tau) \in L_{1,\varepsilon}[R]$$

and

$$\Im[g(\tau)] = G(\varpi).$$

Then,

$$\frac{1}{\sqrt{\Gamma(1+\varepsilon)}} \int_{-\infty}^{\infty} |g(\tau)|^2 (d\tau)^\varepsilon = \frac{1}{\sqrt{\Gamma(1+\varepsilon)}} \int_{-\infty}^{\infty} |G(\varpi)|^2 (d\varpi)^\varepsilon. \tag{3.149}$$

Proof. By setting

$$g_1(\tau) = g_2(\tau) = g(\tau)$$

in (3.144), we conclude that

$$g_1(\tau)\overline{g_2(\tau)} = g(\tau)\overline{g(\tau)} = |g(\tau)|^2 \tag{3.150}$$

and

$$G_1(\varpi)\overline{G_2(\varpi)} = G(\varpi)\overline{G(\varpi)} = |G(\varpi)|^2. \tag{3.151}$$

We thus obtain the desired result (3.149). □

3.3 Applications to signal analysis

3.3.1 The analogous distributions defined on Cantor sets

In light of Definition 3.5, we define the function $\delta_\varepsilon(\tau)$, which is called the analogous Dirac (Dirac-like function) distribution via the local fractional integral operator (also called the local fractional Dirac function).

Definition 3.8. The analogous Dirac distribution via the local fractional integral operator is defined by

$$\begin{aligned}\psi(0) &= \frac{1}{\Gamma(1+\varepsilon)}\int_{-\infty}^{\infty}\delta_{\varepsilon}(\tau)\,\psi(\tau)\,(d\tau)^{\varepsilon}\\ &= (\delta_{\varepsilon} * \psi)(\tau)\\ &= \delta_{\varepsilon}(\tau) * \psi(\tau),\end{aligned} \tag{3.152}$$

and it has the following properties:

(a) $\delta_{\varepsilon}(\tau) \geq 0$, for $\tau \in R$;
(b) $\delta_{\varepsilon}(\tau) = 0$, for $\tau \neq 0$; and
(c) $\frac{1}{\Gamma(1+\varepsilon)}\int_{-\infty}^{\infty}\delta_{\varepsilon}(\tau)\,(d\tau)^{\varepsilon} = 1$.

It follows that

$$\frac{1}{\Gamma(1+\varepsilon)}\int_{-\infty}^{\infty}\delta_{\varepsilon}(\tau - t)\,\psi(\tau)\,(d\tau)^{\varepsilon} = \psi(t), \tag{3.153}$$

$$\frac{1}{\Gamma(1+\varepsilon)}\int_{-\infty}^{\infty}\delta_{\varepsilon}(\tau)\,\psi^{(k\varepsilon)}(\tau)\,(d\tau)^{\varepsilon} = \psi^{(k\varepsilon)}(0), \tag{3.154}$$

and

$$\frac{1}{\Gamma(1+\varepsilon)}\int_{-\infty}^{\infty}\delta_{\varepsilon}(\tau - t)\,\psi^{(k\varepsilon)}(\tau)\,(d\tau)^{\varepsilon} = \psi^{(k\varepsilon)}(t), \tag{3.155}$$

where $k \in N$.

According to the earlier works [1–27], for $\mu \in R$, we have

$$\frac{1}{\Gamma(1+\varepsilon)}\int_{-\infty}^{\infty}\frac{1}{\frac{(4\pi\mu)^{\varepsilon/2}}{\Gamma(1+\varepsilon)}}E_{\varepsilon}\left(-\frac{\tau^{2\varepsilon}}{(4\mu)^{\varepsilon}}\right)(d\tau)^{\varepsilon} = 1, \tag{3.156}$$

from which we obtain the analogous Dirac distribution as follows:

$$\delta_{\varepsilon}(\tau) = \lim_{\mu\to 0}\frac{1}{\frac{(4\pi\mu)^{\varepsilon/2}}{\Gamma(1+\varepsilon)}}E_{\varepsilon}\left(-\frac{\tau^{2\varepsilon}}{(4\mu)^{\varepsilon}}\right). \tag{3.157}$$

For $\mu \in R$, we also have

$$\Omega_{\varepsilon}(\mu,\tau) = \frac{1}{\frac{(4\pi\mu)^{\varepsilon/2}}{\Gamma(1+\varepsilon)}}E_{\varepsilon}\left(-\frac{\tau^{2\varepsilon}}{(4\mu)^{\varepsilon}}\right)\quad\left(\mu = \frac{1}{4\pi}\right), \tag{3.158}$$

which yields

$$\Omega_{\varepsilon}\left(\frac{1}{4\pi},\tau\right) = \Gamma(1+\varepsilon)\,E_{\varepsilon}\left(-\pi^{\varepsilon}\tau^{2\varepsilon}\right). \tag{3.159}$$

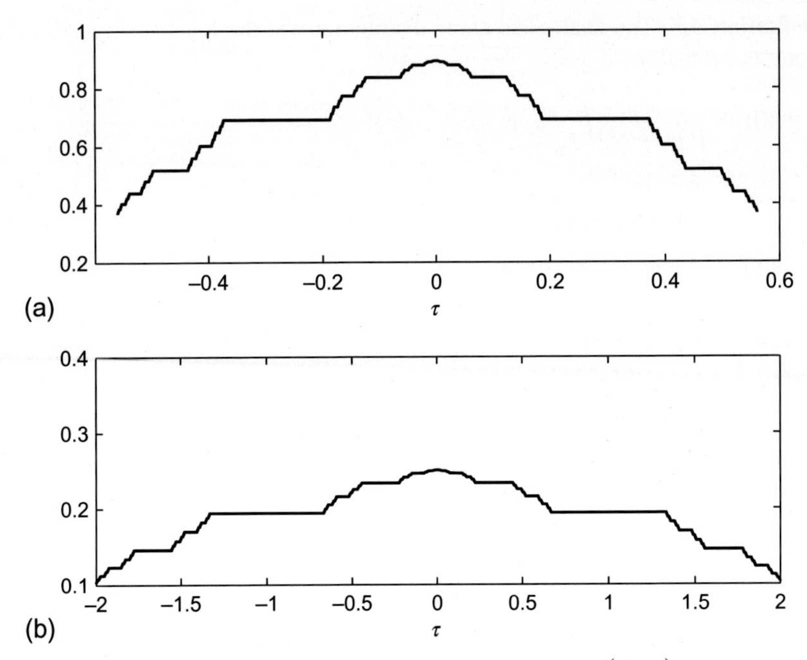

Figure 3.1 The plots of a family of good kernels: (a) the plot of $\Omega_\varepsilon\left(\frac{1}{4\pi},\tau\right)$ with fractal dimension $\varepsilon = \ln 2/\ln 3$ and (b) the plot of $\Omega_\varepsilon(1,\tau)$ with fractal dimension $\varepsilon = \ln 2/\ln 3$.

In this case, we get

$$\frac{1}{\Gamma(1+\varepsilon)}\int_{-\infty}^{\infty}\Gamma(1+\varepsilon)E_\varepsilon\left(-\pi^\varepsilon\tau^{2\varepsilon}\right)(\mathrm{d}\tau)^\varepsilon = 1. \tag{3.160}$$

We also call $\Omega_\varepsilon\left(\frac{1}{4\pi},\tau\right)$ a good kernel and $\Omega_\varepsilon(\mu,\tau)$ a family of good kernels as $\mu \to 0$. The graphs of the functions $\Omega_\varepsilon\left(\frac{1}{4\pi},\tau\right)$ and $\Omega_\varepsilon(1,\tau)$ are shown in Figure 3.1.

Theorem 3.17. *If*

$$\Omega_\varepsilon(\tau) = \Gamma(1+\varepsilon)E_\varepsilon\left(-\pi^\varepsilon\tau^{2\varepsilon}\right), \tag{3.161}$$

then,

$$\Omega_\varepsilon(\varpi) = \widehat{\Omega_\varepsilon(\varpi)}. \tag{3.162}$$

Proof. We define

$$X(\widehat{\varpi) = \Omega_\varepsilon}(\varpi) = \frac{1}{\sqrt{\Gamma(1+\varepsilon)}}\int_{-\infty}^{\infty}\Omega_\varepsilon(\tau)E_\varepsilon\left[-(2\pi\,\mathrm{i})^\varepsilon\tau^\varepsilon\varpi^\varepsilon\right](\mathrm{d}\tau)^\varepsilon \tag{3.163}$$

and find that

$$\widehat{X(0)} = \widehat{\Omega_\varepsilon}(0) = \sqrt{\Gamma(1+\varepsilon)}.$$

We now have

$$\Omega_\varepsilon^{(\varepsilon)}(\tau) = -(2\pi)^\varepsilon \tau^\varepsilon \Omega_\varepsilon(\tau), \tag{3.164}$$

so that

$$\begin{aligned}\widehat{X^{(\varepsilon)}(\varpi)} &= \left[\frac{1}{\sqrt{\Gamma(1+\varepsilon)}}\int_{-\infty}^{\infty}\Omega_\varepsilon(\tau)\left(-(2\pi\,\mathrm{i})^\varepsilon\tau^\varepsilon\right)E_\varepsilon\left[-(2\pi\,\mathrm{i})^\varepsilon\tau^\varepsilon\varpi^\varepsilon\right](\mathrm{d}\tau)^\varepsilon\right]\\ &= \mathrm{i}^\varepsilon\left[\frac{1}{\sqrt{\Gamma(1+\varepsilon)}}\int_{-\infty}^{\infty}\Omega_\varepsilon^{(\varepsilon)}(\tau)E_\varepsilon\left[-(2\pi\,\mathrm{i})^\varepsilon\tau^\varepsilon\varpi^\varepsilon\right](\mathrm{d}\tau)^\varepsilon\right].\end{aligned} \tag{3.165}$$

Using (3.165) together with (3.118), we conclude that

$$X^{(\varepsilon)}(\varpi) = \mathrm{i}^\varepsilon(2\pi\,\mathrm{i}\varpi)^\varepsilon\widehat{\Omega_\varepsilon(\varpi)} = -(2\pi\varpi)^\varepsilon\,\Omega_\varepsilon(\varpi), \tag{3.166}$$

which, for any $2\pi\varpi$, yields

$$\widehat{\Omega_\varepsilon(\varpi)} = \Omega_\varepsilon(\varpi). \tag{3.167}$$

Now, for any good kernel $\Omega_\varepsilon(\mu,\tau)$, we can write

$$\Omega_\varepsilon(\mu,\tau)*\psi(\tau)-\psi(\tau) = \frac{1}{\Gamma(1+\varepsilon)}\int_{-\infty}^{\infty}\delta_\varepsilon(\tau)\left[\psi(t-\tau)-\psi(t)\right](\mathrm{d}\tau)^\varepsilon. \tag{3.168}$$

In this case, we have

$$\begin{aligned}\theta(\tau) &= \frac{1}{(2\pi)^\varepsilon}\int_{-\infty}^{\infty}E_\varepsilon\left(\mathrm{i}^\varepsilon\tau^\varepsilon\rho^\varepsilon\right)\left[\frac{1}{\Gamma(1+\varepsilon)}\int_{-\infty}^{\infty}\theta(t)E_\varepsilon\left(-\mathrm{i}^\varepsilon t^\varepsilon\rho^\varepsilon\right)(\mathrm{d}t)^\varepsilon\right](\mathrm{d}\rho)^\varepsilon\\ &= \frac{1}{\Gamma(1+\varepsilon)}\int_{-\infty}^{\infty}\theta(t)\left[\frac{1}{(2\pi)^\varepsilon}\int_{-\infty}^{\infty}E_\varepsilon\left(\mathrm{i}^\varepsilon(\tau-t)^\varepsilon\rho^\varepsilon\right)(\mathrm{d}\rho)^\varepsilon\right](\mathrm{d}t)^\varepsilon\\ &= \frac{1}{\Gamma(1+\varepsilon)}\int_{-\infty}^{\infty}\theta(t)\,\delta_\varepsilon(\tau-t)(\mathrm{d}t)^\varepsilon,\end{aligned} \tag{3.169}$$

where

$$\delta_\varepsilon(\tau-t) = \frac{1}{(2\pi)^\varepsilon}\int_{-\infty}^{\infty}E_\varepsilon\left(\mathrm{i}^\varepsilon(\tau-t)^\varepsilon\rho^\varepsilon\right)(\mathrm{d}\rho)^\varepsilon \tag{3.170}$$

and

$$\begin{aligned}g(\tau) &= \frac{1}{\sqrt{\Gamma(1+\varepsilon)}}\int_{-\infty}^{\infty}\left[\frac{\int_{-\infty}^{\infty}g(t)E_\varepsilon\left[-(2\pi\,\mathrm{i})^\varepsilon t^\varepsilon\upsilon^\varepsilon\right](\mathrm{d}t)^\varepsilon}{\sqrt{\Gamma(1+\varepsilon)}}\right]E_\varepsilon\\ &\quad\left[(2\pi\,\mathrm{i})^\varepsilon\tau^\varepsilon\upsilon^\varepsilon\right](\mathrm{d}\upsilon)^\varepsilon\\ &= \frac{1}{\sqrt{\Gamma(1+\varepsilon)}}\int_{-\infty}^{\infty}g(t)\left\{\frac{\int_{-\infty}^{\infty}E_\varepsilon\left[(2\pi\,\mathrm{i})^\varepsilon(\tau-t)^\varepsilon\upsilon^\varepsilon\right](\mathrm{d}\upsilon)^\varepsilon}{\sqrt{\Gamma(1+\varepsilon)}}\right\}(\mathrm{d}t)^\varepsilon\\ &= \frac{1}{\sqrt{\Gamma(1+\varepsilon)}}\int_{-\infty}^{\infty}g(t)\left(\frac{\delta_\varepsilon(\tau-t)}{\sqrt{\Gamma(1+\varepsilon)}}\right)(\mathrm{d}t)^\varepsilon.\end{aligned} \tag{3.171}$$

Here,

$$\delta_\varepsilon(\tau - t) = \int_{-\infty}^{\infty} E_\varepsilon\left[(2\pi\, \mathrm{i})^\varepsilon (\tau - t)^\varepsilon \upsilon^\varepsilon\right] (\mathrm{d}\upsilon)^\varepsilon. \tag{3.172}$$

□

In fact, by virtue of (3.169) and (3.171), we find from (3.170) and (3.172) that

$$\delta_\varepsilon(\tau) = \frac{1}{(2\pi)^\varepsilon} \int_{-\infty}^{\infty} E_\varepsilon\left(\mathrm{i}^\varepsilon \tau^\varepsilon \rho^\varepsilon\right) (\mathrm{d}\rho)^\varepsilon \tag{3.173}$$

and

$$\delta_\varepsilon(\tau) = \int_{-\infty}^{\infty} E_\varepsilon\left[(2\pi\, \mathrm{i})^\varepsilon \tau^\varepsilon \upsilon^\varepsilon\right] (\mathrm{d}\upsilon)^\varepsilon, \tag{3.174}$$

where $t = 0$.

In a similar way, we have

$$\begin{aligned}
\Theta(\omega) &= \frac{1}{\Gamma(1+\varepsilon)} \int_{-\infty}^{\infty} E_\varepsilon\left(-\mathrm{i}^\varepsilon \rho^\varepsilon \omega^\varepsilon\right) \left[\frac{1}{(2\pi)^\varepsilon} \int_{-\infty}^{\infty} \Theta(t) E_\varepsilon\left(\mathrm{i}^\varepsilon \rho^\varepsilon t^\varepsilon\right) (\mathrm{d}t)^\varepsilon\right] (\mathrm{d}\rho)^\varepsilon \\
&= \frac{1}{(2\pi)^\varepsilon} \int_{-\infty}^{\infty} \Theta(t) \left[\frac{1}{\Gamma(1+\varepsilon)} \int_{-\infty}^{\infty} E_\varepsilon\left(-\mathrm{i}^\varepsilon \rho^\varepsilon (\omega - t)^\varepsilon\right) (\mathrm{d}\rho)^\varepsilon\right] (\mathrm{d}t)^\varepsilon \\
&= \frac{1}{(2\pi)^\varepsilon} \int_{-\infty}^{\infty} \Theta(t) \left[\frac{(2\pi)^\varepsilon}{\Gamma(1+\varepsilon)} \delta_\varepsilon(\omega - t)\right] (\mathrm{d}t)^\varepsilon,
\end{aligned} \tag{3.175}$$

where

$$\frac{(2\pi)^\varepsilon}{\Gamma(1+\varepsilon)} \delta_\varepsilon(\omega - t) = \frac{1}{\Gamma(1+\varepsilon)} \int_{-\infty}^{\infty} E_\varepsilon\left(-\mathrm{i}^\varepsilon \rho^\varepsilon (\omega - t)^\varepsilon\right) (\mathrm{d}\rho)^\varepsilon \tag{3.176}$$

and

$$\begin{aligned}
G(\varpi) &= \frac{1}{\sqrt{\Gamma(1+\varepsilon)}} \int_{-\infty}^{\infty} \left\{\frac{\int_{-\infty}^{\infty} G(\rho) E_\varepsilon\left[(2\pi\, \mathrm{i})^\varepsilon t^\varepsilon \rho^\varepsilon\right] (\mathrm{d}\rho)^\varepsilon}{\sqrt{\Gamma(1+\varepsilon)}}\right\} E_\varepsilon\left[-(2\pi\, \mathrm{i})^\varepsilon t^\varepsilon \varpi^\varepsilon\right] (\mathrm{d}t)^\varepsilon \\
&= \frac{1}{\sqrt{\Gamma(1+\varepsilon)}} \int_{-\infty}^{\infty} G(\rho) \left\{\frac{\int_{-\infty}^{\infty} E_\varepsilon\left[-(2\pi\, \mathrm{i})^\varepsilon t^\varepsilon (\varpi - \rho)^\varepsilon\right] (\mathrm{d}t)^\varepsilon}{\sqrt{\Gamma(1+\varepsilon)}}\right\} (\mathrm{d}\rho)^\varepsilon \\
&= \frac{1}{\sqrt{\Gamma(1+\varepsilon)}} \int_{-\infty}^{\infty} G(\rho) \left\{\frac{1}{\sqrt{\Gamma(1+\varepsilon)}} \delta_\varepsilon(\varpi - \rho)\right\} (\mathrm{d}\rho)^\varepsilon,
\end{aligned} \tag{3.177}$$

so that

$$\delta_\varepsilon(\varpi - \rho) = \int_{-\infty}^{\infty} E_\varepsilon\left[-(2\pi\, \mathrm{i})^\varepsilon t^\varepsilon (\varpi - \rho)^\varepsilon\right] (\mathrm{d}t)^\varepsilon. \tag{3.178}$$

In fact, in light of (3.175) and (3.177), we find by using (3.176) and (3.178) that

$$\frac{(2\pi)^\varepsilon}{\Gamma(1+\varepsilon)} \delta_\varepsilon(\omega) = \frac{1}{\Gamma(1+\varepsilon)} \int_{-\infty}^{\infty} E_\varepsilon\left(-\mathrm{i}^\varepsilon \rho^\varepsilon \omega^\varepsilon\right) (\mathrm{d}\rho)^\varepsilon \tag{3.179}$$

and

$$\delta_\varepsilon(\varpi) = \int_{-\infty}^{\infty} E_\varepsilon\left[-(2\pi\ \mathrm{i})^\varepsilon t^\varepsilon \varpi^\varepsilon\right](\mathrm{d}t)^\varepsilon, \tag{3.180}$$

so that $t = 0$ and $\rho = 0$. Therefore, we show the following results:

$$\frac{1}{\Gamma(1+\varepsilon)} \int_{-\infty}^{\infty} E_\varepsilon\left(-\mathrm{i}^\varepsilon \omega^\varepsilon \tau^\varepsilon\right)(\mathrm{d}\tau)^\varepsilon = \frac{(2\pi)^\varepsilon}{\Gamma(1+\varepsilon)} \delta_\varepsilon(\omega), \tag{3.181}$$

$$\frac{1}{(2\pi)^\varepsilon} \int_{-\infty}^{\infty} E_\varepsilon\left(\mathrm{i}^\varepsilon \tau^\varepsilon \omega^\varepsilon\right)(\mathrm{d}\omega)^\varepsilon = \delta_\varepsilon(\tau), \tag{3.182}$$

$$\frac{1}{\sqrt{\Gamma(1+\varepsilon)}} \int_{-\infty}^{\infty} E_\varepsilon\left[-(2\pi\ \mathrm{i})^\varepsilon \varpi^\varepsilon \tau^\varepsilon\right](\mathrm{d}\tau)^\varepsilon = \frac{\delta_\varepsilon(\varpi)}{\sqrt{\Gamma(1+\varepsilon)}}, \tag{3.183}$$

and

$$\frac{1}{\sqrt{\Gamma(1+\varepsilon)}} \int_{-\infty}^{\infty} E_\varepsilon\left[(2\pi\ \mathrm{i})^\varepsilon \tau^\varepsilon \varpi^\varepsilon\right](\mathrm{d}\varpi)^\varepsilon = \frac{\delta_\varepsilon(\tau)}{\sqrt{\Gamma(1+\varepsilon)}}. \tag{3.184}$$

Definition 3.9. Let $H_\varepsilon(\tau)$ be the Heaviside function defined on Cantor sets as follows:

$$H_\varepsilon(\tau) = \begin{cases} 0, & \text{if } x < 0, \\ 1, & \text{if } x \geq 0, \end{cases} \tag{3.185}$$

with a distribution given by the formula:

$$\frac{1}{\Gamma(1+\varepsilon)} \int_{-\infty}^{\infty} H_\varepsilon(\tau)\psi(\tau)(\mathrm{d}\tau)^\varepsilon = \frac{1}{\Gamma(1+\varepsilon)} \int_{0}^{\infty} \psi(\tau)(\mathrm{d}\tau)^\varepsilon. \tag{3.186}$$

Definition 3.10. The local fractional derivative of the analogous Dirac distribution, denoted by $u_\varepsilon(\tau)$, is defined by

$$u_\varepsilon(\tau) = \frac{\mathrm{d}^\varepsilon \delta_\varepsilon(\tau)}{\mathrm{d}\tau^\varepsilon} \tag{3.187}$$

with the following properties:

$$\frac{1}{\Gamma(1+\varepsilon)} \int_{-\infty}^{\infty} u_\varepsilon(\tau)\psi(\tau)(\mathrm{d}\tau)^\varepsilon = -\psi^{(\varepsilon)}(0). \tag{3.188}$$

Definition 3.11. The analogous rectangular pulse defined on Cantor sets, denoted by $\mathrm{rect}_\varepsilon(\tau)$, is defined by

$$\mathrm{rect}_\varepsilon(\tau) = \begin{cases} 1 & \text{if } |\tau| \leq \frac{1}{2}, \\ 0 & \text{if } |\tau| > \frac{1}{2}. \end{cases} \tag{3.189}$$

For finding the local fractional Fourier transform of the analogous rectangular pulse, it is observed that

$$\begin{aligned}\Im\left[\mathrm{rect}_\varepsilon(\tau)\right] &= \left(\Im\,\mathrm{rect}_\varepsilon\right)(\omega)\\ &= \frac{1}{\Gamma(1+\varepsilon)}\int_{-\infty}^{\infty}\mathrm{rect}_\varepsilon(\tau)\,E_\varepsilon\left(-i^\varepsilon\tau^\varepsilon\omega^\varepsilon\right)(d\tau)^\varepsilon\\ &= \frac{1}{\Gamma(1+\varepsilon)}\int_{-1/2}^{1/2}E_\varepsilon\left(-i^\varepsilon\tau^\varepsilon\omega^\varepsilon\right)(d\tau)^\varepsilon\\ &= \frac{E_\varepsilon\left(i^\varepsilon\left(\frac{\omega}{2}\right)^\varepsilon\right)-E_\varepsilon\left(-i^\varepsilon\left(\frac{\omega}{2}\right)^\varepsilon\right)}{i\omega}\\ &= \frac{2\sin_\varepsilon\left(\frac{\omega}{2}\right)^\varepsilon}{\omega^\varepsilon}.\end{aligned} \tag{3.190}$$

We thus find that

$$\lim_{\omega\to 0}\frac{2\sin_\varepsilon\left(\frac{\omega}{2}\right)^\varepsilon}{\omega^\varepsilon}=\lim_{\omega\to 0}\frac{\frac{\partial^\varepsilon}{\partial\omega^\varepsilon}2\sin_\varepsilon\left(\frac{\omega}{2}\right)^\varepsilon}{\frac{\partial^\varepsilon\omega^\varepsilon}{\partial\omega^\varepsilon}}=\frac{2^{1+\varepsilon}}{\Gamma(1+\varepsilon)} \tag{3.191}$$

and

$$\lim_{\omega\to\infty}\frac{2\sin_\varepsilon\left(\frac{\omega}{2}\right)^\varepsilon}{\omega^\varepsilon}=0. \tag{3.192}$$

Hence, we get

$$\left(\Im\,\mathrm{rect}_\varepsilon\right)(0)=\frac{2^{1+\varepsilon}}{\Gamma(1+\varepsilon)}. \tag{3.193}$$

In Figure 3.2, $\mathrm{rect}_\varepsilon(\tau)$ and $\left(\Im\,\mathrm{rect}_\varepsilon\right)(\omega)$ are sketched.

From (3.191), we have

$$\lim_{\omega\to 0}\frac{\sin_\varepsilon\left(\frac{\omega}{2}\right)^\varepsilon}{\frac{\left(\frac{\omega}{2}\right)^\varepsilon}{\Gamma(1+\varepsilon)}}=1 \tag{3.194}$$

or

$$\lim_{\omega\to 0}\frac{\sin_\varepsilon\left(\frac{\omega}{2}\right)^\varepsilon}{\left(\frac{\omega}{2}\right)^\varepsilon}=\frac{1}{\Gamma(1+\varepsilon)}. \tag{3.195}$$

When $\omega=1$, we obtain

$$\lim_{\omega\to 0}\frac{\sin\frac{\omega}{2}}{\frac{\omega}{2}}=1. \tag{3.196}$$

Definition 3.12. The analogous triangle function defined on Cantor sets, denoted by $\mathrm{trig}_\varepsilon(\tau)$ is defined as

$$\mathrm{trig}_\varepsilon(\tau)=\begin{cases}\frac{(1-|\tau|^\varepsilon)}{\Gamma(1+\varepsilon)} & \text{if } |\tau|\le 1,\\ 0 & \text{if } |\tau|>1.\end{cases} \tag{3.197}$$

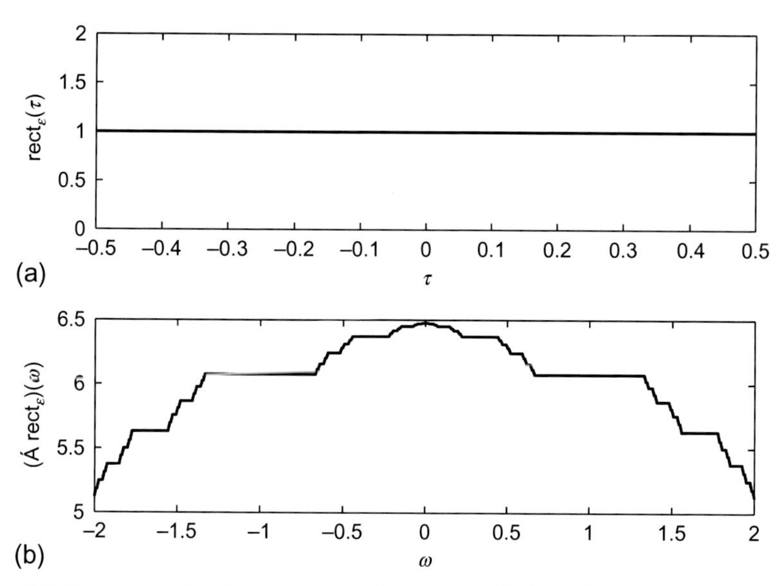

Figure 3.2 The graphs of analogous rectangular pulse and its local fractional Fourier transform: (a) the graph of $\text{rect}_\varepsilon(\tau)$ and (b) the graph of $(\Im\, \text{rect}_\varepsilon)(\omega)$.

Find the local fractional Fourier transform of analogous triangle function, namely,

$$\begin{aligned}
\Im\left[\text{triang}_\varepsilon(\tau)\right] &= \left(\Im\, \text{triang}_\varepsilon\right)(\omega) \\
&= \frac{1}{\Gamma(1+\varepsilon)}\int_{-\infty}^{\infty} \text{triang}_\varepsilon(\tau)\, E_\varepsilon\left(-\mathrm{i}^\varepsilon \tau^\varepsilon \omega^\varepsilon\right)(\mathrm{d}\tau)^\varepsilon \\
&= \frac{1}{\Gamma(1+\varepsilon)}\int_{-1}^{1} \frac{\left(1-|\tau|^\varepsilon\right)}{\Gamma(1+\varepsilon)} E_\varepsilon\left(-\mathrm{i}^\varepsilon \tau^\varepsilon \omega^\varepsilon\right)(\mathrm{d}\tau)^\varepsilon \\
&= \frac{1}{\Gamma(1+\varepsilon)}\int_{-1}^{1} \frac{\left(1-|\tau|^\varepsilon\right)}{\Gamma(1+\varepsilon)} \left[\cos_\varepsilon\left(\tau^\varepsilon \omega^\varepsilon\right) - \mathrm{i}^\varepsilon \sin_\varepsilon\left(\tau^\varepsilon \omega^\varepsilon\right)\right](\mathrm{d}\tau)^\varepsilon \\
&= \frac{1}{\Gamma(1+\varepsilon)}\int_{-1}^{1} \frac{\left(1-|\tau|^\varepsilon\right)}{\Gamma(1+\varepsilon)} \cos_\varepsilon\left(\tau^\varepsilon \omega^\varepsilon\right)(\mathrm{d}\tau)^\varepsilon \\
&= \frac{2}{\Gamma(1+\varepsilon)}\int_{0}^{1} \frac{\left(1-\tau^\varepsilon\right)}{\Gamma(1+\varepsilon)} \cos_\varepsilon\left(\tau^\varepsilon \omega^\varepsilon\right)(\mathrm{d}\tau)^\varepsilon \\
&= \frac{2}{\omega^\varepsilon}\left[\frac{\left(1-\tau^\varepsilon\right)}{\Gamma(1+\varepsilon)} \sin_\varepsilon\left(\tau^\varepsilon \omega^\varepsilon\right)\right]_0^1 + \frac{2}{\omega^\varepsilon}\left[\frac{1}{\Gamma(1+\varepsilon)}\int_0^1 \sin_\varepsilon\left(\tau^\varepsilon \omega^\varepsilon\right)(\mathrm{d}\tau)^\varepsilon\right] \\
&= \frac{2}{\omega^\varepsilon}\left[\frac{1}{\Gamma(1+\varepsilon)}\int_0^1 \sin_\varepsilon\left(\tau^\varepsilon \omega^\varepsilon\right)(\mathrm{d}\tau)^\varepsilon\right] \\
&= \frac{2}{\omega^{2\varepsilon}}\left[1-\cos_\varepsilon\left(\omega^\varepsilon\right)\right] \\
&= \frac{4\sin_\varepsilon^2\left(\frac{\omega}{2}\right)^\varepsilon}{\omega^{2\varepsilon}}.
\end{aligned} \tag{3.198}$$

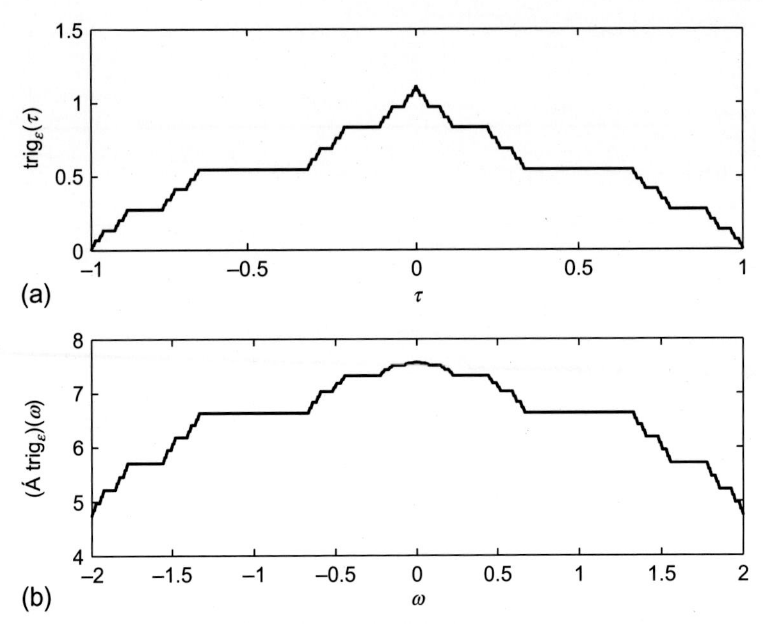

Figure 3.3 The graphs of the analogous triangle function and its local fractional Fourier transform: (a) the plot of $\text{triang}_\varepsilon(\tau)$ and (b) the plot of $\left(\Im\,\text{triang}_\varepsilon\right)(\omega)$.

Using (3.128), we find that

$$\left(\Im\,\text{triang}_\varepsilon\right)(0) = \lim_{\omega\to 0}\frac{4\sin_\varepsilon^2\left(\frac{\omega}{2}\right)^\varepsilon}{\omega^{2\varepsilon}} = \lim_{\omega\to 0}\frac{2}{\Gamma(1+2\varepsilon)} \tag{3.199}$$

and

$$\left(\Im\,\text{triang}_\varepsilon\right)(\infty) = \lim_{\omega\to \infty}\frac{4\sin_\varepsilon^2\left(\frac{\omega}{2}\right)^\varepsilon}{\omega^{2\varepsilon}} = 0. \tag{3.200}$$

In Figure 3.3, the graphs of $\text{triang}_\varepsilon(\tau)$ and $\left(\Im\,\text{triang}_\varepsilon\right)(\omega)$ are drawn.

Definition 3.13. The two-sided Mittag–Leffler distribution defined on Cantor sets with the positive parameter a, denoted by $M_{a.\varepsilon}(\tau)$, is defined by

$$M_{a.\varepsilon}(\tau) = \begin{cases} E_\varepsilon\left(-(a\tau)^\varepsilon\right), & \tau > 0, \\ -E_\varepsilon\left((a\tau)^\varepsilon\right), & \tau < 0. \end{cases} \tag{3.201}$$

We see that

$$\begin{aligned}\Im\left[M_{a.\varepsilon}(\tau)\right] &= \left(\Im M_{a.\varepsilon}\right)(\omega) \\ &= \frac{1}{\Gamma(1+\varepsilon)}\int_{-\infty}^{\infty} M_{a.\varepsilon}(\tau)\,E_\varepsilon\left(-i^\varepsilon\tau^\varepsilon\omega^\varepsilon\right)(d\tau)^\varepsilon\end{aligned}$$

$$= \frac{1}{\Gamma(1+\varepsilon)} \int_0^\infty E_\varepsilon\left(-\tau^\varepsilon\left(\mathrm{i}^\varepsilon\omega^\varepsilon + a^\varepsilon\right)\right)(\mathrm{d}\tau)^\varepsilon$$
$$- \frac{1}{\Gamma(1+\varepsilon)} \int_{-\infty}^0 E_\varepsilon\left(\tau^\varepsilon\left(a^\varepsilon - \mathrm{i}^\varepsilon\omega^\varepsilon\right)\right)(\mathrm{d}\tau)^\varepsilon$$
$$= \frac{1}{\mathrm{i}^\varepsilon\omega^\varepsilon + a^\varepsilon} - \frac{1}{a^\varepsilon - \mathrm{i}^\varepsilon\omega^\varepsilon}$$
$$= \frac{1}{\mathrm{i}^\varepsilon\omega^\varepsilon + a^\varepsilon} - \frac{1}{a^\varepsilon - \mathrm{i}^\varepsilon\omega^\varepsilon}$$
$$= \frac{-2\mathrm{i}^\varepsilon\omega^\varepsilon}{a^{2\varepsilon} + \omega^{2\varepsilon}}, \tag{3.202}$$

which leads us to

$$(\Im M_{a.\varepsilon})(0) = 0 \tag{3.203}$$

and

$$(\Im M_{a.\varepsilon})(\infty) = 0. \tag{3.204}$$

Definition 3.14. The complex distribution via the Mittag–Leffler function defined on Cantor sets with the positive parameter a, denoted by $\Xi_{a.\varepsilon}(\tau)$, is defined by

$$\Xi_{a.\varepsilon}(\tau) = E_\varepsilon\left(-a^\varepsilon |\tau|^\varepsilon\right). \tag{3.205}$$

We thus find that

$$\Im\left[\Xi_{a.\varepsilon}(\tau)\right] = (\Im\Xi_{a.\varepsilon})(\omega)$$
$$= \frac{1}{\Gamma(1+\varepsilon)} \int_{-\infty}^\infty \Xi_{a.\varepsilon}(\tau) E_\varepsilon\left(-\mathrm{i}^\varepsilon\tau^\varepsilon\omega^\varepsilon\right)(\mathrm{d}\tau)^\varepsilon$$
$$= \frac{1}{\Gamma(1+\varepsilon)} \int_0^\infty E_\varepsilon\left(-\tau^\varepsilon\left(\mathrm{i}^\varepsilon\omega^\varepsilon - a^\varepsilon\right)\right)(\mathrm{d}\tau)^\varepsilon$$
$$+ \frac{1}{\Gamma(1+\varepsilon)} \int_{-\infty}^0 E_\varepsilon\left(\tau^\varepsilon\left(-\omega^\varepsilon\mathrm{i}^\varepsilon - a^\varepsilon\right)\right)(\mathrm{d}\tau)^\varepsilon$$
$$= \frac{1}{\mathrm{i}^\varepsilon\omega^\varepsilon - a^\varepsilon} - \frac{1}{\omega^\varepsilon\mathrm{i}^\varepsilon + a^\varepsilon}$$
$$= \frac{2a^\varepsilon}{a^{2\varepsilon} + \omega^{2\varepsilon}}, \tag{3.206}$$

which leads us to

$$(\Im\Xi_{a.\varepsilon})(0) = \frac{2a^\varepsilon}{a^{2\varepsilon}} \tag{3.207}$$

and

$$(\Im\Xi_{a.\varepsilon})(\infty) = 0. \tag{3.208}$$

Definition 3.15. The signum distribution defined on Cantor sets, denoted by $\mathrm{sgn}_\varepsilon(\tau)$, is defined as

$$\mathrm{sgn}_\varepsilon(\tau) = \begin{cases} 1, & \tau \geq 0, \\ -1, & \tau < 0. \end{cases} \tag{3.209}$$

Thus, we have

$$\begin{aligned} \Im\left[\mathrm{sgn}_\varepsilon(\tau)\right] &= \left(\Im\,\mathrm{sgn}_\varepsilon\right)(\omega) \\ &= \frac{1}{\Gamma(1+\varepsilon)}\int_{-\infty}^{\infty} \mathrm{sgn}_\varepsilon(\tau)\, E_\varepsilon\left(-\mathrm{i}^\varepsilon\tau^\varepsilon\omega^\varepsilon\right)(\mathrm{d}\tau)^\varepsilon \\ &= \frac{1}{\Gamma(1+\varepsilon)}\int_{0}^{\infty} E_\varepsilon\left(-\mathrm{i}^\varepsilon\tau^\varepsilon\omega^\varepsilon\right)(\mathrm{d}\tau)^\varepsilon \\ &- \frac{1}{\Gamma(1+\varepsilon)}\int_{-\infty}^{0} E_\varepsilon\left(\mathrm{i}^\varepsilon\tau^\varepsilon(-\omega)^\varepsilon\right)(\mathrm{d}\tau)^\varepsilon \\ &= \frac{2}{\mathrm{i}^\varepsilon\omega^\varepsilon}, \end{aligned} \tag{3.210}$$

which leads to

$$\left(\Im\,\mathrm{sgn}_\varepsilon\right)(0) = \infty \tag{3.211}$$

and

$$\left(\Im\,\Xi_{a.\varepsilon}\right)(\infty) = 0. \tag{3.212}$$

Definition 3.16. The Mittag–Leffler distribution defined on Cantor sets with the squared variable and positive parameter a, denoted by $\Lambda_\varepsilon(\tau)$, is defined as

$$\Lambda_\varepsilon(\tau) = E_\varepsilon\left(-a\tau^{2\varepsilon}\right). \tag{3.213}$$

As a result, by the definition of the local fractional Fourier transform operator, we have

$$\begin{aligned} \Im\left[\Lambda_\varepsilon(\tau)\right] &= \left(\Im\Lambda_\varepsilon\right)(\omega) \\ &= \frac{1}{\Gamma(1+\varepsilon)}\int_{-\infty}^{\infty} \Lambda_\varepsilon E_\varepsilon\left(-\mathrm{i}^\varepsilon\tau^\varepsilon\omega^\varepsilon\right)(\mathrm{d}\tau)^\varepsilon \\ &= \frac{1}{\Gamma(1+\varepsilon)}\int_{-\infty}^{\infty} E_\varepsilon\left(-a\tau^{2\varepsilon} - \mathrm{i}^\varepsilon\tau^\varepsilon\omega^\varepsilon\right)(\mathrm{d}\tau)^\varepsilon \\ &= \frac{2}{\Gamma(1+\varepsilon)}\int_{0}^{\infty} E_\varepsilon\left(-a\tau^{2\varepsilon}\right)\cos_\varepsilon\left(\tau^\varepsilon\omega^\varepsilon\right)(\mathrm{d}\tau)^\varepsilon. \end{aligned} \tag{3.214}$$

In addition, we find that

$$\begin{aligned}\Im^{(\varepsilon)}\left[\Lambda_{\varepsilon}(\tau)\right] &= \frac{-2}{\Gamma(1+\varepsilon)}\int_{0}^{\infty}E_{\varepsilon}\left(-a\tau^{2\varepsilon}\right)\tau^{\varepsilon}\sin_{\varepsilon}\left(\tau^{\varepsilon}\omega^{\varepsilon}\right)(\mathrm{d}\tau)^{\varepsilon} \\ &= \left[\frac{2^{1-\varepsilon}}{a}E_{\varepsilon}\left(-a\tau^{2\varepsilon}\right)\sin_{\varepsilon}\left(\tau^{\varepsilon}\omega^{\varepsilon}\right)\right]_{0}^{\infty} \\ &\quad - \frac{2^{1-\varepsilon}\omega^{\varepsilon}}{a}\left[\frac{1}{\Gamma(1+\varepsilon)}\int_{0}^{\infty}E_{\varepsilon}\left(-a\tau^{2\varepsilon}\right)\cos_{\varepsilon}\left(\tau^{\varepsilon}\omega^{\varepsilon}\right)(\mathrm{d}\tau)^{\varepsilon}\right],\end{aligned} \tag{3.215}$$

which leads us to

$$\left[\frac{2^{1-\varepsilon}}{a}E_{\varepsilon}\left(-a\tau^{2\varepsilon}\right)\sin_{\varepsilon}\left(\tau^{\varepsilon}\omega^{\varepsilon}\right)\right]_{0}^{\infty} = 0 \tag{3.216}$$

and

$$\begin{aligned}\Im^{(\varepsilon)}\left[\Lambda_{\varepsilon}(\tau)\right] &= -\frac{2^{1-\varepsilon}\omega^{\varepsilon}}{a}\left[\frac{1}{\Gamma(1+\varepsilon)}\int_{0}^{\infty}E_{\varepsilon}\left(-a\tau^{2\varepsilon}\right)\cos_{\varepsilon}\left(\tau^{\varepsilon}\omega^{\varepsilon}\right)(\mathrm{d}\tau)^{\varepsilon}\right] \\ &= -\frac{2^{-\varepsilon}\omega^{\varepsilon}}{a}\frac{1}{\Gamma(1+\varepsilon)}\int_{-\infty}^{\infty}E_{\varepsilon}\left(-a\tau^{2\varepsilon}-\mathrm{i}^{\varepsilon}\tau^{\varepsilon}\omega^{\varepsilon}\right)(\mathrm{d}\tau)^{\varepsilon} \\ &= -\frac{2^{-\varepsilon}\omega^{\varepsilon}}{a}\Im\left[\Lambda_{\varepsilon}(\tau)\right].\end{aligned} \tag{3.217}$$

In this case, by making use of (3.217), we have

$$\Im^{(\varepsilon)}\left[\Lambda_{\varepsilon}(\tau)\right] + \frac{2^{-\varepsilon}}{a}\omega^{\varepsilon}\Im\left[\Lambda_{\varepsilon}(\tau)\right] = 0, \tag{3.218}$$

which yields to the nondifferentiable solution in the form

$$\Im\left[\Lambda_{\varepsilon}(\tau)\right] = \left(\Im\Lambda_{\varepsilon}\right)(\omega) = M_{0}E_{\varepsilon}\left(-\frac{2^{-2\varepsilon}}{a}\omega^{2\varepsilon}\right), \tag{3.219}$$

where M_0 can be confirmed by

$$M_{0} = \left(\Im\Lambda_{\varepsilon}\right)(0). \tag{3.220}$$

By using (3.156), we rewrite (3.220) as follows:

$$\begin{aligned}M_{0} &= \left(\Im\Lambda_{\varepsilon}\right)(0) \\ &= \frac{1}{\Gamma(1+\varepsilon)}\int_{\infty}^{\infty}E_{\varepsilon}\left(-a\tau^{2\varepsilon}\right)(\mathrm{d}\tau)^{\varepsilon} \\ &= \frac{\pi^{\varepsilon/2}\sqrt{\frac{1}{a}}}{\Gamma(1+\varepsilon)}.\end{aligned} \tag{3.221}$$

Thus, from (3.219) and (3.221), we obtain

$$\Im\left[E_\varepsilon\left(-a\tau^{2\varepsilon}\right)\right]=\frac{\pi^{\varepsilon/2}\sqrt{\frac{1}{a}}}{\Gamma(1+\varepsilon)}E_\varepsilon\left[-\frac{1}{a}\left(\frac{\omega}{2}\right)^{2\varepsilon}\right]. \tag{3.222}$$

In this case, for $\varepsilon=1$, we get

$$\int_{-\infty}^{\infty}\exp\left(-a\tau^2-\mathrm{i}\tau\omega\right)\mathrm{d}\tau=\sqrt{\frac{\pi}{a}}\exp\left(-\frac{1}{4a}\omega^2\right). \tag{3.223}$$

For $a=1$, we rewrite (3.222) as follows:

$$\Im\left[E_\varepsilon\left(-\tau^{2\varepsilon}\right)\right]=\frac{\pi^{\varepsilon/2}}{\Gamma(1+\varepsilon)}E_\varepsilon\left[-\left(\frac{\omega}{2}\right)^{2\varepsilon}\right]. \tag{3.224}$$

For $a=\pi^\varepsilon$, we obtain

$$\Im\left[E_\varepsilon\left(-\pi^\varepsilon\tau^{2\varepsilon}\right)\right]=\frac{1}{\Gamma(1+\varepsilon)}E_\varepsilon\left[-\frac{1}{\pi^\varepsilon}\left(\frac{\omega}{2}\right)^{2\varepsilon}\right]. \tag{3.225}$$

In fact, we rewrite $H_\varepsilon(\tau)$ as follows:

$$H_\varepsilon(\tau)=\frac{1+\mathrm{sgn}_\varepsilon(\tau)}{2}. \tag{3.226}$$

In this case, we find that

$$\begin{aligned}\Im\left[H_\varepsilon(\tau)\right]&=\left(\Im H_\varepsilon\right)(\omega)\\&=\frac{1}{\Gamma(1+\varepsilon)}\int_{-\infty}^{\infty}H_\varepsilon(\tau)E_\varepsilon\left(-\mathrm{i}^\varepsilon\tau^\varepsilon\omega^\varepsilon\right)(\mathrm{d}\tau)^\varepsilon\\&=\frac{1}{\Gamma(1+\varepsilon)}\int_{-\infty}^{\infty}\left[\frac{1+\mathrm{sgn}_\varepsilon(\tau)}{2}\right]E_\varepsilon\left(-\mathrm{i}^\varepsilon\tau^\varepsilon\omega^\varepsilon\right)(\mathrm{d}\tau)^\varepsilon\\&=\frac{1}{\Gamma(1+\varepsilon)}\int_{-\infty}^{\infty}\frac{1}{2}E_\varepsilon\left(-\mathrm{i}^\varepsilon\tau^\varepsilon\omega^\varepsilon\right)(\mathrm{d}\tau)^\varepsilon\\&\quad+\frac{1}{\Gamma(1+\varepsilon)}\int_{-\infty}^{\infty}\frac{\mathrm{sgn}_\varepsilon(\tau)}{2}E_\varepsilon\left(-\mathrm{i}^\varepsilon\tau^\varepsilon\omega^\varepsilon\right)(\mathrm{d}\tau)^\varepsilon,\end{aligned} \tag{3.227}$$

which, by using (3.181) and (3.210), becomes

$$\Im\left[H_\varepsilon(\tau)\right]=\frac{1}{2}\frac{(2\pi)^\varepsilon}{\Gamma(1+\varepsilon)}\delta_\varepsilon(\omega)+\frac{1}{\mathrm{i}^\varepsilon\omega^\varepsilon}. \tag{3.228}$$

3.3.2 Applications of signal analysis on Cantor sets

Let us consider the signal given by

$$\psi(\tau)=\delta_\varepsilon(\tau-\tau_0). \tag{3.229}$$

Also, let its local fractional Fourier transform be read as follows:

$$\begin{aligned}\Im[\psi(\tau)] &= (\Im\psi)(\omega) \\ &= \frac{1}{\Gamma(1+\varepsilon)}\int_{-\infty}^{\infty}\delta_\varepsilon(\tau-\tau_0)\,E_\varepsilon\left(-\mathrm{i}^\varepsilon\tau^\varepsilon\omega^\varepsilon\right)(\mathrm{d}\tau)^\varepsilon \\ &= E_\varepsilon\left(-\mathrm{i}^\varepsilon\tau_0^\varepsilon\omega^\varepsilon\right).\end{aligned} \tag{3.230}$$

We now find the local fractional Fourier transform of the signal given by

$$\psi(\tau) = \delta_\varepsilon(\tau-\tau_0)\,\theta(\tau).$$

Indeed, by using the local fractional Fourier transform operator, we have

$$\begin{aligned}\Im[\psi(\tau)] &= (\Im\psi)(\omega) \\ &= \frac{1}{\Gamma(1+\varepsilon)}\int_{-\infty}^{\infty}\delta_\varepsilon(\tau-\tau_0)\,\theta(\tau)\,E_\varepsilon\left(-\mathrm{i}^\varepsilon\tau^\varepsilon\omega^\varepsilon\right)(\mathrm{d}\tau)^\varepsilon \\ &= \theta(\tau_0)\,E_\varepsilon\left(-\mathrm{i}^\varepsilon\tau_0^\varepsilon\omega^\varepsilon\right).\end{aligned} \tag{3.231}$$

The next step is to find the local fractional Fourier transform of the signal given by

$$\psi(\tau) = E_\varepsilon\left(\mathrm{i}^\varepsilon\tau^\varepsilon\omega_0^\varepsilon\right).$$

Taking into account the local fractional Fourier transform operator, we have

$$\begin{aligned}\Im[\psi(\tau)] &= (\Im\psi)(\omega) \\ &= \frac{1}{\Gamma(1+\varepsilon)}\int_{-\infty}^{\infty}E_\varepsilon\left(\mathrm{i}^\varepsilon\tau^\varepsilon\omega_0^\varepsilon\right)E_\varepsilon\left(-\mathrm{i}^\varepsilon\tau^\varepsilon\omega^\varepsilon\right)(\mathrm{d}\tau)^\varepsilon \\ &= \frac{1}{\Gamma(1+\varepsilon)}\int_{-\infty}^{\infty}E_\varepsilon\left(-\mathrm{i}^\varepsilon\tau^\varepsilon(\omega-\omega_0)^\varepsilon\right)(\mathrm{d}\tau)^\varepsilon \\ &= \frac{(2\pi)^\varepsilon}{\Gamma(1+\varepsilon)}\delta_\varepsilon(\omega-\omega_0).\end{aligned} \tag{3.232}$$

Now, it is natural to find the local fractional Fourier transform of the signal given by

$$\psi(\tau) = \sin_\varepsilon\left(a^\varepsilon\tau^\varepsilon\right).$$

Indeed we have

$$\begin{aligned}\Im[\psi(\tau)] &= (\Im\psi)(\omega) \\ &= \frac{1}{\Gamma(1+\varepsilon)}\int_{-\infty}^{\infty}\sin_\varepsilon\left(a^\varepsilon\tau^\varepsilon\right)E_\varepsilon\left(-\mathrm{i}^\varepsilon\tau^\varepsilon\omega^\varepsilon\right)(\mathrm{d}\tau)^\varepsilon \\ &= \frac{1}{\Gamma(1+\varepsilon)}\int_{-\infty}^{\infty}\frac{E_\varepsilon(\mathrm{i}^\varepsilon a^\varepsilon\tau^\varepsilon)-E_\varepsilon(-\mathrm{i}^\varepsilon a^\varepsilon\tau^\varepsilon)}{2\mathrm{i}^\varepsilon}E_\varepsilon\left(-\mathrm{i}^\varepsilon\tau^\varepsilon\omega^\varepsilon\right)(\mathrm{d}\tau)^\varepsilon \\ &= \frac{(2\pi)^\varepsilon}{\Gamma(1+\varepsilon)}\frac{[\delta_\varepsilon(\omega+a)-\delta_\varepsilon(\omega-a)]}{2\mathrm{i}^\varepsilon}.\end{aligned} \tag{3.233}$$

In order to find the local fractional Fourier transform of the signal given by

$$\psi(\tau) = \cos_\varepsilon\left(a^\varepsilon \tau^\varepsilon\right),$$

we observe that

$$\begin{aligned}
\Im[\psi(\tau)] &= (\Im\psi)(\omega) \\
&= \frac{1}{\Gamma(1+\varepsilon)} \int_{-\infty}^{\infty} \cos_\varepsilon\left(a^\varepsilon \tau^\varepsilon\right) E_\varepsilon\left(-\mathrm{i}^\varepsilon \tau^\varepsilon \omega^\varepsilon\right) (\mathrm{d}\tau)^\varepsilon \\
&= \frac{1}{\Gamma(1+\varepsilon)} \int_{-\infty}^{\infty} \frac{E_\varepsilon(\mathrm{i}^\varepsilon a^\varepsilon \tau^\varepsilon) + E_\varepsilon(-\mathrm{i}^\varepsilon a^\varepsilon \tau^\varepsilon)}{2} E_\varepsilon\left(-\mathrm{i}^\varepsilon \tau^\varepsilon \omega^\varepsilon\right) (\mathrm{d}\tau)^\varepsilon \\
&= \frac{(2\pi)^\varepsilon}{\Gamma(1+\varepsilon)} \frac{[\delta_\varepsilon(\omega + a) + \delta_\varepsilon(\omega - a)]}{2}.
\end{aligned} \tag{3.234}$$

We now find the local fractional Fourier transform of the signal given by

$$\psi(\tau) = \delta_\varepsilon^{(\varepsilon)}(\tau).$$

In fact, from the definition of the local fractional Fourier transform operator, we have

$$\begin{aligned}
\Im[\psi(\tau)] &= (\Im\psi)(\omega) \\
&= \frac{1}{\Gamma(1+\varepsilon)} \int_{-\infty}^{\infty} \delta_\varepsilon^{(\varepsilon)}(\tau) E_\varepsilon\left(-\mathrm{i}^\varepsilon \tau^\varepsilon \omega^\varepsilon\right) (\mathrm{d}\tau)^\varepsilon \\
&= \mathrm{i}^\varepsilon \omega^\varepsilon \frac{1}{\Gamma(1+\varepsilon)} \int_{-\infty}^{\infty} \delta_\varepsilon(\tau) E_\varepsilon\left(-\mathrm{i}^\varepsilon \tau^\varepsilon \omega^\varepsilon\right) (\mathrm{d}\tau)^\varepsilon \\
&= \mathrm{i}^\varepsilon \omega^\varepsilon.
\end{aligned} \tag{3.235}$$

In order to find the local fractional Fourier transform of the constant signal given by

$$\psi(\tau) = C,$$

we observe from the definition of the local fractional Fourier transform operator that

$$\begin{aligned}
\Im[\psi(\tau)] &= (\Im\psi)(\omega) \\
&= \frac{1}{\Gamma(1+\varepsilon)} \int_{-\infty}^{\infty} C E_\varepsilon\left(-\mathrm{i}^\varepsilon \tau^\varepsilon \omega^\varepsilon\right) (\mathrm{d}\tau)^\varepsilon \\
&= C \frac{1}{\Gamma(1+\varepsilon)} \int_{-\infty}^{\infty} E_\varepsilon\left(-\mathrm{i}^\varepsilon \tau^\varepsilon \omega^\varepsilon\right) (\mathrm{d}\tau)^\varepsilon \\
&= \frac{C(2\pi)^\varepsilon}{\Gamma(1+\varepsilon)} \delta_\varepsilon(\omega).
\end{aligned} \tag{3.236}$$

Find the local fractional Fourier transform of the signal given by

$$\psi(\tau) = \tau^\varepsilon.$$

From the definition of the local fractional Fourier transform operator, we get

$$\begin{aligned}\Im[\psi(\tau)] &= (\Im\psi)(\omega)\\ &= \frac{1}{\Gamma(1+\varepsilon)}\int_{-\infty}^{\infty}\tau^{\varepsilon}E_{\varepsilon}\left(-\mathrm{i}^{\varepsilon}\tau^{\varepsilon}\omega^{\varepsilon}\right)(\mathrm{d}\tau)^{\varepsilon}\\ &= \mathrm{i}^{\varepsilon}\Im^{(\varepsilon)}[1]\\ &= \frac{(2\pi\,\mathrm{i})^{\varepsilon}}{\Gamma(1+\varepsilon)}\delta_{\varepsilon}^{(\varepsilon)}(\omega).\end{aligned} \tag{3.237}$$

Find the local fractional Fourier transform of the signal given by

$$\psi(\tau) = H_{\varepsilon}(\tau)E_{\varepsilon}\left(-\tau^{\varepsilon}a^{\varepsilon}\right).$$

By using the definition of the local fractional Fourier transform operator, we have

$$\begin{aligned}\Im[\psi(\tau)] &= (\Im\psi)(\omega)\\ &= \frac{1}{\Gamma(1+\varepsilon)}\int_{-\infty}^{\infty}\left[H_{\varepsilon}(\tau)E_{\varepsilon}\left(-\tau^{\varepsilon}a^{\varepsilon}\right)\right]E_{\varepsilon}\left(-\mathrm{i}^{\varepsilon}\tau^{\varepsilon}\omega^{\varepsilon}\right)(\mathrm{d}\tau)^{\varepsilon}\\ &= \frac{1}{\Gamma(1+\varepsilon)}\int_{0}^{\infty}E_{\varepsilon}\left(-\tau^{\varepsilon}a^{\varepsilon}\right)E_{\varepsilon}\left(-\mathrm{i}^{\varepsilon}\tau^{\varepsilon}\omega^{\varepsilon}\right)(\mathrm{d}\tau)^{\varepsilon}\\ &= \frac{1}{\Gamma(1+\varepsilon)}\int_{0}^{\infty}E_{\varepsilon}\left(-\tau^{\varepsilon}\left(\omega^{\varepsilon}\mathrm{i}^{\varepsilon}+a^{\varepsilon}\right)\right)(\mathrm{d}\tau)^{\varepsilon}\\ &= \frac{1}{\omega^{\varepsilon}\mathrm{i}^{\varepsilon}+a^{\varepsilon}}.\end{aligned} \tag{3.238}$$

3.4 Solving local fractional differential equations

3.4.1 Applications of local fractional ordinary differential equations

We consider the fractal relaxation equation governed by the local fractional ordinary differential equation in the form

$$\frac{\partial^{\varepsilon}\Phi(\mu)}{\partial\mu^{\varepsilon}} + p\Phi(\mu) = \delta_{\varepsilon}(\mu) \tag{3.239}$$

subject to the initial condition:

$$\Phi(0) = 1. \tag{3.240}$$

Taking the local fractional Fourier transform of (3.239), we first obtain

$$\mathrm{i}^{\varepsilon}\omega^{\varepsilon}(\Im\Phi)(\omega) + p(\Im\Phi)(\omega) = 1, \tag{3.241}$$

which implies that

$$(\Im\Phi)(\omega) = \frac{1}{\mathrm{i}^{\varepsilon}\omega^{\varepsilon}+p}. \tag{3.242}$$

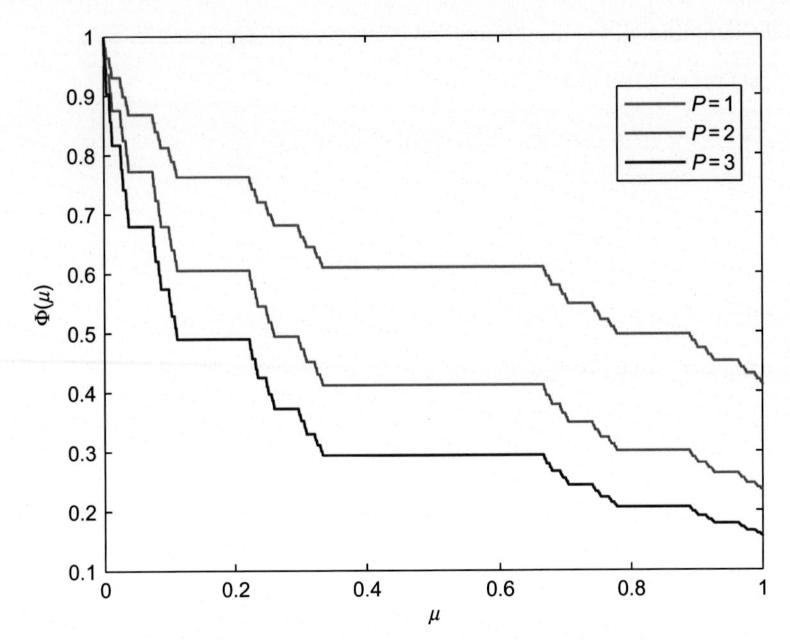

Figure 3.4 The graphs of $\Phi(\mu)$ when $\varepsilon = \ln 2/\ln 3$, $p = 1$, $p = 2$, and $p = 3$.

Now, taking the inverse fractional Fourier transform in (3.242), we have

$$\Phi(\mu) = H_\varepsilon(\tau)\, E_\varepsilon\left(-p\tau^\varepsilon\right).$$

We have drawn the graphs when $p = 1$, $p = 2$, and $p = 3$ in Figure 3.4.

We write the local fractional ordinary differential equation in the form

$$\frac{\partial^\varepsilon \Phi(\mu)}{\partial \mu^\varepsilon} + 2\Phi(\mu) = H_\varepsilon(\tau)\, E_\varepsilon(-\mu) \tag{3.243}$$

subject to the initial condition:

$$\Phi(0) = 0. \tag{3.244}$$

By using the local fractional Fourier transform in (3.243), we have

$$\mathrm{i}^\varepsilon \omega^\varepsilon (\Im\Phi)(\omega) + 2(\Im\Phi)(\omega) = \frac{1}{\mathrm{i}^\varepsilon \omega^\varepsilon + 1}. \tag{3.245}$$

Considering (3.245), we obtain

$$(\Im\Phi)(\omega) = \frac{1}{\mathrm{i}^\varepsilon \omega^\varepsilon + 1} - \frac{1}{2 + \mathrm{i}^\varepsilon \omega^\varepsilon} \tag{3.246}$$

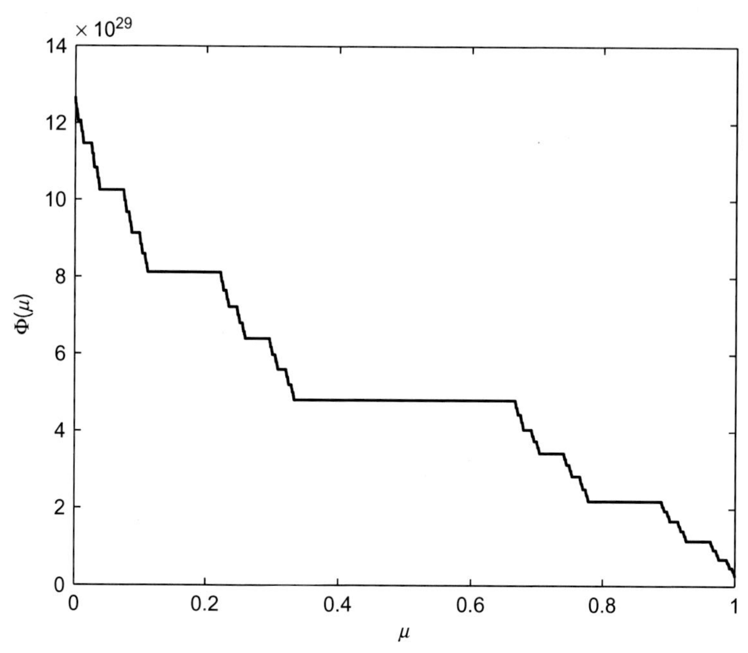

Figure 3.5 The graph of $\Phi(\mu)$ when $\varepsilon = \ln 2/\ln 3$.

such that

$$\Phi(\mu) = H_\varepsilon(\tau)\left[E_\varepsilon(-\mu) - E_\varepsilon(-2\mu)\right]. \tag{3.247}$$

The graph of $\Phi(\mu)$ when $\varepsilon = \ln 2/\ln 3$ is shown in Figure 3.5.

Let us now consider the local fractional ordinary differential equation with the positive parameter p in the form

$$\frac{\partial^{2\varepsilon}\Phi(\mu)}{\partial\mu^{2\varepsilon}} + p\Phi(\mu) = \delta_\varepsilon(\tau) \tag{3.248}$$

subject to initial condition:

$$\Phi(0) = 0. \tag{3.249}$$

Taking the local fractional Fourier transform in (3.248), we get

$$\mathrm{i}^{2\varepsilon}\omega^{2\varepsilon}(\Im\Phi)(\omega) + p(\Im\Phi)(\omega) = 1, \tag{3.250}$$

which yields

$$(\Im\Phi)(\omega) = \frac{1}{p - \omega^{2\varepsilon}}. \tag{3.251}$$

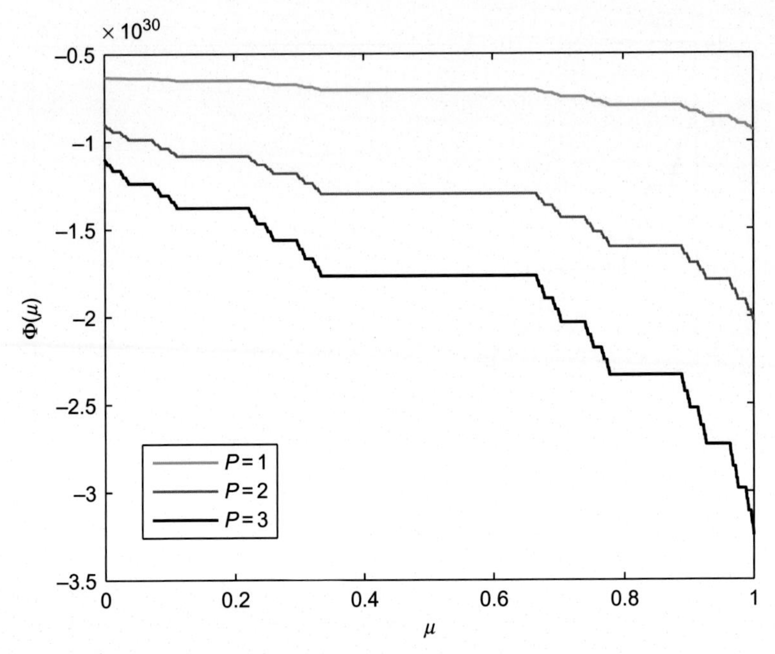

Figure 3.6 The graph of $\Phi(\mu)$ when $\varepsilon = \ln 2/\ln 3$.

Hence, we have

$$(\Im\Phi)(\omega) = \frac{\sqrt{p}}{2}\left[\frac{1}{\sqrt{p}-\omega^{\varepsilon}} + \frac{1}{\sqrt{p}+\omega^{\varepsilon}}\right] \tag{3.252}$$

such that

$$\Phi(\mu) = -\frac{\sqrt{p}}{2} H_{\varepsilon}(\tau)\left[E_{\varepsilon}\left(\sqrt{p}\mu\right) - E_{\varepsilon}\left(-\sqrt{p}\mu\right)\right]. \tag{3.253}$$

The graph for the parameters $p = 1$, $p = 2$, and $p = 3$ is shown in Figure 3.6.

3.4.2 Applications of local fractional partial differential equations

We consider here the local fractional diffusion equation in 1 + 1 fractal dimensional space as follows:

$$\frac{\partial^{\varepsilon}\Phi(\mu,\tau)}{\partial\tau^{\varepsilon}} = k^{2}\frac{\partial^{2\varepsilon}\Phi(\mu,\tau)}{\partial\mu^{2\varepsilon}}, \quad \tau > 0, \ -\infty < \mu < \infty \tag{3.254}$$

subject to the initial condition:

$$\Phi(\mu,0) = \delta_{\varepsilon}(\mu). \tag{3.255}$$

Taking the local fractional Fourier transform with fractal space in (3.254) and (3.255), we get

$$\frac{\partial^{\varepsilon}\Phi(\omega,\tau)}{\partial\tau^{\varepsilon}}=-k^{\varepsilon}\omega^{2\varepsilon}\Phi(\omega,\tau) \tag{3.256}$$

and

$$\Phi(\mu,0)=1. \tag{3.257}$$

Making use of (3.256) and (3.257), we have

$$\Phi(\omega,\tau)=E_{\varepsilon}\left(-k^{\varepsilon}\omega^{2\varepsilon}\tau^{\varepsilon}\right). \tag{3.258}$$

Taking the inverse local fractional Fourier transform in (3.258), we obtain

$$\begin{aligned}\Phi(\mu,\tau)&=\Im^{-1}\left[\Phi(\omega,\tau)\right]\\&=\frac{1}{(2\pi)^{\varepsilon}}\int_{-\infty}^{\infty}\Phi(\omega,\tau)E_{\varepsilon}\left(i^{\varepsilon}\mu^{\varepsilon}\omega^{\varepsilon}\right)(d\omega)^{\varepsilon}\\&=\frac{1}{(2\pi)^{\varepsilon}}\int_{-\infty}^{\infty}E_{\varepsilon}\left(-k^{\varepsilon}\tau^{\varepsilon}\omega^{2\varepsilon}\right)E_{\varepsilon}\left(i^{\varepsilon}\mu^{\varepsilon}\omega^{\varepsilon}\right)(d\omega)^{\varepsilon},\end{aligned} \tag{3.259}$$

which leads us to

$$\Im^{-1}\left\{E_{\varepsilon}\left[-u\omega^{2\varepsilon}\right]\right\}=\frac{1}{\frac{\pi^{\varepsilon/2}\sqrt{4^{\varepsilon}u}}{\Gamma(1+\varepsilon)}}E_{\varepsilon}\left(-\frac{1}{u}\left(\frac{\tau}{2}\right)^{2\varepsilon}\right) \tag{3.260}$$

and

$$\Phi(\mu,\tau)=\frac{\Gamma(1+\varepsilon)}{(4\pi k\tau)^{\varepsilon/2}}E_{\varepsilon}\left(-\frac{\mu^{2\varepsilon}}{(4k\tau)^{\varepsilon}}\right). \tag{3.261}$$

In this case, we obtain the same result as with the local fractional similarity solution [23]. The corresponding solution was discussed in Chapter 1.

Let us consider the local fractional Laplace equation in 1 + 1 fractal dimensional space as follows:

$$\frac{\partial^{2\varepsilon}\Phi(\mu,\eta)}{\partial\mu^{2\varepsilon}}+\frac{\partial^{2\varepsilon}\Phi(\mu,\eta)}{\partial\eta^{2\varepsilon}}=0,\quad \eta>0,\ -\infty<\mu<\infty \tag{3.262}$$

subject to the initial conditions:

$$\Phi(\mu,0)=\delta_{\varepsilon}(\mu) \tag{3.263}$$

and

$$\lim_{\eta\to\infty}\Phi(\mu,\eta)=0. \tag{3.264}$$

Let us take the local fractional Fourier transform with the variable μ in (3.262), (3.263), and (3.264). We then obtain

$$\frac{\partial^{2\varepsilon}\Phi(\omega,\eta)}{\partial\eta^{2\varepsilon}}-\omega^{2\varepsilon}\Phi(\omega,\eta)=0, \tag{3.265}$$

$$\Phi(\omega,0)=1, \tag{3.266}$$

and

$$\lim_{\eta\to\infty}\Phi(\omega,\eta)=0. \tag{3.267}$$

From (3.265), we have the general solution in the form

$$\Phi(\omega,\eta)=P(\omega)E_\varepsilon\left(-\omega^\varepsilon\eta^\varepsilon\right)+Q(\omega)E_\varepsilon\left(\omega^\varepsilon\eta^\varepsilon\right), \tag{3.268}$$

which, in view of (3.267), is rewritten as follows:

$$\Phi(\omega,\eta)=P(\omega)E_\varepsilon\left(-|\omega|^\varepsilon\eta^\varepsilon\right) \tag{3.269}$$

and

$$Q(\omega)=0. \tag{3.270}$$

From (3.266) and (3.269), we have

$$P(\omega)=1 \tag{3.271}$$

such that

$$\Phi(\omega,\eta)=E_\varepsilon\left(-|\omega|^\varepsilon\eta^\varepsilon\right). \tag{3.272}$$

Taking the inverse local fractional Fourier transform in (3.272) yields

$$\begin{aligned}
\Phi(\mu,\eta)&=\Im^{-1}\left[\Phi(\omega,\eta)\right]\\
&=\frac{1}{(2\pi)^\varepsilon}\int_{-\infty}^{\infty}\Phi(\omega,\eta)E_\varepsilon\left(\mathrm{i}^\varepsilon\mu^\varepsilon\omega^\varepsilon\right)(\mathrm{d}\omega)^\varepsilon\\
&=\frac{1}{(2\pi)^\varepsilon}\int_{-\infty}^{\infty}E_\varepsilon\left(-|\omega|^\varepsilon\eta^\varepsilon\right)E_\varepsilon\left(\mathrm{i}^\varepsilon\mu^\varepsilon\omega^\varepsilon\right)(\mathrm{d}\omega)^\varepsilon\\
&=\frac{1}{(2\pi)^\varepsilon}\int_{0}^{\infty}E_\varepsilon\left(-\omega^\varepsilon\eta^\varepsilon\right)E_\varepsilon\left(\mathrm{i}^\varepsilon\mu^\varepsilon\omega^\varepsilon\right)(\mathrm{d}\omega)^\varepsilon\\
&\quad+\frac{1}{(2\pi)^\varepsilon}\int_{-\infty}^{0}E_\varepsilon\left(\omega^\varepsilon\eta^\varepsilon\right)E_\varepsilon\left(\mathrm{i}^\varepsilon\mu^\varepsilon\omega^\varepsilon\right)(\mathrm{d}\omega)^\varepsilon\\
&=\frac{1}{(2\pi)^\varepsilon}\int_{0}^{\infty}E_\varepsilon\left[-\omega^\varepsilon\left(\eta^\varepsilon-\mathrm{i}^\varepsilon\mu^\varepsilon\right)\right](\mathrm{d}\omega)^\varepsilon\\
&\quad+\frac{1}{(2\pi)^\varepsilon}\int_{-\infty}^{0}E_\varepsilon\left[\omega^\varepsilon\left(\eta^\varepsilon+\mathrm{i}^\varepsilon\mu^\varepsilon\right)\right](\mathrm{d}\omega)^\varepsilon\\
&=\frac{\Gamma(1+\varepsilon)}{(2\pi)^\varepsilon}\left(\frac{1}{\eta^\varepsilon-\mathrm{i}^\varepsilon\mu^\varepsilon}+\frac{1}{\eta^\varepsilon+\mathrm{i}^\varepsilon\mu^\varepsilon}\right)\\
&=\frac{\Gamma(1+\varepsilon)}{(2\pi)^\varepsilon}\frac{2\eta^\varepsilon}{\mu^{2\varepsilon}+\eta^{2\varepsilon}}.
\end{aligned} \tag{3.273}$$

We thus conclude that

$$\Omega(\omega,\tau)=\frac{\Gamma(1+\varepsilon)}{(2\pi)^{\varepsilon}}\frac{2\eta^{\varepsilon}}{(\mu-\tau)^{2\varepsilon}+\eta^{2\varepsilon}},\tag{3.274}$$

so that

$$\begin{aligned}\Im\left[\frac{\Gamma(1+\varepsilon)}{(2\pi)^{\varepsilon}}\frac{2\eta^{\varepsilon}}{(\mu-\tau)^{2\varepsilon}+\eta^{2\varepsilon}}\right]&=\frac{1}{(2\pi)^{\varepsilon}}\int_{-\infty}^{\infty}E_{\varepsilon}\left(-|\omega|^{\varepsilon}\eta^{\varepsilon}\right)\frac{(2\pi)^{\varepsilon}}{\Gamma(1+\varepsilon)}\delta_{\varepsilon}(\omega)(d\omega)^{\varepsilon}\\&=1,\end{aligned}\tag{3.275}$$

which leads us to

$$\delta_{\varepsilon}(\tau)=\lim_{\tau\to 0}\Omega(\omega,\tau)=\lim_{\tau\to 0}\frac{\Gamma(1+\varepsilon)}{(2\pi)^{\varepsilon}}\frac{2\eta^{\varepsilon}}{(\mu-\tau)^{2\varepsilon}+\eta^{2\varepsilon}}\tag{3.276}$$

and

$$\frac{1}{\Gamma(1+\varepsilon)}\int_{-\infty}^{\infty}\frac{\Gamma(1+\varepsilon)}{(2\pi)^{\varepsilon}}\frac{2\eta^{\varepsilon}}{(\mu-\tau)^{2\varepsilon}+\eta^{2\varepsilon}}(d\tau)^{\varepsilon}=1,\tag{3.277}$$

respectively.

Local fractional Laplace transform and applications

4

4.1 Introduction

A complex number has the form $z = x + \mathrm{i}y$, where $x, y \in \mathrm{R}$ and i is the imaginary unit that satisfies $\mathrm{i}^2 = -1$. Recently, the number in the fractal dimension space, namely, $x^\varepsilon + \mathrm{i}^\varepsilon y^\varepsilon$, where $x, y \in \mathrm{R}$ and i^ε is the fractal imaginary unit fulfilling $\mathrm{i}^{2\varepsilon} = -1$, which has the relation $z^\varepsilon = x^\varepsilon + \mathrm{i}^\varepsilon y^\varepsilon \in \Re$. Thus, we can derive both the local fractional derivative and integral operators based on the numbers in the fractal dimension space.

The local fractional derivative operator of $\psi(z)$ of order ε $(0 < \varepsilon \leq 1)$ at the point z_0 is defined as [16, 21, 108]

$$
{}_{z_0}D_z^\varepsilon \psi(z_0) = \frac{\Delta^\varepsilon (\psi(z) - \psi(z_0))}{(z - z_0)^\varepsilon}, \tag{4.1}
$$

where $\Delta^\varepsilon (\psi(z) - \psi(z_0)) \cong \Gamma(1+\varepsilon)(\psi(z) - \psi(z_0))$.

The local fractional integral operator of $\Phi(z)$ of order ε $(0 < \varepsilon \leq 1)$ from the point z_p to the point z_q is defined as given below [16, 21, 108]:

$$
\begin{aligned}
{}_{z_p}I_{z_q}^\varepsilon \psi(z) &= \frac{1}{\Gamma(1+\varepsilon)} \lim_{\Delta z_i \to 0} \sum_{i=1}^{k} \Phi(z_i)(\Delta z_i)^\varepsilon \\
&= \frac{1}{\Gamma(1+\varepsilon)} \int_{z_p}^{z_q} \Phi(z)(\mathrm{d}z)^\varepsilon,
\end{aligned} \tag{4.2}
$$

where $\Delta z_i = z_i - z_{i-1}$, $z_0 = z_p$, and $z_n = z_q$.

We recall that the function

$$
\psi(z) = \nu(\mu, \eta) + \mathrm{i}^\varepsilon \upsilon(\mu, \eta) \tag{4.3}
$$

is a local fractional analytic in the region $\Re$ if

$$
\frac{\partial^\varepsilon \nu(\mu,\eta)}{\partial \mu^\varepsilon} - \frac{\partial^\varepsilon \upsilon(\mu,\eta)}{\partial \eta^\varepsilon} = 0 \tag{4.4}
$$

and

$$
\frac{\partial^\varepsilon \upsilon(\mu,\eta)}{\partial \mu^\varepsilon} + \frac{\partial^\varepsilon \nu(\mu,\eta)}{\partial \eta^\varepsilon} = 0. \tag{4.5}
$$

Also, there is

$$
\frac{1}{(2\pi)^\varepsilon \mathrm{i}^\varepsilon} \cdot \frac{1}{\Gamma(1+\varepsilon)} \oint_C \frac{\psi(z)}{(z - z_0)^\varepsilon} (\mathrm{d}z)^\varepsilon = \psi(z_0), \tag{4.6}
$$

Local Fractional Integral Transforms and Their Applications. http://dx.doi.org/10.1016/B978-0-12-804002-7.00004-8

where $\psi(z)$ denotes the local fractional analytic within and on a simple closed contour C and z_0 is any point interior to C.

Generally, we have [16–21]

$$\frac{1}{(2\pi)^{\varepsilon} \mathrm{i}^{\varepsilon}} \cdot \frac{1}{\Gamma(1+\varepsilon)} \oint_C \frac{f(z)}{(z-z_0)^{(n+1)\varepsilon}} (\mathrm{d}z)^{\varepsilon} = f^{(n\varepsilon)}(z_0). \tag{4.7}$$

When $\psi(z) = 1$, from (4.6), we conclude

$$\frac{1}{\Gamma(1+\varepsilon)} \oint_C \frac{(\mathrm{d}z)^{\varepsilon}}{(z-z_0)^{\varepsilon}} = (2\pi)^{\varepsilon} \mathrm{i}^{\varepsilon}, \tag{4.8}$$

where z_0 is any point interior to C.

For C: $|z-z_0| \leq R$, a local fractional Laurent series of $\psi(z)$ has the form [16–21]

$$\psi(z) = \sum_{k=-\infty}^{\infty} \gamma(k, z_0)(z-z_0)^{k\alpha}, \tag{4.9}$$

where

$$\gamma(k, z_0) = \frac{1}{(2\pi)^{\varepsilon} \mathrm{i}^{\varepsilon}} \cdot \frac{1}{\Gamma(1+\varepsilon)} \oint_C \frac{\psi(z)}{(z-z_0)^{(k+1)\varepsilon}} (\mathrm{d}z)^{\varepsilon}. \tag{4.10}$$

We observe that from (4.10), a generalized residue of $f(z)$ at the point $z = z_0$ via the local fractional integral operator, denoted by $\operatorname*{Res}_{z=z_0} \psi(z)$, is [16, 21]

$$\frac{1}{(2\pi)^{\varepsilon} \mathrm{i}^{\varepsilon} \Gamma(1+\varepsilon)} \oint_C \psi(z)(\mathrm{d}z)^{\varepsilon} = \operatorname*{Res}_{z=z_0} \psi(z), \tag{4.11}$$

where C: $|z-z_0| \leq R$.

As a result, we obtain a new transform (Laplace-like transform) based upon the numbers in the fractal dimension space. There is a local fractional Fourier transform of $\Phi(\omega)$ as

$$\Phi(\omega) = \frac{1}{\Gamma(1+\varepsilon)} \int_{-\infty}^{\infty} \phi(\tau) E_{\varepsilon}\left(-\mathrm{i}^{\varepsilon}\tau^{\varepsilon}\omega^{\varepsilon}\right)(\mathrm{d}\tau)^{\varepsilon}, \tag{4.12}$$

which leads to a new transform in the fractal dimension space, namely,

$$\Phi(s) = \frac{1}{\Gamma(1+\varepsilon)} \int_{-\infty}^{\infty} \phi(\tau) E_{\varepsilon}\left(-\tau^{\varepsilon}s^{\varepsilon}\right)(\mathrm{d}\tau)^{\varepsilon}, \tag{4.13}$$

where $s = \mathrm{i}\omega$ and $\omega \to \infty$.

We generalized (4.13) as

$$\Phi(s) = \frac{1}{\Gamma(1+\varepsilon)} \int_{-\infty}^{\infty} \phi(\tau) E_{\varepsilon}\left(-\tau^{\varepsilon}s^{\varepsilon}\right)(\mathrm{d}\tau)^{\varepsilon}, \tag{4.14}$$

where $s = \beta + \mathrm{i}\omega$, $s^{\varepsilon} = \beta^{\varepsilon} + \mathrm{i}^{\varepsilon}\omega^{\varepsilon}$ and $\omega \to \infty$.

We observe that when $\varepsilon = 1$, we get the Laplace transform as [109]

$$\Phi(s) = \int_{-\infty}^{\infty} \phi(\tau) \exp(-\tau s)\,\mathrm{d}\tau, \tag{4.15}$$

where $s = \beta + \mathrm{i}\omega$ and $\omega \to \infty$.

Thus, we have

$$\phi(\tau) = \frac{1}{(2\pi)^{\varepsilon}} \int_{-\infty}^{\infty} \Phi(\omega) E_{\varepsilon}\left(\mathrm{i}^{\varepsilon}\tau^{\varepsilon}\omega^{\varepsilon}\right)(\mathrm{d}\omega)^{\varepsilon} \tag{4.16}$$

such that the inverse formula takes the form

$$\phi(\tau) = \frac{1}{(2\pi)^{\varepsilon}} \int_{-\mathrm{i}\omega}^{\mathrm{i}\omega} \Phi(s) E_{\varepsilon}\left(\tau^{\varepsilon}s^{\varepsilon}\right)(\mathrm{d}s)^{\varepsilon}, \tag{4.17}$$

where $s = \mathrm{i}\omega$ and $\omega \to \infty$.

In a similar manner, (4.17) is generalized in the following form:

$$\phi(\tau) = \frac{1}{(2\pi)^{\varepsilon}} \int_{\beta-\mathrm{i}\omega}^{\beta+\mathrm{i}\omega} \Phi(s) E_{\varepsilon}\left(\tau^{\varepsilon}s^{\varepsilon}\right)(\mathrm{d}s)^{\varepsilon}, \tag{4.18}$$

where $s = \beta + \mathrm{i}\omega$, $s^{\varepsilon} = \beta^{\varepsilon} + \mathrm{i}^{\varepsilon}\omega^{\varepsilon}$ and $\omega \to \infty$.

When $\varepsilon = 1$, by using (4.18), we conclude that the inverse Laplace transform has the form [109]

$$\phi(\tau) = \frac{1}{2\pi} \int_{\beta-\mathrm{i}\omega}^{\beta+\mathrm{i}\omega} \Phi(s) \exp(\tau s)\ \mathrm{d}s, \tag{4.19}$$

where $s = \beta + \mathrm{i}\omega$ and $\omega \to \infty$.

Based on the relation (4.11), we compute (4.18).

In this case, we have

$$\begin{aligned}\phi(\tau) &= \frac{1}{(2\pi)^{\varepsilon}} \int_{\beta-\mathrm{i}\omega}^{\beta+\mathrm{i}\omega} \Phi(s) E_{\varepsilon}\left(\tau^{\varepsilon}s^{\varepsilon}\right)(\mathrm{d}s)^{\varepsilon} \\ &= \frac{1}{(2\pi)^{\varepsilon}} \int_{\beta-\mathrm{i}\omega}^{\beta+\mathrm{i}\omega} \left[\frac{1}{\Gamma(1+\varepsilon)} \int_{-\infty}^{\infty} \phi(\tau) E_{\varepsilon}\left(-\tau^{\varepsilon}s^{\varepsilon}\right)(\mathrm{d}\tau)^{\varepsilon}\right] E_{\varepsilon}\left(\tau^{\varepsilon}s^{\varepsilon}\right)(\mathrm{d}s)^{\varepsilon},\end{aligned} \tag{4.20}$$

where $s = \beta + \mathrm{i}\omega$, $s^{\varepsilon} = \beta^{\varepsilon} + \mathrm{i}^{\varepsilon}\omega^{\varepsilon}$ and $\omega \to \infty$.

In the particular case, we have

$$\begin{aligned}\phi(\tau) &= \frac{1}{(2\pi)^{\varepsilon}} \int_{\beta-\mathrm{i}\omega}^{\beta+\mathrm{i}\omega} \Phi(s) E_{\varepsilon}\left(\tau^{\varepsilon}s^{\varepsilon}\right)(\mathrm{d}s)^{\varepsilon} \\ &= \frac{1}{(2\pi)^{\varepsilon}} \int_{\beta-\mathrm{i}\omega}^{\beta+\mathrm{i}\omega} \left[\frac{1}{\Gamma(1+\varepsilon)} \int_{0}^{\infty} \phi(\tau) E_{\varepsilon}\left(-\tau^{\varepsilon}s^{\varepsilon}\right)(\mathrm{d}\tau)^{\varepsilon}\right] E_{\varepsilon}\left(\tau^{\varepsilon}s^{\varepsilon}\right)(\mathrm{d}s)^{\varepsilon},\end{aligned} \tag{4.21}$$

such that

$$\Phi(s) = \frac{1}{\Gamma(1+\varepsilon)} \int_{0}^{\infty} \phi(\tau) E_{\varepsilon}\left(-\tau^{\varepsilon}s^{\varepsilon}\right)(\mathrm{d}\tau)^{\varepsilon}, \tag{4.22}$$

$$\phi(\tau) = \frac{1}{(2\pi)^{\varepsilon}} \int_{\beta-\mathrm{i}\omega}^{\beta+\mathrm{i}\omega} \Phi(s) E_{\varepsilon}\left(\iota^{\varepsilon}s^{\varepsilon}\right)(\mathrm{d}s)^{\varepsilon}, \tag{4.23}$$

where $s = \beta + \mathrm{i}\omega, s^{\varepsilon} = \beta^{\varepsilon} + \mathrm{i}^{\varepsilon}\omega^{\varepsilon}$ and $\omega \to \infty$.

The structure of this chapter is as follows. In Section 4.2, we present the definitions and properties of the local fractional Laplace transform operators. In Section 4.3, we discuss the application of the local fractional Laplace transform operator to signal analysis. In Section 4.4, the local fractional Laplace transform operator was utilized to solve some local fractional differential equation (ODEs and PDEs).

4.2 Definitions and properties

Below, we introduce the basic definitions of the local fractional Laplace operators and its properties [16, 21, 110–115].

4.2.1 The basic definitions of the local fractional Laplace transform operators

Definition 4.1. Setting $\theta \in L_{1,\varepsilon}[\mathrm{R}_+]$ and $\|\theta\|_{1,\varepsilon} < \infty$, the local fractional Laplace transform operator, denoted by $\mathrm{M}[\theta(\tau)] = \Theta(s)$, is defined as

$$\mathrm{M}[\theta(\tau)] = \Theta(s) = \frac{1}{\Gamma(1+\varepsilon)} \int_0^{\infty} \theta(\tau) E_\varepsilon\left(-\tau^\varepsilon s^\varepsilon\right) (\mathrm{d}\tau)^\varepsilon, \tag{4.24}$$

where M is called the local fractional Laplace transform operator.

Definition 4.2. The inverse local fractional Laplace transform operator, denoted by $\mathrm{M}^{-1}[\Theta(s)] = \theta(\tau)$, is defined as

$$\mathrm{M}^{-1}[\Theta(s)] = \theta(\tau) = \frac{1}{(2\pi)^\varepsilon} \int_{\beta-\mathrm{i}\infty}^{\beta+\mathrm{i}\infty} \Theta(s) E_\varepsilon\left(\tau^\varepsilon s^\varepsilon\right) (\mathrm{d}s)^\varepsilon, \tag{4.25}$$

where M^{-1} is called the inverse local fractional Laplace transform operator.

A sufficient condition for convergence is presented as follows:

$$\frac{1}{\Gamma(1+\varepsilon)} \int_0^{\infty} |\psi(\tau)| (\mathrm{d}\tau)^\varepsilon < K < \infty. \tag{4.26}$$

Definition 4.3. Setting $\theta \in L_{1,\varepsilon}[\mathrm{R}]$ and $\|\theta\|_{1,\varepsilon} < \infty$, the two-sided local fractional Laplace transform operator, denoted by $\mathrm{A}[\theta(\tau)] = \Theta(s)$, is defined as

$$\mathrm{A}[\theta(\tau)] = \Theta(s) = \frac{1}{\Gamma(1+\varepsilon)} \int_{\infty}^{\infty} \theta(\tau) E_\varepsilon\left(-\tau^\varepsilon s^\varepsilon\right) (\mathrm{d}\tau)^\varepsilon, \tag{4.27}$$

where A is called the two-sided local fractional Laplace transform operator.

Definition 4.4. The inverse two-sided local fractional Laplace transform operator, denoted by $\mathrm{A}^{-1}[\Theta(s)] = \theta(\tau)$, is defined as

$$\mathrm{A}^{-1}\left[\Theta\left(s\right)\right]=\theta\left(\tau\right)=\frac{1}{\left(2\pi\right)^{\varepsilon}}\int_{\beta-\mathrm{i}\infty}^{\beta+\mathrm{i}\infty}\Theta\left(s\right)E_{\varepsilon}\left(\tau^{\varepsilon}s^{\varepsilon}\right)\left(\mathrm{d}s\right)^{\varepsilon},\tag{4.28}$$

where A^{-1} is called the inverse two-sided local fractional Laplace transform operator.

A sufficient condition for convergence is presented as

$$\frac{1}{\Gamma\left(1+\varepsilon\right)}\int_{-\infty}^{\infty}\left|\psi\left(\tau\right)\right|\left(\mathrm{d}\tau\right)^{\varepsilon}<K<\infty.\tag{4.29}$$

4.2.2 The properties and theorems for the local fractional Laplace transform operator

Property 21 (Linearity for local fractional Laplace transform operator). *Suppose that* $\theta_1\left(\tau\right),\theta_2\left(\tau\right)\in L_{1,\varepsilon}\left[R_+\right]$, $M\left[\theta_1\left(\tau\right)\right]=\Theta_1\left(s\right)$ and $M\left[\theta_2\left(\tau\right)\right]=\Theta_2\left(s\right)$, *then there is*

$$M\left[a\theta_1\left(\tau\right)\pm b\theta_2\left(\tau\right)\right]=a\Theta_1\left(s\right)\pm b\Theta_2\left(s\right),\tag{4.30}$$

where a and b are constants.

Proof. We have, by definition of the local fractional Laplace transform operator,

$$\begin{aligned}\mathrm{M}\left[a\theta_1\left(\tau\right)\pm b\theta_2\left(\tau\right)\right]&=\frac{1}{\Gamma\left(1+\varepsilon\right)}\int_0^{\infty}\left[a\theta_1\left(\tau\right)\pm b\theta_2\left(\tau\right)\right]E_{\varepsilon}\left(-\tau^{\varepsilon}s^{\varepsilon}\right)\left(\mathrm{d}\tau\right)^{\varepsilon}\\&=a\frac{1}{\Gamma\left(1+\varepsilon\right)}\int_0^{\infty}\left[\theta_1\left(\tau\right)E_{\varepsilon}\left(-\tau^{\varepsilon}s^{\varepsilon}\right)\right]\left(\mathrm{d}\tau\right)^{\varepsilon}\\&\quad\pm b\frac{1}{\Gamma\left(1+\varepsilon\right)}\int_0^{\infty}\left[\theta_2\left(\tau\right)E_{\varepsilon}\left(-\tau^{\varepsilon}s^{\varepsilon}\right)\right]\left(\mathrm{d}\tau\right)^{\varepsilon}\\&=a\Theta_1\left(s\right)\pm b\Theta_2\left(s\right).\end{aligned}\tag{4.31}$$

Thus, the proof is finished. □

Property 22 (Shifting time for local fractional Laplace transform operator). *Suppose that* $\theta\left(\tau\right)\in L_{1,\varepsilon}\left[R_+\right]$, $M\left[\theta\left(\tau\right)\right]=\Theta\left(s\right)$ *and a is a constant, then there is*

$$M\left[\theta\left(\tau-a\right)\right]=E_{\varepsilon}\left(-a^{\varepsilon}s^{\varepsilon}\right)M\left[\theta\left(\tau\right)\right].\tag{4.32}$$

Proof. By using the definition of the local fractional Laplace transform operator, we obtain

$$\begin{aligned}\mathrm{M}\left[\theta\left(\tau-a\right)\right]&=\frac{1}{\Gamma\left(1+\varepsilon\right)}\int_0^{\infty}\theta\left(\tau-a\right)E_{\varepsilon}\left(-\tau^{\varepsilon}s^{\varepsilon}\right)\left(\mathrm{d}\tau\right)^{\varepsilon}\\&=E_{\varepsilon}\left(-a^{\varepsilon}s^{\varepsilon}\right)\frac{1}{\Gamma\left(1+\varepsilon\right)}\int_0^{\infty}\theta\left(\tau-a\right)E_{\varepsilon}\left(-\left(\tau-a\right)^{\varepsilon}s^{\varepsilon}\right)\left(\mathrm{d}\tau\right)^{\varepsilon}\end{aligned}$$

$$= E_\varepsilon\left(-a^\varepsilon s^\varepsilon\right)\Theta(s)$$
$$= E_\varepsilon\left(-a^\varepsilon s^\varepsilon\right)\mathrm{M}\left[\theta(\tau)\right]. \tag{4.33}$$

Therefore, the proof is finished. □

Property 23 (Scaling time for local fractional Laplace transform operator). *Suppose that* $\theta(\tau) \in L_{1,\varepsilon}[\mathrm{R}_+]$, $\mathrm{M}[\theta(\tau)] = \Theta(s)$ *and* a $(a > 0)$ *is a constant, then there is*

$$M\left[\theta(a\tau)\right] = \frac{1}{a^\varepsilon}\Theta\left(\frac{s}{a}\right). \tag{4.34}$$

Proof. With the help of the definition of the local fractional Laplace transform operator, we obtain

$$\begin{aligned}\mathrm{M}\left[\theta(a\tau)\right] &= \frac{1}{\Gamma(1+\varepsilon)}\int_0^\infty \theta(a\tau)E_\varepsilon\left(-\tau^\varepsilon s^\varepsilon\right)(\mathrm{d}\tau)^\varepsilon\\ &= \frac{1}{a^\varepsilon}\frac{1}{\Gamma(1+\varepsilon)}\int_0^\infty \theta(a\tau)E_\varepsilon\left(-\left(\frac{a\tau}{a}\right)^\varepsilon s^\varepsilon\right)(\mathrm{d}a\tau)^\varepsilon\\ &= \frac{1}{a^\varepsilon}\Theta\left(\frac{s}{a}\right).\end{aligned} \tag{4.35}$$

Thus, we completed this proof. □

Property 24 (Translation for local fractional Laplace transform operator). *Suppose that* $\theta(\tau) \in L_{1,\varepsilon}[\mathrm{R}_+]$, $M[\theta(\tau)] = \Theta(s)$ *and* a *is a constant, then there is*

$$M\left[E_\varepsilon\left(a^\varepsilon\tau^\varepsilon\right)\theta(\tau)\right] = \Theta(s-a). \tag{4.36}$$

Proof. By utilizing the definition of the local fractional Laplace transform operator, we conclude that

$$\begin{aligned}M\left[E_\varepsilon\left(-a^\varepsilon\tau^\varepsilon\right)\theta(\tau)\right] &= \frac{1}{\Gamma(1+\varepsilon)}\int_0^\infty \left[\theta(\tau)E_\varepsilon\left(a^\varepsilon\tau^\varepsilon\right)\right]E_\varepsilon\left(-\tau^\varepsilon s^\varepsilon\right)(\mathrm{d}\tau)^\varepsilon\\ &= \frac{1}{\Gamma(1+\varepsilon)}\int_0^\infty \theta(\tau)E_\varepsilon\left(-(s-a)^\varepsilon\tau^\varepsilon\right)(\mathrm{d}\tau)^\varepsilon\\ &= \Theta(s-a).\end{aligned} \tag{4.37}$$

Therefore, we proved our claim. □

Theorem 4.1. *Suppose that* $\theta(\tau) \in L_{1,\varepsilon}[R_+]$, $M[\theta(\tau)] = \Theta(s)$ *and* $\lim_{\tau\to\infty}\theta(\tau) = 0$, *then there is*

$$M\left[\theta^{(\varepsilon)}(\tau)\right] = s^\varepsilon M\left[\theta(\tau)\right] - \theta(0). \tag{4.38}$$

Proof. Once more, by using the definition of the local fractional Laplace transform operator, we obtain

$$\mathrm{M}\left[\theta^{(\varepsilon)}(\tau)\right]=\frac{1}{\Gamma(1+\varepsilon)}\int_0^{\infty}\theta^{(\varepsilon)}(\tau)E_{\varepsilon}\left(-\tau^{\varepsilon}s^{\varepsilon}\right)(\mathrm{d}\tau)^{\varepsilon}. \tag{4.39}$$

Taking the integration by part for local fractional integral operator and using (4.36), we conclude

$$\begin{aligned}\mathrm{M}\left[\theta^{(\varepsilon)}(\tau)\right]&=\frac{1}{\Gamma(1+\varepsilon)}\int_0^{\infty}\theta^{(\varepsilon)}(\tau)E_{\varepsilon}\left(-\tau^{\varepsilon}s^{\varepsilon}\right)(\mathrm{d}\tau)^{\varepsilon}\\&=\left[\theta(\tau)E_{\varepsilon}\left(-\tau^{\varepsilon}s^{\varepsilon}\right)\right]_0^{\infty}+\frac{s^{\varepsilon}}{\Gamma(1+\varepsilon)}\int_0^{\infty}\theta(\tau)E_{\varepsilon}\left(-\tau^{\varepsilon}s^{\varepsilon}\right)(\mathrm{d}\tau)^{\varepsilon},\end{aligned} \tag{4.40}$$

which, using $\lim_{\tau\to\infty}\theta(\tau)=0$, leads to

$$\begin{aligned}\mathrm{M}\left[\theta^{(\varepsilon)}(\tau)\right]&=s^{\varepsilon}\left[\frac{1}{\Gamma(1+\varepsilon)}\int_0^{\infty}\theta(\tau)E_{\varepsilon}\left(-\tau^{\varepsilon}s^{\varepsilon}\right)(\mathrm{d}\tau)^{\varepsilon}\right]-\theta(0)\\&=s^{\varepsilon}\mathrm{M}\left[\theta(\tau)\right]-\theta(0).\end{aligned} \tag{4.41}$$

Thus, we completed the proof. □

We remark that there is

$$\mathrm{M}\left[\theta^{(k\varepsilon)}(\tau)\right]=s^{k\varepsilon}\mathrm{M}\left[\theta(\tau)\right]-s^{(k-1)\varepsilon}\theta(0)-s^{(k-2)\varepsilon}\theta^{(\varepsilon)}(0)-\cdots-\theta^{((k-1)\varepsilon)}(0), \tag{4.42}$$

where $k\in\mathrm{N}$.

Theorem 4.2. *Suppose that* $\theta(\tau)\in L_{1,\varepsilon}[R_+]$, $M[\theta(\tau)]=\Theta(s)$ *and* $\lim_{\tau\to\infty}0I_{\tau}^{(\varepsilon)}\theta(\tau)=0$, *then there is*

$$M\left[0I_{\tau}^{(\varepsilon)}\theta(\tau)\right]=\frac{1}{s^{\varepsilon}}M\left[\theta(\tau)\right]. \tag{4.43}$$

Proof. In accordance with the definition of the local fractional Laplace transform operator, we have

$$\mathrm{M}\left[0I_{\tau}^{(\varepsilon)}\theta(\tau)\right]=\frac{1}{\Gamma(1+\varepsilon)}\int_0^{\infty}\left(0I_{\tau}^{(\varepsilon)}\theta(\tau)\right)E_{\varepsilon}\left(-\tau^{\varepsilon}s^{\varepsilon}\right)(\mathrm{d}\tau)^{\varepsilon}, \tag{4.44}$$

which is, using the integration by part for local fractional integral operator,

$$\begin{aligned}\mathrm{M}\left[0I_{\tau}^{(\varepsilon)}\theta(\tau)\right]&=\left[0I_{\tau}^{(\varepsilon)}\theta(\tau)E_{\varepsilon}\left(-\tau^{\varepsilon}s^{\varepsilon}\right)\right]\Big|_0^{\infty}\\&\quad+\left[\frac{s^{\varepsilon}}{\Gamma(1+\varepsilon)}\int_0^{\infty}\theta(\tau)E_{\varepsilon}\left(-\tau^{\varepsilon}s^{\varepsilon}\right)(\mathrm{d}\tau)^{\varepsilon}\right].\end{aligned} \tag{4.45}$$

Taking $\lim_{\tau\to\infty} {}_0I_\tau^{(\varepsilon)}\theta(\tau)=0$ into account in (4.45), we have the following result:

$$\begin{aligned}\mathrm{M}\left[{}_0I_\tau^{(\varepsilon)}\theta(\tau)\right] &= \left[\frac{s^\varepsilon}{\Gamma(1+\varepsilon)}\int_0^\infty \theta(\tau)E_\varepsilon\left(-\tau^\varepsilon s^\varepsilon\right)(\mathrm{d}\tau)^\varepsilon\right] \\ &= s^\varepsilon \mathrm{M}[\theta(\tau)]. \end{aligned} \tag{4.46}$$

In this way, we finish the proof. □

In a similar manner, by repeating this process, we have

$$\mathrm{M}\left[{}_0I_\tau^{(k\varepsilon)}\theta(\tau)\right] = \frac{1}{s^{k\varepsilon}}\mathrm{M}[\theta(\tau)], \tag{4.47}$$

where

$$\lim_{\tau\to\infty} {}_0I_\tau^{(k\varepsilon)}\theta(\tau)=0. \tag{4.48}$$

Theorem 4.3. *Suppose $\theta(\tau)\in L_{1,\varepsilon}[R_+]$ and $M[\theta(\tau)]=\Theta(s)$, then there is*

$$M\left[\tau^\varepsilon\theta(\tau)\right]=\Theta^{(\varepsilon)}(s). \tag{4.49}$$

Proof. Again, by using the definition of the inverse local fractional Laplace transform operator, we conclude that

$$\mathrm{M}^{-1}\left[\Theta^{(\varepsilon)}(s)\right] = \frac{1}{(2\pi)^\varepsilon}\int_{\beta-\mathrm{i}\infty}^{\beta+\mathrm{i}\infty}\left[\Theta^{(\varepsilon)}(s)\right]E_\varepsilon\left(\tau^\varepsilon s^\varepsilon\right)(\mathrm{d}s)^\varepsilon, \tag{4.50}$$

which becomes, by using the related integrating by part,

$$\begin{aligned}\mathrm{M}^{-1}\left[\Theta^{(\varepsilon)}(s)\right] &= \frac{\Gamma(1+\varepsilon)}{(2\pi)^\varepsilon}\left[\Theta(s)E_\varepsilon\left(\tau^\varepsilon s^\varepsilon\right)\right]_{\beta-\mathrm{i}\infty}^{\beta+\mathrm{i}\infty} \\ &\quad + \tau^\varepsilon\left[\frac{1}{(2\pi)^\varepsilon}\int_{\beta-\mathrm{i}\infty}^{\beta+\mathrm{i}\infty}\Theta(s)E_\varepsilon\left(\tau^\varepsilon s^\varepsilon\right)(\mathrm{d}s)^\varepsilon\right]. \end{aligned} \tag{4.51}$$

Taking $\lim_{|s|\to\infty}\Theta(s)=0$ into account in (4.51), the final result can be seen as

$$\begin{aligned}\mathrm{M}^{-1}\left[\Theta^{(\varepsilon)}(s)\right] &= \tau^\varepsilon\left[\frac{1}{(2\pi)^\varepsilon}\int_{\beta-\mathrm{i}\infty}^{\beta+\mathrm{i}\infty}\Theta(s)E_\varepsilon\left(\tau^\varepsilon s^\varepsilon\right)(\mathrm{d}s)^\varepsilon\right] \\ &= \tau^\varepsilon\theta(\tau). \end{aligned} \tag{4.52}$$

Therefore, we finished the proof. □

Using the sane way of thinking, we have

$$\mathrm{M}\left[\tau^{k\varepsilon}\theta(\tau)\right] = (-1)^{k\varepsilon}\,\Theta^{(k\varepsilon)}(s), \tag{4.53}$$

where

$$\lim_{|s|\to\infty}\Theta^{(k\varepsilon)}(s)=0, \quad k\in\mathrm{N}. \tag{4.54}$$

Definition 4.5. The local fractional convolution of two functions $\theta_1(\tau)$ and $\theta_2(\tau)$ via local fractional integral operator, denoted by $(\theta_1 * \theta_2)(\tau) = \theta_1(\tau) * \theta_2(\tau)$, is defined as

$$\begin{aligned}(\theta_1 * \theta_2)(\tau) &= \theta_1(\tau) * \theta_2(\tau)\\ &= {}_0I_{\infty}^{(\varepsilon)}\left[\theta_1(t)\,\theta_2(\tau - t)\right]\\ &= \frac{1}{\Gamma(1+\varepsilon)}\int_0^{\infty}\theta_1(t)\,\theta_2(\tau - t)\,(\mathrm{d}t)^{\varepsilon}.\end{aligned} \tag{4.55}$$

From the definition of local fractional convolution, we obtain the following properties, namely:

(a) $\theta_1(\tau) * \theta_2(\tau) = \theta_2(\tau) * \theta_1(\tau)$;
(b) $\theta_1(\tau) * (\theta_2(\tau) + \theta_3(\tau)) = \theta_1(\tau) * \theta_2(\tau) + \theta_1(\tau) * \theta_3(\tau)$.

Theorem 4.4. *Suppose that* $\theta_1(\tau), \theta_2(\tau) \in L_{1,\varepsilon}[R_+]$, $M[\theta_1(\tau)] = \Theta_1(s)$ *and* $M[\theta_2(\tau)] = \Theta_2(s)$, *then there is*

$$M[\theta_1(\tau) * \theta_2(\tau)] = \Theta_1(s)\,\Theta_2(s) \tag{4.56}$$

or

$$\theta_1(\tau) * \theta_2(\tau) = M^{-1}\left[\Theta_1(s)\,\Theta_2(s)\right] \tag{4.57}$$

or, equivalently,

$$\frac{1}{\Gamma(1+\varepsilon)}\int_0^{\infty}\theta_1(t)\,\theta_2(\tau - t)\,(dt)^{\varepsilon} = \frac{1}{(2\pi)^{\varepsilon}}\int_{\beta - \mathrm{i}\omega}^{\beta + \mathrm{i}\omega}\Theta_1(s)\,\Theta_2(s)\,E_{\varepsilon}\left(\tau^{\varepsilon}s^{\varepsilon}\right)(ds)^{\varepsilon}. \tag{4.58}$$

Proof. From the definition of the local fractional Laplace transform operator, we conclude that

$$\begin{aligned}\mathrm{M}[\theta_1(\tau) * \theta_2(\tau)] &= \frac{1}{\Gamma(1+\varepsilon)}\int_0^{\infty}E_{\varepsilon}\left(-\tau^{\varepsilon}s^{\varepsilon}\right)(\mathrm{d}\tau)^{\varepsilon}\\ &\quad\times\left[\frac{1}{\Gamma(1+\varepsilon)}\int_0^{\infty}\theta_1(\eta)\,\theta_2(\tau - \eta)\,(\mathrm{d}\eta)^{\varepsilon}\right]\\ &= \frac{1}{\Gamma(1+\varepsilon)}\int_0^{\infty}\theta_1(\eta)\,E_{\varepsilon}\left(-\eta^{\varepsilon}s^{\varepsilon}\right)\\ &\quad\times\left[\frac{1}{\Gamma(1+\varepsilon)}\int_0^{\infty}E_{\varepsilon}\left(-(\tau - \eta)^{\varepsilon}s^{\varepsilon}\right)\theta_2(\tau - \eta)\,(\mathrm{d}\tau)^{\varepsilon}\right](\mathrm{d}\eta)^{\varepsilon}.\end{aligned} \tag{4.59}$$

From

$$\Theta_2(s) = \frac{1}{\Gamma(1+\varepsilon)}\int_0^{\infty}E_{\varepsilon}\left(-(\tau - \eta)^{\varepsilon}s^{\varepsilon}\right)\theta_2(\tau - \eta)\,(\mathrm{d}\tau)^{\varepsilon}, \tag{4.60}$$

we conclude that

$$
\begin{aligned}
\mathrm{M}\left[\theta_1(\tau) * \theta_2(\tau)\right] &= \frac{1}{\Gamma(1+\varepsilon)} \int_0^{\infty} \theta_1(\eta) E_{\varepsilon}\left(-\eta^{\varepsilon} s^{\varepsilon}\right) \Theta_2(s) (\mathrm{d}\eta)^{\varepsilon} \\
&= \Theta_2(s) \frac{1}{\Gamma(1+\varepsilon)} \int_0^{\infty} \theta_1(\eta) E_{\varepsilon}\left(-\eta^{\varepsilon} s^{\varepsilon}\right) (\mathrm{d}\eta)^{\varepsilon} \\
&= \Theta_1(s) \Theta_2(s).
\end{aligned} \tag{4.61}
$$

Thus, the desired result is proved. □

Theorem 4.5. *Suppose that* $\theta_1(\tau), \theta_2(\tau) \in L_{1,\varepsilon}[R_+]$, $M[\theta_1(\tau)] = \Theta_1(s)$ *and* $M[\theta_2(\tau)] = \Theta_2(s)$, *then there is*

$$
M\left[\theta_1(\tau)\theta_2(\tau)\right] = \Theta_1(s) * \Theta_2(s) \tag{4.62}
$$

or

$$
\theta_1(\tau)\theta_2(\tau) = M^{-1}\left[\Theta_1(s) * \Theta_2(s)\right] \tag{4.63}
$$

or, equivalently,

$$
\begin{aligned}
&\frac{1}{\Gamma(1+\varepsilon)} \int_0^{\infty} \theta_1(\tau)\theta_2(\tau) E_{\varepsilon}\left(-\tau^{\varepsilon} s^{\varepsilon}\right) (d\tau)^{\varepsilon} \\
&= \frac{1}{(2\pi)^{\varepsilon}} \int_{\beta-\mathrm{i}\omega}^{\beta-\mathrm{i}\omega} \Theta_1(\tilde{s}) \Theta_2(s-\tilde{s}) (d\tilde{s})^{\varepsilon}.
\end{aligned} \tag{4.64}
$$

Proof. In addition, we conclude that

$$
\begin{aligned}
\mathrm{M}^{-1}\left[\Theta_1(s) * \Theta_2(s)\right] &= \frac{1}{(2\pi)^{\varepsilon}} \int_{\beta-\mathrm{i}\infty}^{\beta+\mathrm{i}\infty} \left[\frac{1}{(2\pi)^{\varepsilon}} \int_{\beta-\mathrm{i}\infty}^{\beta+\mathrm{i}\infty} \Theta_1(\tilde{s}) \Theta_2(s-\tilde{s}) (\mathrm{d}\tilde{s})^{\varepsilon}\right] \\
&\quad \times E_{\varepsilon}\left(\tau^{\varepsilon} s^{\varepsilon}\right) (\mathrm{d}s)^{\varepsilon} \\
&= \frac{1}{(2\pi)^{\varepsilon}} \int_{\beta-\mathrm{i}\infty}^{\beta+\mathrm{i}\infty} E_{\varepsilon}\left(\tau^{\varepsilon} \tilde{s}^{\varepsilon}\right) \Theta_1(\tilde{s}) \\
&\quad \times \left[\frac{1}{(2\pi)^{\varepsilon}} \int_{\beta-\mathrm{i}\infty}^{\beta+\mathrm{i}\infty} \Theta_2(s-\tilde{s}) E_{\varepsilon}\left(\tau^{\varepsilon}(s-\tilde{s})^{\varepsilon}\right) (\mathrm{d}s)^{\varepsilon}\right] (\mathrm{d}\tilde{s})^{\varepsilon}.
\end{aligned} \tag{4.65}
$$

Thus, we have

$$
\theta_2(\tau) = \frac{1}{(2\pi)^{\varepsilon}} \int_{\beta-\mathrm{i}\infty}^{\beta+\mathrm{i}\infty} \Theta_2(s-\tilde{s}) E_{\varepsilon}\left(\tau^{\varepsilon}(s-\tilde{s})^{\varepsilon}\right) (\mathrm{d}s)^{\varepsilon} \tag{4.66}
$$

such that (4.65) is expressed as

$$
\begin{aligned}
\mathrm{M}^{-1}\left[\Theta_1(s) * \Theta_2(s)\right] &= \frac{1}{(2\pi)^{\varepsilon}} \int_{\beta-\mathrm{i}\infty}^{\beta+\mathrm{i}\infty} E_{\varepsilon}\left(\tau^{\varepsilon}\tilde{s}^{\varepsilon}\right) \Theta_1(\tilde{s})\, \theta_2(\tau)\, (\mathrm{d}\tilde{s})^{\varepsilon} \\
&= \left[\frac{1}{(2\pi)^{\varepsilon}} \int_{\beta-\mathrm{i}\infty}^{\beta+\mathrm{i}\infty} E_{\varepsilon}\left(\tau^{\varepsilon}\tilde{s}^{\varepsilon}\right) \Theta_1(\tilde{s})\, (\mathrm{d}\tilde{s})^{\varepsilon}\right] \theta_2(\tau) \\
&= \theta_1(\tau)\, \theta_2(\tau).
\end{aligned} \tag{4.67}
$$

Therefore, we completed the proof. □

Theorem 4.6 (Convolution theorem for local fractional Laplace transform operator)**.** *Suppose $\theta_1(\tau), \theta_2(\tau) \in L_{1,\varepsilon}[R_+]$, $M[\theta_1(\tau)] = \Theta_1(s)$, and $M[\theta_2(\tau)] = \Theta_2(s)$, then there is*

$$
\frac{1}{\Gamma(1+\varepsilon)} \int_0^{\infty} \theta_1(\tau)\, \overline{\theta_2(\tau)}\, (d\tau)^{\varepsilon} = \frac{1}{(2\pi)^{\varepsilon}} \int_{\beta-i\infty}^{\beta+i\infty} \Theta_1(s)\, \overline{\Theta_2(s)}\, (ds)^{\varepsilon}. \tag{4.68}
$$

Proof. We consider that

$$
\begin{aligned}
\overline{\theta_2(\tau)} &= \overline{\frac{1}{(2\pi)^{\varepsilon}} \int_{\beta-\mathrm{i}\infty}^{\beta+\mathrm{i}\infty} \Theta_2(s)\, E_{\varepsilon}\left(\tau^{\varepsilon}s^{\varepsilon}\right) (\mathrm{d}s)^{\varepsilon}} \\
&= \frac{1}{(2\pi)^{\varepsilon}} \int_{\beta-\mathrm{i}\infty}^{\beta+\mathrm{i}\infty} \overline{\Theta_2(s)} E_{\varepsilon}\left(-\tau^{\varepsilon}s^{\varepsilon}\right) (\mathrm{d}s)^{\varepsilon}.
\end{aligned} \tag{4.69}
$$

In this case, from (4.68), we write

$$
\begin{aligned}
&\frac{1}{\Gamma(1+\varepsilon)} \int_0^{\infty} \theta_1(\tau)\, \overline{\theta_2(\tau)}\, (\mathrm{d}\tau)^{\varepsilon} \\
&= \frac{1}{\Gamma(1+\varepsilon)} \int_0^{\infty} \theta_1(\tau) \left[\frac{1}{(2\pi)^{\varepsilon}} \int_{\beta-\mathrm{i}\infty}^{\beta+\mathrm{i}\infty} \overline{\Theta_2(s)} E_{\varepsilon}\left(-\tau^{\varepsilon}s^{\varepsilon}\right) (\mathrm{d}s)^{\varepsilon}\right] (\mathrm{d}\tau)^{\varepsilon} \\
&= \left[\frac{1}{(2\pi)^{\varepsilon}} \int_{\beta-\mathrm{i}\infty}^{\beta+\mathrm{i}\infty} \overline{\Theta_2(s)} \left[\frac{1}{\Gamma(1+\varepsilon)} \int_0^{\infty} \theta_1(\tau)\, E_{\varepsilon}\left(-\tau^{\varepsilon}s^{\varepsilon}\right) (\mathrm{d}\tau)^{\varepsilon}\right] (\mathrm{d}s)^{\varepsilon}\right].
\end{aligned} \tag{4.70}
$$

Therefore, observing

$$
\Theta_1(s) = \frac{1}{\Gamma(1+\varepsilon)} \int_0^{\infty} \theta_1(\tau)\, E_{\varepsilon}\left(-\tau^{\varepsilon}s^{\varepsilon}\right) (\mathrm{d}\tau)^{\varepsilon}, \tag{4.71}
$$

we clearly say that

$$
\frac{1}{\Gamma(1+\varepsilon)} \int_0^{\infty} \theta_1(\tau)\, \overline{\theta_2(\tau)}\, (\mathrm{d}\tau)^{\varepsilon} = \frac{1}{(2\pi)^{\varepsilon}} \int_{\beta-\mathrm{i}\infty}^{\beta+\mathrm{i}\infty} \Theta_1(s)\, \overline{\Theta_2(s)}\, (\mathrm{d}s)^{\varepsilon}. \tag{4.72}
$$

Thus, the proof is completed. □

Theorem 4.7 (The initial value theorem for local fractional Laplace transform operator)**.** *Suppose that* $M[\theta(\tau)] = \Theta(s)$ *and* $M\left[\theta^{(\varepsilon)}(\tau)\right] = s^{\varepsilon} M[\theta(\tau)] - \theta(0_{-})$, *then there is*

$$\lim_{\tau \to 0_{+}} \theta(\tau) = \theta(0_{+}) = \lim_{s \to \infty} s^{\varepsilon} M[\theta(\tau)]. \tag{4.73}$$

Proof. Due to

$$\lim_{s \to \infty} E_{\varepsilon}\left(-s^{\varepsilon}\tau^{\varepsilon}\right) = 0, \tag{4.74}$$

it follows that

$$\begin{aligned} &\lim_{s \to \infty} \frac{1}{\Gamma(1+\varepsilon)} \int_{0_{+}}^{\infty} \frac{d^{\varepsilon}\theta(\tau)}{d\tau^{\varepsilon}} E_{\varepsilon}\left(-s^{\varepsilon}\tau^{\varepsilon}\right)(d\tau)^{\varepsilon} \\ &= \frac{1}{\Gamma(1+\varepsilon)} \int_{0_{+}}^{\infty} \frac{d^{\varepsilon}\theta(\tau)}{d\tau^{\varepsilon}} \left(\lim_{s \to \infty} E_{\varepsilon}\left(-s^{\varepsilon}\tau^{\varepsilon}\right)\right)(d\tau)^{\varepsilon} \\ &= 0. \end{aligned} \tag{4.75}$$

Thus, we have

$$\begin{aligned} \mathrm{M}\left[\theta^{(\varepsilon)}(\tau)\right] &= s^{\varepsilon}\mathrm{M}[\theta(\tau)] - \theta(0_{-}) \\ &= \lim_{s \to \infty} \frac{1}{\Gamma(1+\varepsilon)} \int_{0_{-}}^{\infty} \frac{d^{\varepsilon}\theta(\tau)}{d\tau^{\varepsilon}} E_{\varepsilon}\left(-s^{\varepsilon}\tau^{\varepsilon}\right)(d\tau)^{\varepsilon}. \end{aligned} \tag{4.76}$$

In this case, we conclude

$$\frac{1}{\Gamma(1+\varepsilon)} \int_{0_{-}}^{0+} \frac{d^{\varepsilon}\theta(\tau)}{d\tau^{\varepsilon}} E_{\varepsilon}\left(-s^{\varepsilon}\tau^{\varepsilon}\right)(d\tau)^{\varepsilon} = \theta(0+) - \theta(0-), \tag{4.77}$$

which leads to

$$\begin{aligned} &\frac{1}{\Gamma(1+\varepsilon)} \int_{0-}^{\infty} \frac{d^{\varepsilon}\theta(\tau)}{d\tau^{\varepsilon}} E_{\varepsilon}\left(-s^{\varepsilon}\tau^{\varepsilon}\right)(d\tau)^{\varepsilon} \\ &= \frac{1}{\Gamma(1+\varepsilon)} \int_{0-}^{0+} \frac{d^{\varepsilon}\theta(\tau)}{d\tau^{\varepsilon}} E_{\varepsilon}\left(-s^{\varepsilon}\tau^{\varepsilon}\right)(d\tau)^{\varepsilon} \\ &\quad + \frac{1}{\Gamma(1+\varepsilon)} \int_{0+}^{\infty} \frac{d^{\varepsilon}\theta(\tau)}{d\tau^{\varepsilon}} E_{\varepsilon}\left(-s^{\varepsilon}\tau^{\varepsilon}\right)(d\tau)^{\varepsilon} \\ &= \theta(0+) - \theta(0-) + \frac{1}{\Gamma(1+\varepsilon)} \int_{0+}^{\infty} \frac{d^{\varepsilon}\theta(\tau)}{d\tau^{\varepsilon}} E_{\varepsilon}\left(-s^{\varepsilon}\tau^{\varepsilon}\right)(d\tau)^{\varepsilon} \\ &= \theta(0+) - \theta(0-). \end{aligned} \tag{4.78}$$

Thus, we conclude

$$\lim_{s \to \infty} \mathrm{M}\left[\theta^{(\varepsilon)}(\tau)\right] = \lim_{s \to \infty}\left[s^{\varepsilon}\mathrm{M}[\theta(\tau)] - \theta(0_{-})\right] \tag{4.79}$$

such that

$$\begin{aligned}
\lim_{s\to\infty} s^{\varepsilon}\mathrm{M}\left[\theta\left(\tau\right)\right] &= \lim_{s\to\infty}\left\{\mathrm{M}\left[\theta^{(\varepsilon)}\left(\tau\right)\right]+\theta\left(0_{-}\right)\right\} \\
&= \lim_{s\to\infty}\left[\frac{1}{\Gamma\left(1+\varepsilon\right)}\int_{0_{-}}^{\infty}\frac{\mathrm{d}^{\varepsilon}\theta\left(\tau\right)}{\mathrm{d}\tau^{\varepsilon}}E_{\varepsilon}\left(-s^{\varepsilon}\tau^{\varepsilon}\right)\left(\mathrm{d}\tau\right)^{\varepsilon}+\theta\left(0_{-}\right)\right] \\
&= \lim_{s\to\infty}\left[\theta\left(0_{+}\right)+\frac{1}{\Gamma\left(1+\varepsilon\right)}\int_{0_{+}}^{\infty}\frac{\mathrm{d}^{\varepsilon}\theta\left(\tau\right)}{\mathrm{d}\tau^{\varepsilon}}E_{\varepsilon}\left(-s^{\varepsilon}\tau^{\varepsilon}\right)\left(\mathrm{d}\tau\right)^{\varepsilon}\right] \\
&= \theta\left(0_{+}\right) \\
&= \lim_{\tau\to 0_{+}}\theta\left(\tau\right).
\end{aligned} \tag{4.80}$$

Thus, we obtained the result. □

Theorem 4.8 (The final value theorem for local fractional Laplace transform operator)**.** *Suppose that* $M\left[\theta\left(\tau\right)\right]=\Theta\left(s\right)$ *and* $M\left[\theta^{(\varepsilon)}\left(\tau\right)\right]=s^{\varepsilon}M\left[\theta\left(\tau\right)\right]-\theta\left(0_{-}\right)$*, then there is*

$$\lim_{\tau\to\infty}\theta(\tau)=\theta(+\infty)=\lim_{s\to 0}s^{\varepsilon}M\left[\theta\left(\tau\right)\right]. \tag{4.81}$$

Proof. We consider

$$\lim_{s\to 0}\mathrm{M}\left[\theta^{(\varepsilon)}\left(\tau\right)\right]=\lim_{s\to 0}\left[s^{\varepsilon}\mathrm{M}\left[\theta\left(\tau\right)\right]-\theta\left(0_{-}\right)\right], \tag{4.82}$$

which leads to

$$\lim_{s\to 0}s^{\varepsilon}\mathrm{M}\left[\theta\left(\tau\right)\right]=\lim_{s\to 0}\left[\mathrm{M}\left[\theta^{(\varepsilon)}\left(\tau\right)\right]+\theta\left(0_{-}\right)\right]. \tag{4.83}$$

In this case, we may easily put it in the form

$$\begin{aligned}
\lim_{s\to 0}\mathrm{M}\left[\theta^{(\varepsilon)}\left(\tau\right)\right] &= \lim_{s\to 0}\left[\frac{1}{\Gamma\left(1+\varepsilon\right)}\int_{0_{-}}^{\infty}\frac{\mathrm{d}^{\varepsilon}\theta\left(\tau\right)}{\mathrm{d}\tau^{\varepsilon}}E_{\varepsilon}\left(-s^{\varepsilon}\tau^{\varepsilon}\right)\left(\mathrm{d}\tau\right)^{\varepsilon}\right] \\
&= \frac{1}{\Gamma\left(1+\varepsilon\right)}\int_{0_{-}}^{\infty}\frac{\mathrm{d}^{\varepsilon}\theta\left(\tau\right)}{\mathrm{d}\tau^{\varepsilon}}\left[\lim_{s\to 0}E_{\varepsilon}\left(-s^{\varepsilon}\tau^{\varepsilon}\right)\right]\left(\mathrm{d}\tau\right)^{\varepsilon} \\
&= \frac{1}{\Gamma\left(1+\varepsilon\right)}\int_{0_{-}}^{\infty}\frac{\mathrm{d}^{\varepsilon}\theta\left(\tau\right)}{\mathrm{d}\tau^{\varepsilon}}\left(\mathrm{d}\tau\right)^{\varepsilon} \\
&= \theta\left(\infty\right)-\theta\left(0_{-}\right).
\end{aligned} \tag{4.84}$$

Thus, making use of (4.83), we have the following result:

$$\begin{aligned}
\lim_{s\to 0}s^{\varepsilon}\mathrm{M}\left[\theta\left(\tau\right)\right] &= \lim_{s\to 0}\left[\mathrm{M}\left[\theta^{(\varepsilon)}\left(\tau\right)\right]+\theta\left(0_{-}\right)\right] \\
&= \lim_{s\to 0}\theta\left(\infty\right) \\
&= \theta\left(\infty\right).
\end{aligned} \tag{4.85}$$

Therefore, the proof of this theorem is reported. □

For the tables of the local fractional Laplace transform operators, the reader can see Appendix F.

4.3 Applications to signal analysis

We consider the signal defined on Cantor sets [16–21].

Let us take the local fractional Laplace transform of the signal defined as

$$\theta(\tau) = 1, \quad \tau > 0. \tag{4.86}$$

We observe that, by using the definition of the local fractional Laplace transform operator, we have

$$\begin{aligned} \mathrm{M}[1] &= \frac{1}{\Gamma(1+\varepsilon)} \int_0^\infty E_\varepsilon\left(-\tau^\varepsilon s^\varepsilon\right)(\mathrm{d}\tau)^\varepsilon \\ &= \frac{1}{s^\varepsilon}. \end{aligned} \tag{4.87}$$

We determine the signal defined on Cantor sets, which is given as

$$\theta(\tau) = E_\varepsilon\left(a^\varepsilon \tau^\varepsilon\right), \quad \tau > 0. \tag{4.88}$$

With the help of the local fractional Laplace transform operator, we report the result:

$$\begin{aligned} \mathrm{M}\left[E_\varepsilon\left(a^\varepsilon \tau^\varepsilon\right)\right] &= \frac{1}{\Gamma(1+\varepsilon)} \int_0^\infty E_\varepsilon\left(a^\varepsilon \tau^\varepsilon\right) E_\varepsilon\left(-\tau^\varepsilon s^\varepsilon\right)(\mathrm{d}\tau)^\varepsilon \\ &= \frac{1}{\Gamma(1+\varepsilon)} \int_0^\infty E_\varepsilon\left(-\tau^\varepsilon (s-a)^\varepsilon\right)(\mathrm{d}\tau)^\varepsilon \\ &= \frac{1}{(s-a)^\varepsilon} \\ &= \frac{1}{s^\varepsilon - a^\varepsilon}. \end{aligned} \tag{4.89}$$

Taking $a^\varepsilon = \mu^\varepsilon + \mathrm{i}^\varepsilon \eta^\varepsilon$ in (4.88), where μ and η are constants, it gives

$$\mathrm{M}\left[E_\varepsilon\left(\left(\mu^\varepsilon + \mathrm{i}^\varepsilon \eta^\varepsilon\right)\tau^\varepsilon\right)\right] = \frac{1}{s^\varepsilon - \left(\mu^\varepsilon + \mathrm{i}^\varepsilon \eta^\varepsilon\right)}. \tag{4.90}$$

Taking $a^\varepsilon = \mu^\varepsilon - \mathrm{i}^\varepsilon \eta^\varepsilon$ in (4.88), we may create

$$\mathrm{M}\left[E_\varepsilon\left(\left(\mu^\varepsilon - \mathrm{i}^\varepsilon \eta^\varepsilon\right)\tau^\varepsilon\right)\right] = \frac{1}{s^\varepsilon - \left(\mu^\varepsilon - \mathrm{i}^\varepsilon \eta^\varepsilon\right)}. \tag{4.91}$$

Taking $\mu = 0$ in (4.90), the final result reads as

$$\mathrm{M}\left[E_\varepsilon\left(\mathrm{i}^\varepsilon \eta^\varepsilon \tau^\varepsilon\right)\right] = \frac{1}{s^\varepsilon - \mathrm{i}^\varepsilon \eta^\varepsilon}. \tag{4.92}$$

Taking $\mu = 0$ in (4.91), this may be added in the form

$$\mathrm{M}\left[E_\varepsilon\left(-\mathrm{i}^\varepsilon \eta^\varepsilon \tau^\varepsilon\right)\right] = \frac{1}{s^\varepsilon + \mathrm{i}^\varepsilon \eta^\varepsilon}. \tag{4.93}$$

We report the local fractional Laplace transform of the signal on Cantor denoted by

$$\theta(\tau) = \cos_\varepsilon\left(\eta^\varepsilon \tau^\varepsilon\right), \quad \tau > 0. \tag{4.94}$$

From (4.92) and (4.93), we conclude that

$$\begin{aligned} \mathrm{M}\left[\cos_\varepsilon\left(\eta^\varepsilon \tau^\varepsilon\right)\right] &= \frac{1}{\Gamma(1+\varepsilon)} \int_0^\infty \cos_\varepsilon\left(\eta^\varepsilon \tau^\varepsilon\right) E_\varepsilon\left(-\tau^\varepsilon s^\varepsilon\right) (\mathrm{d}\tau)^\varepsilon \\ &= \frac{1}{\Gamma(1+\varepsilon)} \int_0^\infty \left[\frac{E_\varepsilon\left(\mathrm{i}^\varepsilon \eta^\varepsilon \tau^\varepsilon\right) + E_\varepsilon\left(-\mathrm{i}^\varepsilon \eta^\varepsilon \tau^\varepsilon\right)}{2}\right] E_\varepsilon\left(-\tau^\varepsilon s^\varepsilon\right) (\mathrm{d}\tau)^\varepsilon \\ &= \frac{\frac{1}{s^\varepsilon - \mathrm{i}^\varepsilon \eta^\varepsilon} + \frac{1}{s^\varepsilon + \mathrm{i}^\varepsilon \eta^\varepsilon}}{2} \\ &= \frac{s^\varepsilon}{s^{2\varepsilon} + \eta^{2\varepsilon}}. \end{aligned} \tag{4.95}$$

We determine the local fractional Laplace transform of the signal on Cantor as

$$\theta(\tau) = \sin_\varepsilon\left(\eta^\varepsilon \tau^\varepsilon\right), \quad \tau > 0. \tag{4.96}$$

Utilizing both (4.92) and (4.93), we see that

$$\begin{aligned} \mathrm{M}\left[\sin_\varepsilon\left(\eta^\varepsilon \tau^\varepsilon\right)\right] &= \frac{1}{\Gamma(1+\varepsilon)} \int_0^\infty \sin_\varepsilon\left(\eta^\varepsilon \tau^\varepsilon\right) E_\varepsilon\left(-\tau^\varepsilon s^\varepsilon\right) (\mathrm{d}\tau)^\varepsilon \\ &= \frac{1}{\Gamma(1+\varepsilon)} \int_0^\infty \left[\frac{E_\varepsilon\left(\mathrm{i}^\varepsilon \eta^\varepsilon \tau^\varepsilon\right) - E_\varepsilon\left(-\mathrm{i}^\varepsilon \eta^\varepsilon \tau^\varepsilon\right)}{2\mathrm{i}^\varepsilon}\right] E_\varepsilon\left(-\tau^\varepsilon s^\varepsilon\right) (\mathrm{d}\tau)^\varepsilon \\ &= \frac{\frac{1}{s^\varepsilon - \mathrm{i}^\varepsilon \eta^\varepsilon} - \frac{1}{s^\varepsilon + \mathrm{i}^\varepsilon \eta^\varepsilon}}{2\mathrm{i}^\varepsilon} \\ &= \frac{\eta^\varepsilon}{s^{2\varepsilon} + \eta^{2\varepsilon}}. \end{aligned} \tag{4.97}$$

We find the local fractional Laplace transform of the signal on Cantor, namely,

$$\theta(\tau) = \cosh_\varepsilon\left(\eta^\varepsilon \tau^\varepsilon\right), \quad \tau > 0. \tag{4.98}$$

By using the formula (4.89), we found that

$$\begin{aligned} \mathrm{M}\left[\cosh_\varepsilon\left(\eta^\varepsilon \tau^\varepsilon\right)\right] &= \frac{1}{\Gamma(1+\varepsilon)} \int_0^\infty \cosh_\varepsilon\left(\eta^\varepsilon \tau^\varepsilon\right) E_\varepsilon\left(-\tau^\varepsilon s^\varepsilon\right) (\mathrm{d}\tau)^\varepsilon \\ &= \frac{1}{\Gamma(1+\varepsilon)} \int_0^\infty \left[\frac{E_\varepsilon\left(\eta^\varepsilon \tau^\varepsilon\right) + E_\varepsilon\left(-\eta^\varepsilon \tau^\varepsilon\right)}{2}\right] E_\varepsilon\left(-\tau^\varepsilon s^\varepsilon\right) (\mathrm{d}\tau)^\varepsilon \\ &= \frac{\frac{1}{s^\varepsilon - \eta^\varepsilon} + \frac{1}{s^\varepsilon + \eta^\varepsilon}}{2} \\ &= \frac{s^\varepsilon}{s^{2\varepsilon} - \eta^{2\varepsilon}}. \end{aligned} \tag{4.99}$$

We show the local fractional Laplace transform of the signal on Cantor

$$\theta(\tau) = \sinh_\varepsilon\left(\eta^\varepsilon \tau^\varepsilon\right), \quad \tau > 0. \tag{4.100}$$

With the help of formula (4.89), we present the final result as

$$\begin{aligned}
M\left[\sinh_\varepsilon\left(\eta^\varepsilon\tau^\varepsilon\right)\right] &= \frac{1}{\Gamma(1+\varepsilon)}\int_0^\infty \sinh_\varepsilon\left(\eta^\varepsilon\tau^\varepsilon\right)E_\varepsilon\left(-\tau^\varepsilon s^\varepsilon\right)(d\tau)^\varepsilon \\
&= \frac{1}{\Gamma(1+\varepsilon)}\int_0^\infty \left[\frac{E_\varepsilon\left(\eta^\varepsilon\tau^\varepsilon\right)-E_\varepsilon\left(-\eta^\varepsilon\tau^\varepsilon\right)}{2}\right]E_\varepsilon\left(-\tau^\varepsilon s^\varepsilon\right)(d\tau)^\varepsilon \\
&= \frac{\frac{1}{s^\varepsilon-\eta^\varepsilon}-\frac{1}{s^\varepsilon+\eta^\varepsilon}}{2} \\
&= \frac{\eta^\varepsilon}{s^{2\varepsilon}-\eta^{2\varepsilon}}. \qquad (4.101)
\end{aligned}$$

We compute the local fractional Laplace transform of the signal on Cantor as

$$\theta(\tau) = \frac{\tau^{k\varepsilon}}{\Gamma(1+k\varepsilon)}, \quad \tau > 0, \ k \in \mathrm{N}. \qquad (4.102)$$

With the help of the definition of the local fractional Laplace transform, we have

$$\begin{aligned}
M\left[\frac{\tau^{k\varepsilon}}{\Gamma(1+k\varepsilon)}\right] &= \frac{1}{\Gamma(1+\varepsilon)}\int_0^\infty \frac{\tau^{k\varepsilon}}{\Gamma(1+k\varepsilon)}E_\varepsilon\left(-\tau^\varepsilon s^\varepsilon\right)(d\tau)^\varepsilon \\
&= \frac{1}{s^\varepsilon}\frac{1}{\Gamma(1+\varepsilon)}\int_0^\infty \frac{\tau^{(k-1)\varepsilon}}{\Gamma(1+(k-1)\varepsilon)}E_\varepsilon\left(-\tau^\varepsilon s^\varepsilon\right)(d\tau)^\varepsilon \\
&= \frac{1}{s^{2\varepsilon}}\frac{1}{\Gamma(1+\varepsilon)}\int_0^\infty \frac{\tau^{(k-2)\varepsilon}}{\Gamma(1+(k-2)\varepsilon)}E_\varepsilon\left(-\tau^\varepsilon s^\varepsilon\right)(d\tau)^\varepsilon \\
&= \frac{1}{s^{\varepsilon(k+1)}}. \qquad (4.103)
\end{aligned}$$

We give the signal defined on Cantor sets by the following expression:

$$\theta(\tau) = \frac{\tau^{k\varepsilon}}{\Gamma(1+k\varepsilon)}E_\varepsilon\left(a^\varepsilon\tau^\varepsilon\right), \quad \tau > 0, \ k \in \mathrm{N}. \qquad (4.104)$$

Its graphs with different parameters a and k are depicted in Figure 4.1.

With the help of (4.36) and (4.103), we conclude that

$$\begin{aligned}
M\left[\frac{\tau^{k\varepsilon}}{\Gamma(1+k\varepsilon)}E_\varepsilon\left(a^\varepsilon\tau^\varepsilon\right)\right] &= \frac{1}{\Gamma(1+\varepsilon)}\int_0^\infty \left[\frac{\tau^{k\varepsilon}}{\Gamma(1+k\varepsilon)}E_\varepsilon\left(a^\varepsilon\tau^\varepsilon\right)\right] \\
&\quad E_\varepsilon\left(-\tau^\varepsilon s^\varepsilon\right)(d\tau)^\varepsilon \\
&= \frac{1}{(s-a)^{(k+1)\varepsilon}}. \qquad (4.105)
\end{aligned}$$

We write the local fractional Laplace transform of the signal on Cantor sets as

$$\theta(\tau) = E_\varepsilon\left(a^\varepsilon\tau^\varepsilon\right)\cos_\varepsilon\left(\eta^\varepsilon\tau^\varepsilon\right), \quad \tau > 0. \qquad (4.106)$$

The corresponding graphs, with different parameters a and η, are shown in Figure 4.2.

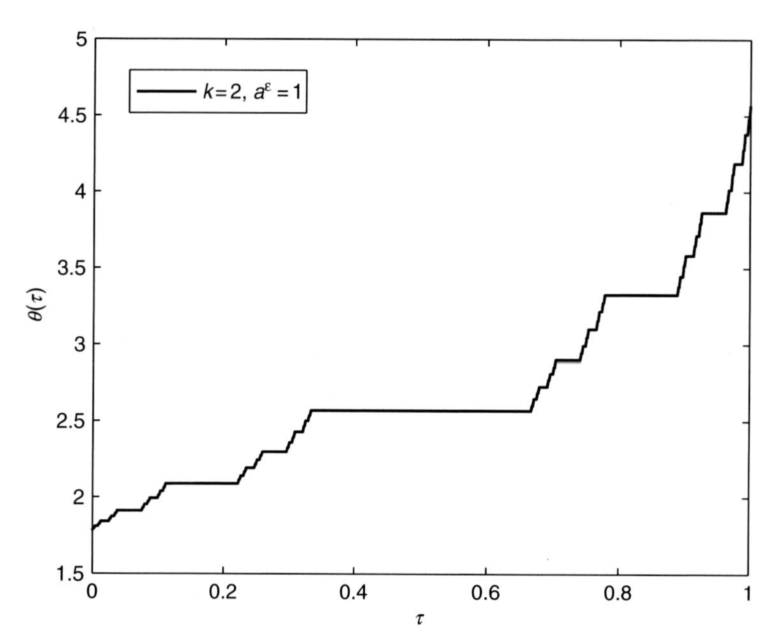

Figure 4.1 The graph of $\theta(\tau)$ when $\varepsilon = \ln 2/\ln 3$.

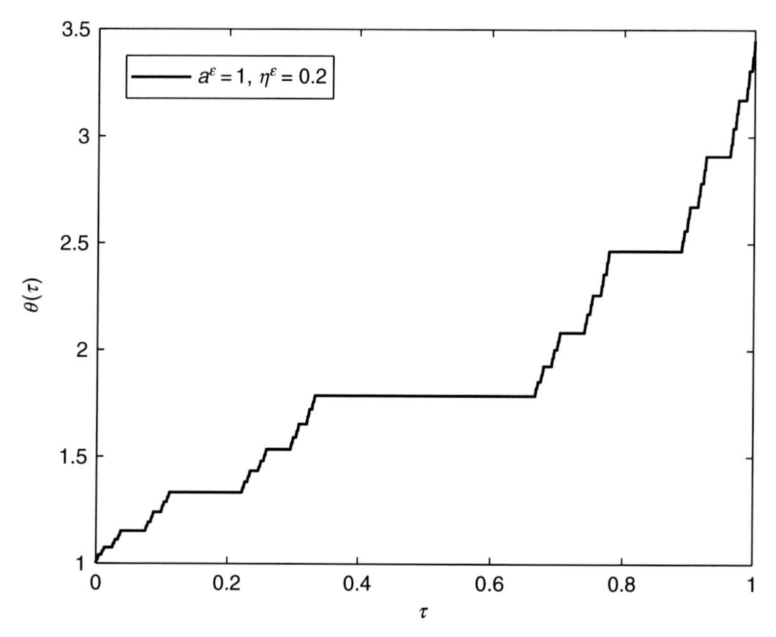

Figure 4.2 The graph of $\theta(\tau)$ when $\varepsilon = \ln 2/\ln 3$.

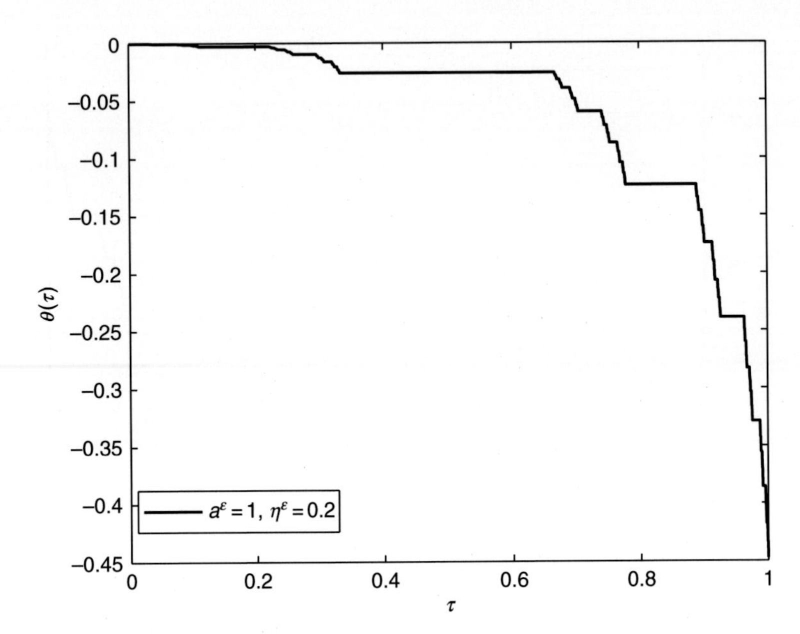

Figure 4.3 The graph of $\theta(\tau)$ when $\varepsilon = \ln 2/\ln 3$.

From formulas (4.90) and (4.91), we conclude that

$$\begin{aligned}
&\mathrm{M}\left[E_{\varepsilon}\left(a^{\varepsilon}\tau^{\varepsilon}\right)\cos_{\varepsilon}\left(\eta^{\varepsilon}\tau^{\varepsilon}\right)\right] \\
&= \frac{1}{\Gamma(1+\varepsilon)}\int_{0}^{\infty}\left[E_{\varepsilon}\left(a^{\varepsilon}\tau^{\varepsilon}\right)\cos_{\varepsilon}\left(\eta^{\varepsilon}\tau^{\varepsilon}\right)\right]E_{\varepsilon}\left(-\tau^{\varepsilon}s^{\varepsilon}\right)(\mathrm{d}\tau)^{\varepsilon} \\
&= \frac{1}{\Gamma(1+\varepsilon)}\int_{0}^{\infty}E_{\varepsilon}\left(a^{\varepsilon}\tau^{\varepsilon}\right)\left[\frac{E_{\varepsilon}\left(\mathrm{i}^{\varepsilon}\eta^{\varepsilon}\tau^{\varepsilon}\right)+E_{\varepsilon}\left(-\mathrm{i}^{\varepsilon}\eta^{\varepsilon}\tau^{\varepsilon}\right)}{2}\right]E_{\varepsilon}\left(-\tau^{\varepsilon}s^{\varepsilon}\right)(\mathrm{d}\tau)^{\varepsilon} \\
&= \frac{\frac{1}{\left((s-a)^{\varepsilon}-\mathrm{i}^{\varepsilon}\eta^{\varepsilon}\right)}+\frac{1}{\left((s-a)^{\varepsilon}+\mathrm{i}^{\varepsilon}\eta^{\varepsilon}\right)}}{2} = \frac{(s-a)^{2\varepsilon}}{(s-a)^{2\varepsilon}+\eta^{2\varepsilon}}.
\end{aligned} \tag{4.107}$$

We consider the local fractional Laplace transform of the signal on Cantor sets as

$$\theta(\tau) = E_{\varepsilon}\left(a^{\varepsilon}\tau^{\varepsilon}\right)\sin_{\varepsilon}\left(\eta^{\varepsilon}\tau^{\varepsilon}\right), \quad \tau > 0. \tag{4.108}$$

Its graphs with different parameters a and η are presented in Figure 4.3.
Using (4.92) and (4.93), it implies that

$$\begin{aligned}
\mathrm{M}\left[E_{\varepsilon}\left(a^{\varepsilon}\tau^{\varepsilon}\right)\sin_{\varepsilon}\left(\eta^{\varepsilon}\tau^{\varepsilon}\right)\right] &= \frac{1}{\Gamma(1+\varepsilon)}\int_{0}^{\infty}\left[E_{\varepsilon}\left(a^{\varepsilon}\tau^{\varepsilon}\right)\sin_{\varepsilon}\left(\eta^{\varepsilon}\tau^{\varepsilon}\right)\right]E_{\varepsilon}\left(-\tau^{\varepsilon}s^{\varepsilon}\right)(\mathrm{d}\tau)^{\varepsilon} \\
&= \frac{1}{\Gamma(1+\varepsilon)}\int_{0}^{\infty}E_{\varepsilon}\left(a^{\varepsilon}\tau^{\varepsilon}\right)
\end{aligned}$$

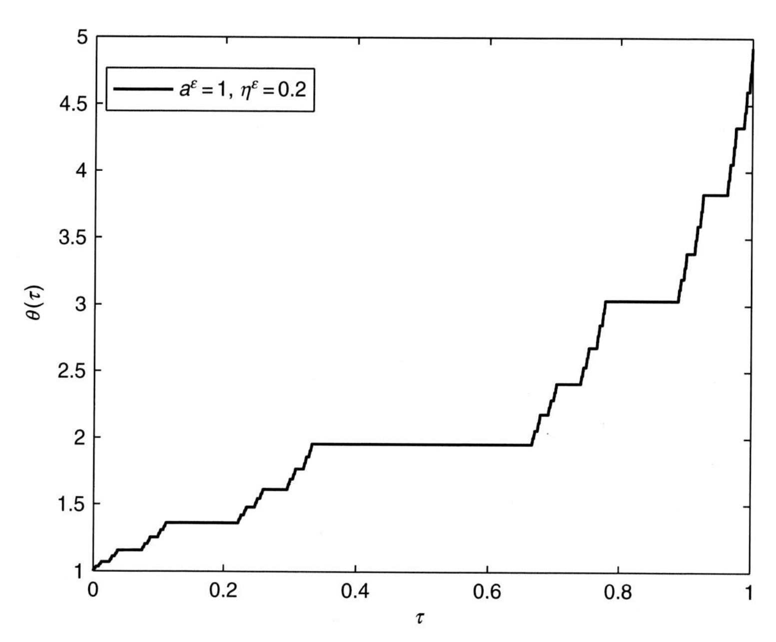

Figure 4.4 The graph of $\theta(\tau)$ when $\varepsilon = \ln 2/\ln 3$.

$$\times \left[\frac{E_\varepsilon\left(\mathrm{i}^\varepsilon\eta^\varepsilon\tau^\varepsilon\right) - E_\varepsilon\left(-\mathrm{i}^\varepsilon\eta^\varepsilon\tau^\varepsilon\right)}{2\mathrm{i}^\varepsilon}\right] E_\varepsilon\left(-\tau^\varepsilon s^\varepsilon\right)(\mathrm{d}\tau)^\varepsilon$$
$$= \frac{\frac{1}{(s-a)^\varepsilon - \mathrm{i}^\varepsilon\eta^\varepsilon} - \frac{1}{(s-a)^\varepsilon + \mathrm{i}^\varepsilon\eta^\varepsilon}}{2\mathrm{i}^\varepsilon}$$
$$= \frac{\eta^\varepsilon}{(s-a)^{2\varepsilon} + \eta.^{2\varepsilon}}. \tag{4.109}$$

We treat the local fractional Laplace transform of the signal on Cantor sets as

$$\theta(\tau) = E_\varepsilon\left(a^\varepsilon\tau^\varepsilon\right)\cosh_\varepsilon\left(\eta^\varepsilon\tau^\varepsilon\right), \quad \tau > 0. \tag{4.110}$$

As we did before, the corresponding graphs for different parameters of a and η are plotted in Figure 4.4.

In view of (4.36) and (4.100), the final result is given below:

$$\mathrm{M}\left[E_\varepsilon\left(a^\varepsilon\tau^\varepsilon\right)\cosh_\varepsilon\left(\eta^\varepsilon\tau^\varepsilon\right)\right] = \frac{1}{\Gamma(1+\varepsilon)}\int_0^\infty \left[E_\varepsilon\left(a^\varepsilon\tau^\varepsilon\right)\cosh_\varepsilon\left(\eta^\varepsilon\tau^\varepsilon\right)\right] E_\varepsilon\left(-\tau^\varepsilon s^\varepsilon\right)(\mathrm{d}\tau)^\varepsilon$$
$$= \frac{1}{\Gamma(1+\varepsilon)}\int_0^\infty E_\varepsilon\left(a^\varepsilon\tau^\varepsilon\right)\left[\frac{E_\varepsilon\left(\eta^\varepsilon\tau^\varepsilon\right) + E_\varepsilon\left(-\eta^\varepsilon\tau^\varepsilon\right)}{2}\right]$$
$$E_\varepsilon\left(-\tau^\varepsilon s^\varepsilon\right)(\mathrm{d}\tau)^\varepsilon$$
$$= \frac{\frac{1}{(s-a)^\varepsilon - \eta^\varepsilon} + \frac{1}{(s-a)^\varepsilon + \eta^\varepsilon}}{2} = \frac{(s-a)^\varepsilon}{(s-a)^{2\varepsilon} - \eta^{2\varepsilon}}. \tag{4.111}$$

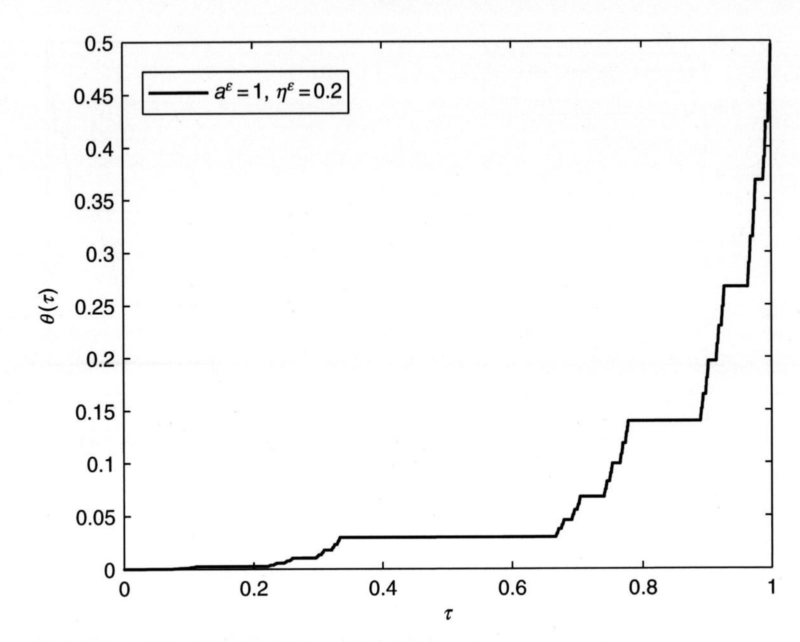

Figure 4.5 The graph of $\theta(\tau)$ when $\varepsilon = \ln 2/\ln 3$.

Let us compute the local fractional Laplace transform of the signal on Cantor sets defined by

$$\theta(\tau) = E_{\varepsilon}\left(a^{\varepsilon}\tau^{\varepsilon}\right)\sinh_{\varepsilon}\left(\eta^{\varepsilon}\tau^{\varepsilon}\right), \quad \tau > 0. \tag{4.112}$$

In Figure 4.5, its graphs with different parameters a and η are illustrated. Adopting (4.36) and (4.101), we conclude that

$$\begin{aligned}
\mathrm{M}\left[E_{\varepsilon}\left(a^{\varepsilon}\tau^{\varepsilon}\right)\sinh_{\varepsilon}\left(\eta^{\varepsilon}\tau^{\varepsilon}\right)\right] &= \frac{1}{\Gamma(1+\varepsilon)}\int_0^{\infty}\left[E_{\varepsilon}\left(a^{\varepsilon}\tau^{\varepsilon}\right)\sinh_{\varepsilon}\left(\eta^{\varepsilon}\tau^{\varepsilon}\right)\right]E_{\varepsilon}\left(-\tau^{\varepsilon}s^{\varepsilon}\right)(\mathrm{d}\tau)^{\varepsilon} \\
&= \frac{1}{\Gamma(1+\varepsilon)}\int_0^{\infty}E_{\varepsilon}\left(a^{\varepsilon}\tau^{\varepsilon}\right)\left[\frac{E_{\varepsilon}\left(\eta^{\varepsilon}\tau^{\varepsilon}\right)-E_{\varepsilon}\left(-\eta^{\varepsilon}\tau^{\varepsilon}\right)}{2}\right] \\
&\quad E_{\varepsilon}\left(-\tau^{\varepsilon}s^{\varepsilon}\right)(\mathrm{d}\tau)^{\varepsilon} \\
&= \frac{\dfrac{1}{(s-a)^{\varepsilon}-\eta^{\varepsilon}}-\dfrac{1}{(s-a)^{\varepsilon}+\eta^{\varepsilon}}}{2} = \frac{\eta^{\varepsilon}}{(s-a)^{2\varepsilon}-\eta.^{2\varepsilon}}.
\end{aligned} \tag{4.113}$$

We present the local fractional Laplace transform of the signal on Cantor sets as

$$\theta(\tau) = E_{\varepsilon}\left(-a\tau^{\varepsilon}\right) - E_{\varepsilon}\left(-b\tau^{\varepsilon}\right), \quad \tau > 0. \tag{4.114}$$

Thus, from the definition of the local fractional Laplace transform operator, we conclude

$$\begin{aligned}\mathrm{M}\left[E_\varepsilon\left(-a\tau^\varepsilon\right)-E_\varepsilon\left(-b\tau^\varepsilon\right)\right]&=\frac{1}{\Gamma(1+\varepsilon)}\int_0^\infty\left[E_\varepsilon\left(-a\tau^\varepsilon\right)-E_\varepsilon\left(-b\tau^\varepsilon\right)\right]\\&\quad E_\varepsilon\left(-\tau^\varepsilon s^\varepsilon\right)(\mathrm{d}\tau)^\varepsilon\\&=\frac{\dfrac{1}{s^\varepsilon+a}-\dfrac{1}{s^\varepsilon+b}}{2}\\&=\frac{b-a}{\left(s^\varepsilon+a\right)\left(s^\varepsilon+b\right)}.\end{aligned}\tag{4.115}$$

We take the local fractional Laplace transform of the signal on Cantor sets defined by

$$\theta(\tau)=E_\varepsilon\left(-a\tau^\varepsilon\right)+\frac{a\tau^\varepsilon}{\Gamma(1+\varepsilon)}-1,\quad \tau>0.\tag{4.116}$$

From the definition of the local fractional Laplace transform operator, it follows that

$$\begin{aligned}&\mathrm{M}\left[E_\varepsilon\left(-a\tau^\varepsilon\right)+\frac{a\tau^\varepsilon}{\Gamma(1+\varepsilon)}-1\right]\\&=\frac{1}{\Gamma(1+\varepsilon)}\int_0^\infty\left[E_\varepsilon\left(-a\tau^\varepsilon\right)+\frac{a\tau^\varepsilon}{\Gamma(1+\varepsilon)}-1\right]E_\varepsilon\left(-\tau^\varepsilon s^\varepsilon\right)(\mathrm{d}\tau)^\varepsilon\\&=\frac{1}{s^\varepsilon+a}+\frac{a}{s^{2\varepsilon}}-\frac{1}{s^\varepsilon}\\&=\frac{a^2}{\left(s^\varepsilon+a\right)s^{2\varepsilon}}.\end{aligned}\tag{4.117}$$

We consider the local fractional Laplace transform of the signal on Cantor sets defined by

$$\theta(\tau)=\frac{1}{\eta^2}\frac{\tau^\varepsilon}{\Gamma(1+\varepsilon)}-\frac{1}{\eta^3}\sin_\varepsilon\left(\eta\tau^\varepsilon\right),\quad \tau>0.\tag{4.118}$$

Thus, we obtain from the definition of the local fractional Laplace transform operator

$$\begin{aligned}&\mathrm{M}\left[\frac{1}{\eta^2}\frac{\tau^\varepsilon}{\Gamma(1+\varepsilon)}-\frac{1}{\eta^3}\sin_\varepsilon\left(\eta\tau^\varepsilon\right)\right]\\&=\frac{1}{\Gamma(1+\varepsilon)}\int_0^\infty\left[\frac{1}{\eta^2}\frac{\tau^\varepsilon}{\Gamma(1+\varepsilon)}-\frac{1}{\eta^3}\sin_\varepsilon\left(\eta\tau^\varepsilon\right)\right]E_\varepsilon\left(-\tau^\varepsilon s^\varepsilon\right)(\mathrm{d}\tau)^\varepsilon\\&=\frac{1}{\eta^2}\left(\frac{1}{s^{2\varepsilon}}-\frac{1}{s^{2\varepsilon}+\eta^2}\right),\\&=\frac{1}{s^{2\varepsilon}\left(s^{2\varepsilon}+\eta^2\right)}.\end{aligned}\tag{4.119}$$

Now, let us consider the local fractional Laplace transform of the signal on Cantor sets as

$$\theta(\tau) = \frac{1}{\eta^2 - \mu^2}\left(\frac{\sin_\varepsilon(\mu\tau^\varepsilon)}{\mu} - \frac{\sin_\varepsilon(\eta\tau^\varepsilon)}{\eta}\right), \quad \tau > 0. \tag{4.120}$$

By using the definition of the local fractional Laplace transform operator, the final result is written as

$$\begin{aligned}
&\mathrm{M}\left[\frac{1}{\eta^2 - \mu^2}\left(\frac{\sin_\varepsilon(\mu\tau^\varepsilon)}{\mu} - \frac{\sin_\varepsilon(\eta\tau^\varepsilon)}{\eta}\right)\right] \\
&= \frac{1}{\Gamma(1+\varepsilon)}\int_0^\infty \left[\frac{1}{\eta^2 - \mu^2}\left(\frac{\sin_\varepsilon(\mu\tau^\varepsilon)}{\mu} - \frac{\sin_\varepsilon(\eta\tau^\varepsilon)}{\eta}\right)\right] E_\varepsilon\left(-\tau^\varepsilon s^\varepsilon\right)(d\tau)^\varepsilon \\
&= \frac{1}{\eta^2 - \mu^2}\left[\frac{1}{s^{2\varepsilon} + \mu^2} - \frac{1}{s^{2\varepsilon} + \eta^2}\right] \\
&= \frac{1}{\left(s^{2\varepsilon} + \mu^2\right)\left(s^{2\varepsilon} + \eta^2\right)}.
\end{aligned} \tag{4.121}$$

We find the local fractional Laplace transform of the signal on Cantor sets by the following expression:

$$\theta(\tau) = \frac{\mu - (\mu - \eta)E_\varepsilon(-\eta\tau^\varepsilon)}{\eta}, \quad \tau > 0. \tag{4.122}$$

We have

$$\begin{aligned}
\mathrm{M}\left[\frac{\mu - \mu E_\varepsilon(-\eta\tau^\varepsilon)}{\eta}\right] &= \frac{1}{\Gamma(1+\varepsilon)}\int_0^\infty \left[\frac{\mu - \mu E_\varepsilon(-\eta\tau^\varepsilon)}{\eta}\right] E_\varepsilon\left(-\tau^\varepsilon s^\varepsilon\right)(d\tau)^\varepsilon \\
&= \frac{\mu}{\eta}\frac{1}{s^\varepsilon} - \frac{\mu}{\eta}\frac{1}{s^\varepsilon + \eta} \\
&= \frac{\mu\eta}{s^\varepsilon\left(s^\varepsilon + \eta\right)}.
\end{aligned} \tag{4.123}$$

We handle the local fractional Laplace transform of the signal on Cantor sets as

$$\theta(\tau) = \left(1 + \frac{\eta^\varepsilon\tau^\varepsilon}{\Gamma(1+\varepsilon)}\right)E_\varepsilon\left(\eta^\varepsilon\tau^\varepsilon\right), \quad \tau > 0. \tag{4.124}$$

Taking the local fractional Laplace transform of (4.124), we conclude that

$$\begin{aligned}
&\mathrm{M}\left[\left(1 + \frac{\eta^\varepsilon\tau^\varepsilon}{\Gamma(1+\varepsilon)}\right)E_\varepsilon\left(\eta^\varepsilon\tau^\varepsilon\right)\right] \\
&= \frac{1}{\Gamma(1+\varepsilon)}\int_0^\infty \left[\left(1 + \frac{\eta^\varepsilon\tau^\varepsilon}{\Gamma(1+\varepsilon)}\right)E_\varepsilon\left(\eta^\varepsilon\tau^\varepsilon\right)\right] E_\varepsilon\left(-\tau^\varepsilon s^\varepsilon\right)(d\tau)^\varepsilon \\
&= \frac{1}{s^\varepsilon - \eta^\varepsilon} + \frac{\eta^\varepsilon}{(s-\eta)^{2\varepsilon}} \\
&= \frac{s^\varepsilon}{(s-\eta)^{2\varepsilon}}.
\end{aligned} \tag{4.125}$$

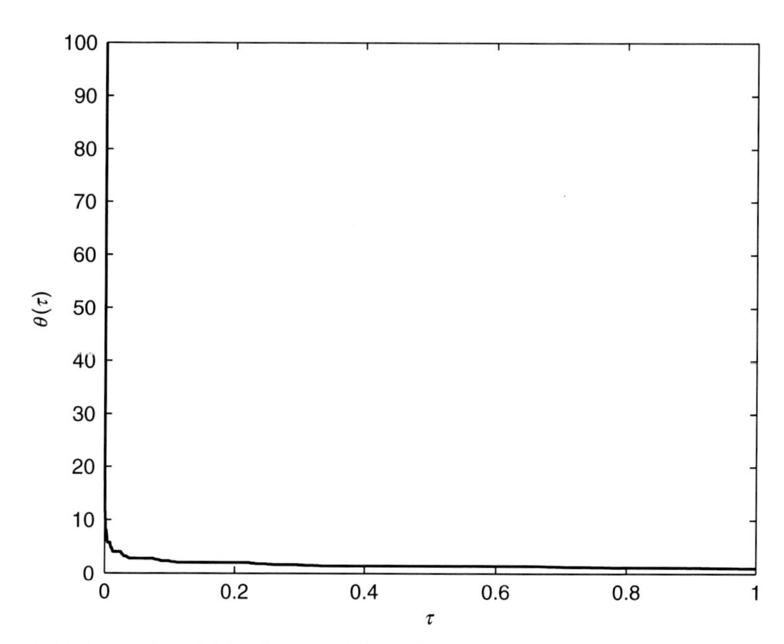

Figure 4.6 The graph of $\theta(\tau)$ when $\varepsilon = \ln 2/\ln 3$.

We give the local fractional Laplace transform of the signal, which is a local fractional Dirac function:

$$\theta(\tau) = \delta_\varepsilon(\tau), \quad \tau > 0. \tag{4.126}$$

Similarly, we have

$$\begin{aligned} \mathrm{M}[\delta_\varepsilon(\tau)] &= \frac{1}{\Gamma(1+\varepsilon)} \int_0^\infty \delta_\varepsilon(\tau) E_\varepsilon\left(-\tau^\varepsilon s^\varepsilon\right) (\mathrm{d}\tau)^\varepsilon \\ &= 1. \end{aligned} \tag{4.127}$$

We consider the local fractional Laplace transform of the signal, which is given as

$$\theta(\tau) = \tau^{-\varepsilon/2}, \quad \tau > 0, \tag{4.128}$$

and we depict its graph in Figure 4.6.

It easily follows that

$$\begin{aligned} \mathrm{M}\left[\tau^{-\varepsilon/2}\right] &= \frac{1}{\Gamma(1+\varepsilon)} \int_0^\infty \tau^{-\varepsilon/2} E_\varepsilon\left(-\tau^\varepsilon s^\varepsilon\right) (\mathrm{d}\tau)^\varepsilon \\ &= \frac{\Gamma(1+2\varepsilon)}{\Gamma^2(1+\varepsilon)} \left[s^{-\varepsilon/2} \left(\frac{1}{\Gamma(1+\varepsilon)} \int_0^\infty E_\varepsilon\left(-\tau^{2\varepsilon}\right) (\mathrm{d}\tau)^\varepsilon \right) \right] \\ &= \frac{\Gamma(1+2\varepsilon)}{2\Gamma^3(1+\varepsilon)} \left(\frac{\pi}{s}\right)^{\varepsilon/2}. \end{aligned} \tag{4.129}$$

Find the local fractional Laplace transform of the analogous rectangular pulse, defined by

$$\mathrm{rect}_{\varepsilon}(\tau,\tau_0,\tau_1)=\begin{cases}0, & \tau_0\leq\tau,\\ 1, & \tau_0<\tau\leq\tau_1,\\ 0, & \tau>\tau_1,\end{cases} \tag{4.130}$$

that is to say,

$$\begin{aligned}\mathrm{M}\left[\mathrm{rect}_{\varepsilon}(\tau,\tau_0,\tau_1)\right]&=\frac{1}{\Gamma(1+\varepsilon)}\int_0^{\infty}\mathrm{rect}_{\varepsilon}(\tau,\tau_0,\tau_1)\,E_{\varepsilon}\left(-\tau^{\varepsilon}s^{\varepsilon}\right)(\mathrm{d}\tau)^{\varepsilon}\\ &=\frac{1}{\Gamma(1+\varepsilon)}\int_{\tau_0}^{\tau_1}E_{\varepsilon}\left(-\tau^{\varepsilon}s^{\varepsilon}\right)(\mathrm{d}\tau)^{\varepsilon}\\ &=\frac{1}{s^{\varepsilon}}\left[E_{\varepsilon}\left(-\tau_0^{\varepsilon}s^{\varepsilon}\right)-E_{\varepsilon}\left(-\tau_1^{\varepsilon}s^{\varepsilon}\right)\right],\end{aligned} \tag{4.131}$$

where $0<\tau_0<\tau_1$.

For some illustrative examples, the reader can see Appendix F.

4.4 Solving local fractional differential equations

We now present the local fractional Laplace transform operators to obtain the nondifferentiable solution of local ordinary and partial differential equations.

4.4.1 Applications of local fractional ordinary differential equations

The potential applications of the local fractional Laplace transform operators are particularly effective for linear local fractional ODEs with constant coefficients. Now, we illustrate the methods with the following examples:

Let us consider the local fractional ordinary differential equation

$$\frac{\mathrm{d}^{\varepsilon}\theta(\mu)}{\mathrm{d}\mu^{\varepsilon}}+a\theta(\mu)=b\sin_{\varepsilon}\left(c\mu^{\varepsilon}\right), \tag{4.132}$$

subject to initial value condition

$$\theta(0)=1, \tag{4.133}$$

where a, b, and c are constants.

Taking the local fractional Laplace transform in (4.131), it gives

$$s^{\varepsilon}\Theta(s)-1+a\Theta(s)=\frac{bc}{s^{2\varepsilon}+c^2}. \tag{4.134}$$

In this case, we can rearrange (4.133) to obtain

$$\begin{aligned}\Theta(s) &= \frac{bc}{(s^{\varepsilon}+a)\left(s^{2\varepsilon}+c^{2}\right)}+\frac{1}{s^{\varepsilon}+a}\\ &= \left(\frac{bc}{a^{2}+c^{2}}+1\right)\frac{1}{s^{\varepsilon}+a}-\frac{bc}{\left(a^{2}+c^{2}\right)}\left(\frac{s^{\varepsilon}}{s^{2\varepsilon}+c^{2}}\right)+\frac{ab}{\left(a^{2}+c^{2}\right)}\left(\frac{c}{s^{2\varepsilon}+c^{2}}\right),\end{aligned} \tag{4.135}$$

which leads to

$$\theta(\mu) = \left(\frac{bc}{a^{2}+c^{2}}+1\right)E_{\varepsilon}\left(-a\mu^{\varepsilon}\right)-\frac{bc}{\left(a^{2}+c^{2}\right)}\cos_{\varepsilon}\left(c\mu^{\varepsilon}\right)+\frac{ab}{\left(a^{2}+c^{2}\right)}\sin_{\varepsilon}\left(c\mu^{\varepsilon}\right). \tag{4.136}$$

We consider the local fractional ordinary differential equation

$$\frac{d^{\varepsilon}\theta(\mu)}{d\mu^{\varepsilon}}+a\theta(\mu) = b\cos_{\varepsilon}\left(c\mu^{\varepsilon}\right), \tag{4.137}$$

with the initial value condition

$$\theta(0) = 1, \tag{4.138}$$

such that a, b, and c are constants.

Similarly, making the local fractional Laplace transform in (4.136), we conclude

$$s^{\varepsilon}\Theta(s)-1+a\Theta(s) = \frac{bs^{\varepsilon}}{s^{2\varepsilon}+c^{2}}. \tag{4.139}$$

We may reformulate (4.138) as

$$\begin{aligned}\Theta(s) &= \frac{bs^{\varepsilon}}{(s^{\varepsilon}+a)\left(s^{2\varepsilon}+c^{2}\right)}+\frac{1}{s^{\varepsilon}+a}\\ &= \left(1-\frac{ab}{c^{2}+a^{2}}\right)\frac{1}{s^{\varepsilon}+a}+\frac{ab}{c^{2}+a^{2}}\left(\frac{s^{\varepsilon}}{s^{2\varepsilon}+c^{2}}\right)\\ &\quad+\left(\frac{b-\frac{a^{2}b}{c^{2}+a^{2}}}{c}\right)\left(\frac{c}{s^{2\varepsilon}+c^{2}}\right).\end{aligned} \tag{4.140}$$

Thus, it follows that

$$\begin{aligned}\theta(\mu) &= \left(1-\frac{ab}{c^{2}+a^{2}}\right)E_{\varepsilon}\left(-a\mu^{\varepsilon}\right)+\frac{ab}{c^{2}+a^{2}}\cos_{\varepsilon}\left(c\mu^{\varepsilon}\right)\\ &\quad+\left(\frac{b-\frac{a^{2}b}{c^{2}+a^{2}}}{c}\right)\sin_{\varepsilon}\left(c\mu^{\varepsilon}\right).\end{aligned} \tag{4.141}$$

The local fractional ordinary differential equation takes the form

$$\frac{d^{\varepsilon}\theta(\mu)}{d\mu^{\varepsilon}}+a\theta(\mu)=bE_{\varepsilon}\left(c\mu^{\varepsilon}\right), \tag{4.142}$$

subject to the initial value condition

$$\theta(0)=1, \tag{4.143}$$

where a, b, and c are constants.

Let us take the local fractional Laplace transform in (4.141). That is to say,

$$s^{\varepsilon}\Theta(s)-1+a\Theta(s)=\frac{b}{s^{\varepsilon}-c}. \tag{4.144}$$

Thus, we have

$$\Theta(s)=\frac{b}{(s^{\varepsilon}-c)(s^{\varepsilon}+a)}+\frac{1}{s^{\varepsilon}+a}. \tag{4.145}$$

Therefore, we have the following result:

$$\theta(\mu)=E_{\varepsilon}\left(-a\mu^{\varepsilon}\right)+\frac{b}{a+c}\left[E_{\varepsilon}\left(a\mu^{\varepsilon}\right)+E_{\varepsilon}\left(-c\mu^{\varepsilon}\right)\right]. \tag{4.146}$$

We consider that the local fractional ordinary differential equation

$$\frac{d^{2\varepsilon}\theta(\mu)}{d\mu^{2\varepsilon}}+a^{2}\theta(\mu)=b\sin_{\varepsilon}\left(c\mu^{\varepsilon}\right), \tag{4.147}$$

subject to the initial value conditions

$$\theta(0)=0, \tag{4.148}$$

$$\frac{d^{\varepsilon}\theta(0)}{d\mu^{\varepsilon}}=0, \tag{4.149}$$

where a, b, and c are constants.

Taking the local fractional Laplace transform in (4.146), we conclude that

$$s^{2\varepsilon}\Theta(s)+a^{2}\Theta(s)=\frac{b}{s^{2\varepsilon}+c^{2}}, \tag{4.150}$$

which leads to

$$\Theta(s)=\frac{b}{\left(s^{2\varepsilon}+c^{2}\right)\left(s^{2\varepsilon}+a^{2}\right)}. \tag{4.151}$$

Thus, from (4.121), this may be added in the form

$$\theta(\mu)=\frac{b}{a^{2}-c^{2}}\left(\frac{\sin_{\varepsilon}\left(c\mu^{\varepsilon}\right)}{c}-\frac{\sin_{\varepsilon}\left(a\mu^{\varepsilon}\right)}{a}\right), \tag{4.152}$$

where $a\neq c$.

For $a = c$, we found that

$$\Theta(s) = \frac{b}{\left(s^{2\varepsilon} + c^2\right)^2} = \frac{b}{2c^2}\left[\frac{1}{s^{2\varepsilon} + c^2} - \frac{s^{2\varepsilon} - c^2}{\left(s^{2\varepsilon} + c^2\right)^2}\right], \tag{4.153}$$

where

$$\mathrm{M}\left[\frac{\mu^{\varepsilon}}{\Gamma(1+\varepsilon)} \cos_{\varepsilon}\left(\zeta \mu^{\varepsilon}\right)\right] = \frac{s^{2\varepsilon} - \zeta^2}{\left(s^{2\varepsilon} + \zeta^{2\varepsilon}\right)^2}, \tag{4.154}$$

such that

$$\theta(\mu) = \frac{b}{2c^2}\left[\frac{\sin_{\varepsilon}(c\mu^{\varepsilon})}{c} - \frac{\mu^{\varepsilon}}{\Gamma(1+\alpha)} \cos_{\varepsilon}\left(c\mu^{\varepsilon}\right)\right]. \tag{4.155}$$

We investigate the local fractional ordinary differential equation

$$\frac{\mathrm{d}^{2\varepsilon}\theta(\mu)}{\mathrm{d}\mu^{2\varepsilon}} + a^2\theta(\mu) = b\cos_{\varepsilon}\left(c\mu^{\varepsilon}\right), \tag{4.156}$$

in the presence of the initial value conditions

$$\theta(0) = 0, \tag{4.157}$$

$$\frac{\mathrm{d}^{\varepsilon}\theta(0)}{\mathrm{d}\mu^{\varepsilon}} = 0, \tag{4.158}$$

where a, b, and c are constants.

Similarly, we present

$$s^{2\varepsilon}\Theta(s) + a^2\Theta(s) = \frac{bs^{\varepsilon}}{s^{2\varepsilon} + c^2}, \tag{4.159}$$

which becomes

$$\Theta(s) = \frac{bs^{\varepsilon}}{\left(s^{2\varepsilon} + c^2\right)\left(s^{2\varepsilon} + a^2\right)}. \tag{4.160}$$

For $a \neq c$, we have

$$\mathrm{M}\left[\frac{1}{a^2 - b^2}\left(\cos_{\varepsilon}\left(b\mu^{\varepsilon}\right) - \cos_{\varepsilon}\left(a\mu^{\varepsilon}\right)\right)\right] = \frac{s^{\varepsilon}}{\left(s^{2\varepsilon} + a^2\right)\left(s^{2\varepsilon} + b^2\right)}, \tag{4.161}$$

such that

$$\theta(\mu) = \frac{b}{a^2 - c^2}\left(\cos_{\varepsilon}\left(c\mu^{\varepsilon}\right) - \cos_{\varepsilon}\left(a\mu^{\varepsilon}\right)\right). \tag{4.162}$$

In case $a = c$, we have

$$\Theta(s) = \frac{bs^{\varepsilon}}{\left(s^{2\varepsilon} + a^2\right)^2} = \frac{b}{a}\frac{as^{\varepsilon}}{\left(s^{2\varepsilon} + a^2\right)^2} \tag{4.163}$$

in such a way that

$$\theta(\mu) = \frac{b}{a} \frac{\mu^{\varepsilon}}{\Gamma(1+\varepsilon)} \sin_{\varepsilon}(a\mu^{\varepsilon}), \tag{4.164}$$

where

$$\mathrm{M}\left[\frac{\mu^{\varepsilon}}{\Gamma(1+\varepsilon)} \sin_{\varepsilon}(\zeta\mu^{\varepsilon})\right] = \frac{\zeta s^{\varepsilon}}{\left(s^{2\varepsilon}+\zeta^{2\varepsilon}\right)^{2}}. \tag{4.165}$$

Now, let us discuss the local fractional ordinary differential equation

$$\frac{d^{2\varepsilon}\theta(\mu)}{d\mu^{2\varepsilon}} - a^2\theta(\mu) = b\sin_{\varepsilon}(c\mu^{\varepsilon}) \tag{4.166}$$

equipped with the initial value conditions

$$\theta(0) = 0, \tag{4.167}$$

$$\frac{d^{\varepsilon}\theta(0)}{d\mu^{\varepsilon}} = 0, \tag{4.168}$$

such that a, b, and c are constants.

After calculating the local fractional Laplace transform in (4.166), we conclude

$$s^{2\varepsilon}\Theta(s) - a^2\Theta(s) = \frac{b}{s^{2\varepsilon}+c^2}, \tag{4.169}$$

which is rearranged in the form

$$\Theta(s) = \frac{b}{\left(s^{2\varepsilon}+c^2\right)\left(s^{2\varepsilon}-a^2\right)} = \frac{b}{c^2+a^2}\left[\frac{1}{s^{2\varepsilon}-a^2} - \frac{1}{s^{2\varepsilon}+c^2}\right]. \tag{4.170}$$

Thus, from (4.170), it results that

$$\theta(\mu) = \frac{b}{c^2+a^2}\left[\sinh_{\varepsilon}(a\mu^{\varepsilon}) - \sin_{\varepsilon}(c\mu^{\varepsilon})\right]. \tag{4.171}$$

Let us consider the local fractional ordinary differential equation

$$\frac{d^{2\varepsilon}\theta(\mu)}{d\mu^{2\varepsilon}} - a^2\theta(\mu) = b\cos_{\varepsilon}(c\mu^{\varepsilon}) \tag{4.172}$$

subjected to the initial value conditions

$$\theta(0) = 0, \tag{4.173}$$

$$\frac{d^{\varepsilon}\theta(0)}{d\mu^{\varepsilon}} = 0, \tag{4.174}$$

with a, b, and c being constants.

In this case, we have

$$s^{2\varepsilon}\Theta(s) - a^2\Theta(s) = \frac{bs^{\varepsilon}}{s^{2\varepsilon}+c^2}, \tag{4.175}$$

which becomes

$$\Theta(s) = \frac{bs^{\varepsilon}}{\left(s^{2\varepsilon}+c^{2}\right)\left(s^{2\varepsilon}-a^{2}\right)} = \frac{b}{c^{2}+a^{2}}\left[\frac{s^{\varepsilon}}{s^{2\varepsilon}-a^{2}} - \frac{s^{\varepsilon}}{s^{2\varepsilon}+c^{2}}\right]. \tag{4.176}$$

Thus, we write that

$$\theta(\mu) = \frac{b}{c^{2}+a^{2}}\left[\cosh_{\varepsilon}\left(a\mu^{\varepsilon}\right) - \cos_{\varepsilon}\left(c\mu^{\varepsilon}\right)\right]. \tag{4.177}$$

The next step is to find the solution of the local fractional ordinary differential equation

$$\frac{d^{2\varepsilon}\theta(\mu)}{d\mu^{2\varepsilon}} + a^{2}\theta(\mu) = 0 \tag{4.178}$$

in the presence of the initial value conditions

$$\theta(0) = b, \tag{4.179}$$

$$\frac{d^{\varepsilon}\theta(0)}{d\mu^{\varepsilon}} = c, \tag{4.180}$$

where $a, b,$ and c are constants. Using (4.42), we note that

$$s^{2\varepsilon}\Theta(s) - s^{\varepsilon}b - c + a^{2}\Theta(s) = 0, \tag{4.181}$$

which leads us to

$$\Theta(s) = \frac{s^{\varepsilon}b + c}{s^{2\varepsilon}+a^{2}}. \tag{4.182}$$

Thus, we get

$$\theta(\mu) = b\cos_{\varepsilon}\left(a\mu^{\varepsilon}\right) + \frac{c}{a}\sin_{\varepsilon}\left(a\mu^{\varepsilon}\right). \tag{4.183}$$

Below, we present the solution of the local fractional ordinary differential equation

$$\frac{d^{2\varepsilon}\theta(\mu)}{d\mu^{2\varepsilon}} - a^{2}\theta(\mu) = 0 \tag{4.184}$$

subjected to the initial value conditions

$$\theta(0) = b, \tag{4.185}$$

$$\frac{d^{\varepsilon}\theta(0)}{d\mu^{\varepsilon}} = c, \tag{4.186}$$

where $a, b,$ and c are constants.

Adopting (4.42), we observe that

$$s^{2\varepsilon}\Theta(s) - s^{\varepsilon}b - c + a^{2}\Theta(s) = 0, \tag{4.187}$$

which can be written as

$$\Theta(s) = \frac{s^{\varepsilon}b + c}{s^{2\varepsilon}-a^{2}}. \tag{4.188}$$

Thus, the final result becomes

$$\theta(\mu) = b\cosh_\varepsilon\left(a\mu^\varepsilon\right) + \frac{c}{a}\sinh_\varepsilon\left(a\mu^\varepsilon\right). \tag{4.189}$$

4.4.2 Applications of local fractional partial differential equations

We start with the local fractional partial differential equations

$$\frac{\partial^\varepsilon \theta(\mu,\tau)}{\partial \mu^\varepsilon} + \frac{\partial^\varepsilon \theta(\mu,\tau)}{\partial \tau^\varepsilon} + \theta(\mu,\tau) = \cos_\varepsilon\left(\tau^\varepsilon\right), \tag{4.190}$$

subject to the initial-boundary conditions

$$\theta(\mu,0) = 1, \tag{4.191}$$

$$\theta(0,\tau) = 0. \tag{4.192}$$

Taking the local fractional Laplace transform in (4.190), we find that

$$\frac{\partial^\varepsilon \theta(\mu,s)}{\partial \mu^\varepsilon} + \left(s^\varepsilon + 1\right)\theta(\mu,s) - 1 = \frac{s^\varepsilon}{s^{2\varepsilon}+1} \tag{4.193}$$

and

$$\theta(0,s) = 0. \tag{4.194}$$

In this case, (4.173) gives the following result:

$$\begin{aligned}
\theta(\mu,s) &= \frac{\frac{s^\varepsilon}{s^{2\varepsilon}+1}+1}{s^\varepsilon+1}\left(1 - E_\varepsilon\left(-\left(s^\varepsilon+1\right)\mu^\varepsilon\right)\right) \\
&= \left\{\frac{1}{2}s^\varepsilon\left(\frac{1}{s^{2\varepsilon}-1} + \frac{1}{s^{2\varepsilon}+1}\right) - \frac{1}{2}\left(\frac{1}{s^{2\varepsilon}-1} - \frac{1}{s^{2\varepsilon}+1}\right)\right\} \\
&\quad \left(1 - E_\varepsilon\left(-\left(s^\varepsilon+1\right)\mu^\varepsilon\right)\right) \\
&= \left\{\frac{1}{2}s^\varepsilon\left(\frac{1}{s^{2\varepsilon}-1} + \frac{1}{s^{2\varepsilon}+1}\right) - \frac{1}{2}\left(\frac{1}{s^{2\varepsilon}-1} - \frac{1}{s^{2\varepsilon}+1}\right)\right\} \\
&\quad \left(1 - E_\varepsilon\left(-s^\varepsilon\mu^\varepsilon\right)E_\varepsilon\left(-\mu^\varepsilon\right)\right),
\end{aligned} \tag{4.195}$$

which reduces to

$$\begin{aligned}
\theta(\mu,\tau) = \frac{1}{2}&\left\{\cosh_\varepsilon\left(\tau^\varepsilon\right) + \cos_\varepsilon\left(\tau^\varepsilon\right) - \sinh_\varepsilon\left(\tau^\varepsilon\right) + \sin_\varepsilon\left(\tau^\varepsilon\right)\right. \\
&- E_\varepsilon\left(-\mu^\varepsilon\right)H_\varepsilon(\tau-\mu)\left[\cosh_\varepsilon\left((\tau-\mu)^\varepsilon\right) + \cos_\varepsilon\left((\tau-\mu)^\varepsilon\right)\right) \\
&\left.- E_\varepsilon\left(-\mu^\varepsilon\right)H_\varepsilon(\tau-\mu)\left[\sin_\varepsilon\left((\tau-\mu)^\varepsilon\right) - \sinh_\varepsilon\left((\tau-\mu)^\varepsilon\right)\right]\right]\right\}.
\end{aligned} \tag{4.196}$$

The local fractional wave equation for the fractal vibrating string has the form

$$\frac{\partial^{2\varepsilon}\theta(\mu,\tau)}{\partial\tau^{2\varepsilon}} = a^{2\varepsilon}\frac{\partial^{2\varepsilon}\theta(\mu,\tau)}{\partial\mu^{2\varepsilon}}, \tag{4.197}$$

with the initial-boundary conditions

$$\theta(\mu,0) = 0, \tag{4.198}$$

$$\frac{\partial^{\varepsilon}\theta(\mu,0)}{\partial\tau^{\varepsilon}} = 0, \tag{4.199}$$

$$\theta(0,\tau) = \sin_{\varepsilon}\left(\tau^{\varepsilon}\right), \tag{4.200}$$

$$|\theta(\mu,\tau)| < \Pi, \tag{4.201}$$

where the constant a^{ε} denotes the speed of fractal wave travels.

Taking the local fractional Laplace transform in (4.197), it gives

$$a^{2\varepsilon}\frac{\partial^{2\varepsilon}\theta(\mu,s)}{\partial\mu^{2\varepsilon}} - s^{2\varepsilon}\theta(\mu,s) = 0, \tag{4.202}$$

which is rewritten in the form

$$\frac{\partial^{2\varepsilon}\theta(\mu,s)}{\partial\mu^{2\varepsilon}} - \frac{s^{2\varepsilon}}{a^{2\varepsilon}}\theta(\mu,s) = 0, \tag{4.203}$$

where

$$\theta(0,s) = \frac{1}{s^{2\varepsilon}+1}. \tag{4.204}$$

We observe that the general solution of (4.203) is

$$\theta(\mu,s) = \zeta_1(s)E_{\varepsilon}\left(s^{\varepsilon}\mu^{\varepsilon}/a^{\varepsilon}\right) + \zeta_2(s)E_{\varepsilon}\left(-s^{\varepsilon}\mu^{\varepsilon}/a^{\varepsilon}\right), \tag{4.205}$$

which, from (4.201), leads to

$$\zeta_1(s) = 0 \tag{4.206}$$

or

$$\theta(\mu,s) = \zeta_2(s)E_{\varepsilon}\left(-s^{\varepsilon}\mu^{\varepsilon}/a^{\varepsilon}\right). \tag{4.207}$$

From (4.204), we see that (4.207) can be rewritten as

$$\theta(\mu,s) = \frac{1}{s^{2\varepsilon}+1}E_{\varepsilon}\left(-s^{\varepsilon}\mu^{\varepsilon}/a^{\varepsilon}\right). \tag{4.208}$$

Thus, the inverse local fractional Laplace transform of (4.208) reduces to

$$\theta(\mu,\tau) = H_{\varepsilon}(\tau-\mu/a)\sin_{\varepsilon}\left[(\tau-\mu/a)^{\varepsilon}\right]. \tag{4.209}$$

Let us consider now the local fractional diffusion equation in 1 + 1 fractal dimensional space, namely,

$$\frac{\partial^{\varepsilon}\theta(\mu,\tau)}{\partial\tau^{\varepsilon}} - k^{2\varepsilon}\frac{\partial^{2\varepsilon}\theta(\mu,\tau)}{\partial\mu^{2\varepsilon}} = 0, \tag{4.210}$$

subject to the initial-boundary conditions

$$\theta(0,\tau) = \delta_{\varepsilon}(\tau), \tag{4.211}$$

$$\frac{\partial^{\varepsilon}\theta(\mu,0)}{\partial\mu^{\varepsilon}} = 0, \tag{4.212}$$

$$|\theta(\mu,\tau)| < \Pi. \tag{4.213}$$

Taking the local fractional Laplace transform in (4.210), we have

$$s^{\varepsilon}\theta(\mu,s) - k^{2\varepsilon}\frac{\partial^{2\varepsilon}\theta(\mu,s)}{\partial\mu^{2\varepsilon}} = 0. \tag{4.214}$$

From (4.211), we write

$$\theta(\mu,s) = \zeta_1(s)E_{\varepsilon}\left(s^{\varepsilon/2}\mu^{\varepsilon}/k^{\varepsilon}\right) + \zeta_2(s)E_{\varepsilon}\left(-s^{\varepsilon/2}\mu^{\varepsilon}/k^{\varepsilon}\right). \tag{4.215}$$

From (4.213), it follows that

$$\theta(\mu,s) = \zeta_2(s)E_{\varepsilon}\left(-s^{\varepsilon/2}\mu^{\varepsilon}/k^{\varepsilon}\right), \tag{4.216}$$

$$\zeta_1(s) = 0. \tag{4.217}$$

From (4.211), $\theta(0,s) = 1$, we conclude that

$$\zeta_2(s) = 1. \tag{4.218}$$

We now observe from (4.216) and (4.218) that

$$\theta(\mu,s) = E_{\varepsilon}\left(-s^{\varepsilon/2}\mu^{\varepsilon}/k^{\varepsilon}\right). \tag{4.219}$$

Thus, we have the final result:

$$\theta(\mu,\tau) = \frac{\Gamma(1+\varepsilon)}{\sqrt{4^{\varepsilon}\pi^{\varepsilon}k^{\varepsilon}}}E_{\varepsilon}\left(-\mu^{2\varepsilon}/4^{\varepsilon}\pi^{\varepsilon}k^{\varepsilon}\right). \tag{4.220}$$

Coupling the local fractional Laplace transform with analytic methods

5.1 Introduction

Many challenging problems, such as vibrating strings, traffic flow, mass, and heat transfer in fractal dimensional time-space, have opened new frontiers in physics, mathematics, and engineering applications. The local fractional partial differential equations were used to investigate some anomalous, still unsolved nondifferential phenomena in nature. The local fractional Fourier and Laplace transform operators handle these types of equations (e.g., Chapters 3 and 4). Also, there are analytic and numerical methods to deal with the local fractional partial differential equations. This chapter presents the variational iteration, decomposition methods, and the coupling methods of the Laplace transform with them within the local fractional operators.

In order to clearly illustrate the analytic methods, we consider the local fractional partial differential equation in a local fractional operator form, which is given by

$$L_{\varepsilon}^{(n)}\Phi + R_{\varepsilon}\Phi = 0, \tag{5.1}$$

where $L_{\varepsilon}^{(n)}$ is linear local fractional operators of $n\varepsilon$ order and R_{ε} is the linear local fractional operators of order less than $L_{\varepsilon}^{(n)}$.

The structure of the chapter is given below. In Section 5.2, the variational iteration method of local fractional operator is presented. Section 5.3 gives the decomposition method of local fractional operator. In Section 5.4, the coupling Laplace transform with variational iteration method of local fractional operator is given. Section 5.5 is devoted to the coupling Laplace transform with decomposition method of local fractional operator.

5.2 Variational iteration method of the local fractional operator

In this section, the idea of the variational iteration method of local fractional operator is briefly introduced. The variational iteration method, proposed by He, was used to find the approximate solutions for the linear partial differential equations [116, 117]. The variational iteration method of the local fractional operator was employed to solve the local fractional partial differential equations [75, 104, 118–125].

Local Fractional Integral Transforms and Their Applications. http://dx.doi.org/10.1016/B978-0-12-804002-7.00005-X

As a starting point, we consider the local fractional variational iteration algorithm, namely,

$$\Phi_{n+1}(\tau) = \Phi_n(\tau) + \frac{1}{\Gamma(1+\varepsilon)} \int_0^{\tau} \lambda \left\{ L_{\varepsilon}^{(n)} \Phi_n(\xi) + R_{\varepsilon} \Phi_n(\xi) \right\} (d\xi)^{\varepsilon}. \quad (5.2)$$

The local fractional correction functional is written as

$$\Phi_{n+1}(\tau) = \Phi_n(\tau) + \frac{1}{\Gamma(1+\varepsilon)} \int_0^{\tau} \lambda \left\{ \left\{ L_{\varepsilon}^{(n)} \Phi_n(\xi) + R_{\varepsilon} \widehat{\Phi}_n(\xi) \right\} \right\} (d\xi)^{\varepsilon}, \quad (5.3)$$

where $\widehat{\Phi}_n$ is considered as a restricted local fractional variation, that is, $\delta^{\varepsilon} \widehat{\Phi}_n = 0$ (for more details, the reader can read the references [75, 104, 118–125] and the references therein).

After identifying the multiplier, namely,

$$\lambda = (-1)^n \frac{(\xi - \tau)^{(n-1)\varepsilon}}{\Gamma(1+(n-1)\varepsilon)}, \quad (5.4)$$

we have

$$\begin{aligned} \Phi_{n+1}(\tau) &= \Phi_n(\tau) + \frac{1}{\Gamma(1+\varepsilon)} \int_0^{\tau} \frac{(-1)^n (\xi - \tau)^{(n-1)\varepsilon}}{\Gamma(1+(n-1)\varepsilon)} \\ &\quad \left\{ \left\{ L_{\varepsilon}^{(n)} \Phi_n(\xi) + R_{\varepsilon} \widehat{\Phi}_n(\xi) \right\} \right\} (d\xi)^{\varepsilon} \\ &= \Phi_n(\tau) + {}_0I_{\tau}^{(\varepsilon)} \left\{ \frac{(-1)^n (\xi - \tau)^{(n-1)\varepsilon}}{\Gamma(1+(n-1)\varepsilon)} \left[L_{\varepsilon}^{(n)} \Phi_n(\xi) + R_{\varepsilon} \widehat{\Phi}_n(\xi) \right] \right\}. \end{aligned} \quad (5.5)$$

Finally, the nondifferentiable solution of (5.1) admits

$$\Phi(\tau) = \lim_{n \to \infty} \Phi_n(\tau). \quad (5.6)$$

Let us consider now the local fractional partial differential equation

$$\frac{\partial^{2\varepsilon} \Phi(\mu,\tau)}{\partial \tau^{2\varepsilon}} - \frac{\partial^{2\varepsilon} \Phi(\mu,\tau)}{\partial \mu^{2\varepsilon}} - \frac{\partial^{3\varepsilon} \Phi(\mu,\tau)}{\partial \mu^{\varepsilon} \partial \tau^{2\varepsilon}} = 0, \quad (5.7)$$

subjected to the initial-boundary condition

$$\Phi(\mu,0) = \frac{\mu^{2\varepsilon}}{\Gamma(1+2\varepsilon)}, \quad 0 \le \mu \le 1, \quad (5.8)$$

$$\frac{\partial^{\varepsilon} \Phi(\mu,0)}{\partial \tau^{\varepsilon}} = 0, \quad 0 \le \mu \le 1, \quad (5.9)$$

$$\Phi(l,\tau) = \Phi(0,\tau) = 0, \quad \tau > 0, \quad (5.10)$$

$$\frac{\partial^{\varepsilon} \Phi(l,\tau)}{\partial \mu^{\varepsilon}} = \frac{\partial^{\varepsilon} \Phi(0,\tau)}{\partial \mu^{\varepsilon}} = 0, \quad \tau > 0. \quad (5.11)$$

We structure the local fractional variational iteration algorithm as

$$\Phi_{n+1}(\mu,\tau)=\Phi_n(\mu,\tau)+{}_0I_\tau^{(\varepsilon)}\left\{\frac{(\xi-\tau)^\varepsilon}{\Gamma(1+\varepsilon)}\left[\frac{\partial^{2\varepsilon}\Phi_n(\mu,\tau)}{\partial\tau^{2\varepsilon}}\right.\right.$$
$$\left.\left.-\frac{\partial^{2\varepsilon}\Phi_n(\mu,\tau)}{\partial\mu^{2\varepsilon}}-\frac{\partial^{3\varepsilon}\Phi_n(\mu,\tau)}{\partial\mu^{\varepsilon}\partial\tau^{2\varepsilon}}\right]\right\}, \tag{5.12}$$

where

$$\Phi_0(\mu,\tau)=\frac{\mu^{2\varepsilon}}{\Gamma(1+2\varepsilon)}. \tag{5.13}$$

With the help of (5.12) and (5.13), we obtain the following approximations:

$$\begin{aligned}\Phi_1(\mu,\tau)&=\Phi_0(\mu,\tau)+{}_0I_\tau^{(\varepsilon)}\left\{\frac{(\xi-\tau)^\varepsilon}{\Gamma(1+\varepsilon)}\left[\frac{\partial^{2\varepsilon}\Phi_0(\mu,\tau)}{\partial\tau^{2\varepsilon}}\right.\right.\\&\quad\left.\left.-\frac{\partial^{2\varepsilon}\Phi_0(\mu,\tau)}{\partial\mu^{2\varepsilon}}-\frac{\partial^{3\varepsilon}\Phi_0(\mu,\tau)}{\partial\mu^{\varepsilon}\partial\tau^{2\varepsilon}}\right]\right\}\\&=\frac{\mu^{2\varepsilon}}{\Gamma(1+2\varepsilon)}+{}_0I_\tau^{(\varepsilon)}\left\{\frac{(\xi-\tau)^\varepsilon}{\Gamma(1+\varepsilon)}\right\}\\&=\frac{\mu^{2\varepsilon}}{\Gamma(1+2\varepsilon)}+\frac{\tau^{2\varepsilon}}{\Gamma(1+2\varepsilon)},\end{aligned} \tag{5.14}$$

$$\begin{aligned}\Phi_2(\mu,\tau)&=\Phi_1(\mu,\tau)+{}_0I_\tau^{(\varepsilon)}\left\{\frac{(\xi-\tau)^\varepsilon}{\Gamma(1+\varepsilon)}\left[\frac{\partial^{2\varepsilon}\Phi_1(\mu,\tau)}{\partial\tau^{2\varepsilon}}\right.\right.\\&\quad\left.\left.-\frac{\partial^{2\varepsilon}\Phi_1(\mu,\tau)}{\partial\mu^{2\varepsilon}}-\frac{\partial^{3\varepsilon}\Phi_1(\mu,\tau)}{\partial\mu^{\varepsilon}\partial\tau^{2\varepsilon}}\right]\right\}\\&=\frac{\mu^{2\varepsilon}}{\Gamma(1+2\varepsilon)}+\frac{\tau^{2\varepsilon}}{\Gamma(1+2\varepsilon)},\end{aligned} \tag{5.15}$$

$$\begin{aligned}\Phi_3(\mu,\tau)&=\Phi_2(\mu,\tau)+{}_0I_\tau^{(\varepsilon)}\left\{\frac{(\xi-\tau)^\varepsilon}{\Gamma(1+\varepsilon)}\left[\frac{\partial^{2\varepsilon}\Phi_2(\mu,\tau)}{\partial\tau^{2\varepsilon}}\right.\right.\\&\quad\left.\left.-\frac{\partial^{2\varepsilon}\Phi_2(\mu,\tau)}{\partial\mu^{2\varepsilon}}-\frac{\partial^{3\varepsilon}\Phi_2(\mu,\tau)}{\partial\mu^{\varepsilon}\partial\tau^{2\varepsilon}}\right]\right\}\\&=\frac{\mu^{2\varepsilon}}{\Gamma(1+2\varepsilon)}+\frac{\tau^{2\varepsilon}}{\Gamma(1+2\varepsilon)},\end{aligned} \tag{5.16}$$

$$\vdots$$

$$\Phi_n(\mu,\tau)=\frac{\mu^{2\varepsilon}}{\Gamma(1+2\varepsilon)}+\frac{\tau^{2\varepsilon}}{\Gamma(1+2\varepsilon)}. \tag{5.17}$$

Thus, we obtain

$$\Phi(\mu,\tau)=\lim_{n\to\infty}\Phi_n(\mu,\tau)=\frac{\mu^{2\varepsilon}}{\Gamma(1+2\varepsilon)}+\frac{\tau^{2\varepsilon}}{\Gamma(1+2\varepsilon)}, \tag{5.18}$$

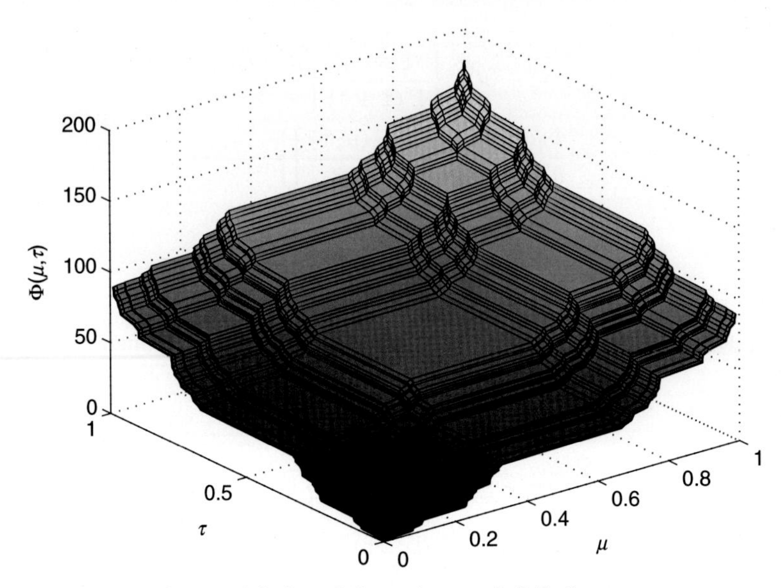

Figure 5.1 The plot of $\Phi(\mu, \tau)$ in fractal dimension $\varepsilon = \ln 2/\ln 3$.

and the related graph is shown in Figure 5.1.

We consider the following local fractional partial differential equation:

$$\frac{\partial^{3\varepsilon}\Phi(\eta,\mu)}{\partial\mu^{3\varepsilon}} - \frac{\partial^{\varepsilon}\Phi(\eta,\mu)}{\partial\mu^{\varepsilon}} = \frac{\partial^{4\varepsilon}\Phi(\eta,\mu)}{\partial\eta^{4\varepsilon}}, \tag{5.19}$$

equipped with the initial value conditions

$$\frac{\partial^{2\varepsilon}\Phi(0,\mu)}{\partial\eta^{2\varepsilon}} = E_\varepsilon(\mu^\varepsilon), \tag{5.20}$$

$$\frac{\partial^{\varepsilon}\Phi(0,\mu)}{\partial\eta^{\varepsilon}} = 0, \tag{5.21}$$

$$\Phi(0,\mu) = 0. \tag{5.22}$$

We structure the local fractional variational iteration algorithm as

$$\Phi_{n+1}(\eta,\mu) = \Phi_n(\eta,\mu) - {}_0I_\tau^{(\varepsilon)}\left\{\frac{(\xi-\tau)^{2\varepsilon}}{\Gamma(1+2\varepsilon)}\left[\frac{\partial^{3\varepsilon}\Phi_n(\eta,\mu)}{\partial\mu^{3\varepsilon}} - \frac{\partial^{\varepsilon}\Phi_n(\eta,\mu)}{\partial\mu^{\varepsilon}} - \frac{\partial^{4\varepsilon}\Phi_n(\eta,\mu)}{\partial\eta^{4\varepsilon}}\right]\right\}, \tag{5.23}$$

where

$$\Phi_0(\eta,\mu) = \frac{\eta^{2\varepsilon}}{\Gamma(1+2\varepsilon)}E_\varepsilon(\mu^\varepsilon). \tag{5.24}$$

Thus, the approximations read as

$$\Phi_1(\eta,\mu) = \Phi_0(\eta,\mu) - 0I_\tau^{(\varepsilon)}\left\{\frac{(\xi-\tau)^{2\varepsilon}}{\Gamma(1+2\varepsilon)}\left[\frac{\partial^{3\varepsilon}\Phi_0(\eta,\mu)}{\partial\mu^{3\varepsilon}} - \frac{\partial^{\varepsilon}\Phi_0(\eta,\mu)}{\partial\mu^{\varepsilon}} - \frac{\partial^{4\varepsilon}\Phi_0(\eta,\mu)}{\partial\eta^{4\varepsilon}}\right]\right\}$$
$$= \frac{\eta^{2\varepsilon}}{\Gamma(1+2\varepsilon)}E_\varepsilon(\mu^\varepsilon) - 0I_\tau^{(\varepsilon)}\left\{\frac{(\xi-\tau)^{2\varepsilon}}{\Gamma(1+2\varepsilon)}\left[\frac{\eta^{2\varepsilon}}{\Gamma(1+2\varepsilon)}E_\varepsilon(\mu^\varepsilon) - \frac{\eta^{2\varepsilon}}{\Gamma(1+2\varepsilon)}E_\varepsilon(\mu^\varepsilon) - 0\right]\right\}$$
$$= \frac{\eta^{2\varepsilon}}{\Gamma(1+2\varepsilon)}E_\varepsilon(\mu^\varepsilon), \tag{5.25}$$

$$\Phi_2(\eta,\mu) = \Phi_1(\eta,\mu) - 0I_\tau^{(\varepsilon)}\left\{\frac{(\xi-\tau)^{2\varepsilon}}{\Gamma(1+2\varepsilon)}\left[\frac{\partial^{3\varepsilon}\Phi_1(\eta,\mu)}{\partial\mu^{3\varepsilon}} - \frac{\partial^{\varepsilon}\Phi_1(\eta,\mu)}{\partial\mu^{\varepsilon}} - \frac{\partial^{4\varepsilon}\Phi_1(\eta,\mu)}{\partial\eta^{4\varepsilon}}\right]\right\}$$
$$= \frac{\eta^{2\varepsilon}}{\Gamma(1+2\varepsilon)}E_\varepsilon(\mu^\varepsilon) - 0I_\tau^{(\varepsilon)}\left\{\frac{(\xi-\tau)^{2\varepsilon}}{\Gamma(1+2\varepsilon)}\left[\frac{\eta^{2\varepsilon}}{\Gamma(1+2\varepsilon)}E_\varepsilon(\mu^\varepsilon) - \frac{\eta^{2\varepsilon}}{\Gamma(1+2\varepsilon)}E_\varepsilon(\mu^\varepsilon) - 0\right]\right\}$$
$$= \frac{\eta^{2\varepsilon}}{\Gamma(1+2\varepsilon)}E_\varepsilon(\mu^\varepsilon), \tag{5.26}$$

$$\Phi_3(\eta,\mu) = \Phi_2(\eta,\mu) - 0I_\tau^{(\varepsilon)}\left\{\frac{(\xi-\tau)^{2\varepsilon}}{\Gamma(1+2\varepsilon)}\left[\frac{\partial^{3\varepsilon}\Phi_2(\eta,\mu)}{\partial\mu^{3\varepsilon}} - \frac{\partial^{\varepsilon}\Phi_2(\eta,\mu)}{\partial\mu^{\varepsilon}} - \frac{\partial^{4\varepsilon}\Phi_2(\eta,\mu)}{\partial\eta^{4\varepsilon}}\right]\right\}$$
$$= \frac{\eta^{2\varepsilon}}{\Gamma(1+2\varepsilon)}E_\varepsilon(\mu^\varepsilon) - 0I_\tau^{(\varepsilon)}\left\{\frac{(\xi-\tau)^{2\varepsilon}}{\Gamma(1+2\varepsilon)}\left[\frac{\eta^{2\varepsilon}}{\Gamma(1+2\varepsilon)}E_\varepsilon(\mu^\varepsilon) - \frac{\eta^{2\varepsilon}}{\Gamma(1+2\varepsilon)}E_\varepsilon(\mu^\varepsilon) - 0\right]\right\}$$

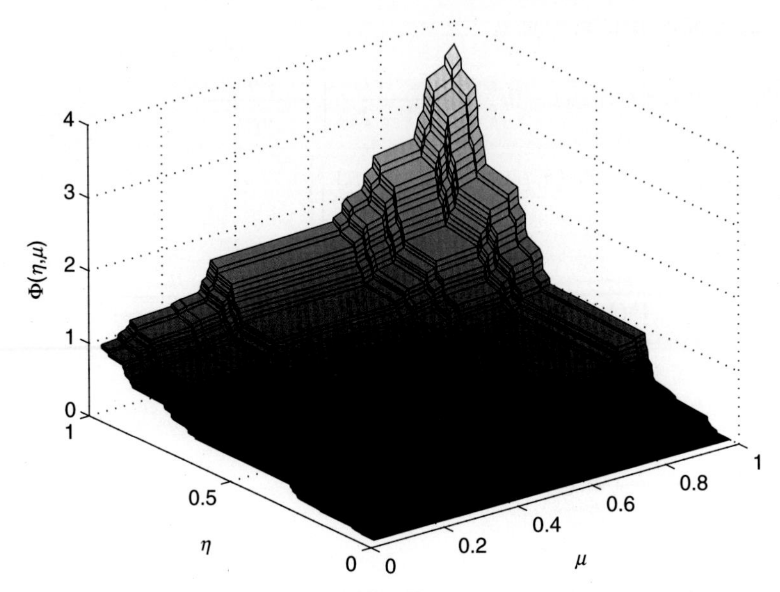

Figure 5.2 The plot of $\Phi(\eta,\mu)$ in fractal dimension $\varepsilon = \ln 2/\ln 3$.

$$= \frac{\eta^{2\varepsilon}}{\Gamma(1+2\varepsilon)} E_{\varepsilon}(\mu^{\varepsilon}), \tag{5.27}$$

$$\vdots$$

$$\Phi_n(\eta,\mu) = \frac{\eta^{2\varepsilon}}{\Gamma(1+2\varepsilon)} E_{\varepsilon}(\mu^{\varepsilon}). \tag{5.28}$$

As a result, we report

$$\Phi(\eta,\mu) = \lim_{n\to\infty} \Phi_n(\eta,\mu) = \frac{\eta^{2\varepsilon}}{\Gamma(1+2\varepsilon)} E_{\varepsilon}(\mu^{\varepsilon}), \tag{5.29}$$

and its graph is shown in Figure 5.2.

5.3 Decomposition method of the local fractional operator

In this section, the idea of the decomposition method of local fractional operator is considered. The decomposition method, proposed by Adomian, was used to find the approximate solutions for linear partial differential equations [126, 127]. Also, the decomposition method of local fractional operator was considered to find the nondifferentiable solutions of the local fractional partial differential equations [125, 128, 129].

When $L_{\varepsilon}^{(n)}$ in (5.1) is a local fractional differential operator of order $n\varepsilon$, we denote it as

$$L_{\varepsilon}^{(n)} = L_{\mu}^{(n\varepsilon)} = \frac{\partial^{n\varepsilon}}{\partial \mu^{n\varepsilon}} \tag{5.30}$$

and

$$R_{\varepsilon}\Phi(\mu) = \sum_{n=0} a_n \frac{\partial^{(n-1)\varepsilon}}{\partial \mu^{(n-1)\varepsilon}}. \tag{5.31}$$

Defining the n-fold local fractional integral operator

$$L_{\varepsilon}^{(-n)}\Phi(\mu) = {}_0I_{\mu}^{(n\varepsilon)}\Phi(\mu), \tag{5.32}$$

we conclude

$$L_{\varepsilon}^{(-n)}L_{\varepsilon}^{(n)}\Phi = L_{\varepsilon}^{(-n)}R_{\varepsilon}\Phi. \tag{5.33}$$

Therefore, (5.33) implies

$$\Phi(\mu) = I(\mu) + L_{\varepsilon}^{(-n)}R_{\varepsilon}\Phi(\mu), \tag{5.34}$$

where $I(\mu)$ is to be determined from the fractal initial conditions.

Hence, the iterative formula is expressed as

$$\Phi_n(\mu) = L_{\varepsilon}^{(-n)}R_{\varepsilon}\Phi_n(\mu), \tag{5.35}$$

where $\Phi_0(\mu) = I(\mu)$.

Thus, for $n \geq 0$, the following recurrence formula reads

$$\begin{cases} \Phi_n(\mu) = L_{\varepsilon}^{(-n)}R_{\varepsilon}\Phi_n(\mu), \\ \Phi_0(\mu) = I(\mu). \end{cases} \tag{5.36}$$

Finally, we obtain the following result:

$$\Phi(\mu) = \lim_{n\to\infty}\sum_{n=0}^{\infty}\Phi_n(\mu). \tag{5.37}$$

The next step is to consider the local fractional partial differential equation:

$$\frac{\partial^{2\varepsilon}\Phi(\mu,\eta)}{\partial \eta^{2\varepsilon}} = 2\frac{\partial^{2\varepsilon}\Phi(\mu,\eta)}{\partial \mu^{2\varepsilon}} - 1, \tag{5.38}$$

subject to the following initial-boundary conditions

$$\Phi(\mu,0) = E_{\varepsilon}(\mu^{\varepsilon}), \quad 0 \leq \mu \leq 1, \tag{5.39}$$

$$\frac{\partial^{\varepsilon}\Phi(\mu,0)}{\partial \mu^{\varepsilon}} = 0, \quad 0 \leq \mu \leq 1, \tag{5.40}$$

$$\Phi(l,\eta) = \Phi(0,\eta) = 0, \quad \eta > 0, \tag{5.41}$$

$$\frac{\partial^{\varepsilon}\Phi(l,\eta)}{\partial\mu^{\varepsilon}}=\frac{\partial^{\varepsilon}\Phi(0,\eta)}{\partial\mu^{\varepsilon}}=0,\quad \eta>0. \tag{5.42}$$

Next, we consider the iterative formula, namely,

$$\Phi_{n+1}(\mu,\eta)=L_{\eta}^{(-2\varepsilon)}\left(2\frac{\partial^{2\varepsilon}\Phi_{n}(\mu,\eta)}{\partial\mu^{2\alpha}}\right), \tag{5.43}$$

together with the initial value

$$\Phi_{0}(\mu,\eta)=E_{\varepsilon}\left(\mu^{\varepsilon}\right)+L_{\eta}^{(-2\varepsilon)}1=E_{\varepsilon}\left(\mu^{\varepsilon}\right)+\frac{\eta^{2\varepsilon}}{\Gamma(1+2\varepsilon)}. \tag{5.44}$$

Using (5.43) and (5.44), the approximations read as

$$\Phi_{1}(\mu,\eta)=L_{\eta}^{(-2\varepsilon)}\left(2\frac{\partial^{2\varepsilon}\Phi_{0}(\mu,\eta)}{\partial\mu^{2\alpha}}\right)=\frac{2\eta^{2\varepsilon}}{\Gamma(1+2\varepsilon)}E_{\varepsilon}\left(\mu^{\varepsilon}\right), \tag{5.45}$$

$$\Phi_{2}(\mu,\eta)=L_{\eta}^{(-2\varepsilon)}\left(2\frac{\partial^{2\varepsilon}\Phi_{1}(\mu,\eta)}{\partial\mu^{2\alpha}}\right)=\frac{4\eta^{4\varepsilon}}{\Gamma(1+4\varepsilon)}E_{\varepsilon}\left(\mu^{\varepsilon}\right), \tag{5.46}$$

$$\Phi_{3}(\mu,\eta)=L_{\eta}^{(-2\varepsilon)}\left(2\frac{\partial^{2\varepsilon}\Phi_{2}(\mu,\eta)}{\partial\mu^{2\alpha}}\right)=\frac{8\eta^{6\varepsilon}}{\Gamma(1+6\varepsilon)}E_{\varepsilon}\left(\mu^{\varepsilon}\right), \tag{5.47}$$

$$\Phi_{4}(\mu,\eta)=L_{\eta}^{(-2\varepsilon)}\left(2\frac{\partial^{2\varepsilon}\Phi_{3}(\mu,\eta)}{\partial\mu^{2\alpha}}\right)=\frac{16\eta^{8\varepsilon}}{\Gamma(1+8\varepsilon)}E_{\varepsilon}\left(\mu^{\varepsilon}\right), \tag{5.48}$$

$$\vdots$$

$$\Phi_{n}(\mu,\eta)=\frac{2^{n}\eta^{2n\varepsilon}}{\Gamma(1+2n\varepsilon)}E_{\varepsilon}\left(\mu^{\varepsilon}\right). \tag{5.49}$$

Finally, the solution containing the nondifferentiable terms is given by

$$\Phi(\mu,\eta)=\lim_{n\to\infty}\sum_{n=0}^{n}\Phi_{n}(\mu,\eta)=E_{\varepsilon}\left(\mu^{\varepsilon}\right)\sum_{n=0}^{n}\frac{2^{n}\eta^{2n\varepsilon}}{\Gamma(1+2n\varepsilon)}+\frac{\eta^{2\varepsilon}}{\Gamma(1+2\varepsilon)}. \tag{5.50}$$

Below, we consider the following local fractional partial differential equation:

$$\frac{\partial^{2\varepsilon}\Phi(\eta,\mu)}{\partial\mu^{2\varepsilon}}+\frac{\partial^{\varepsilon}\Phi(\eta,\mu)}{\partial\mu^{\varepsilon}}+\Phi(\eta,\mu)=\frac{\partial^{5\varepsilon}\Phi(\eta,\mu)}{\partial\eta^{5\varepsilon}}, \tag{5.51}$$

and its initial values read as

$$\Phi(0,\mu)=E_{\varepsilon}(\mu^{\varepsilon}), \tag{5.52}$$

$$\frac{\partial^{\varepsilon}\Phi(0,\mu)}{\partial\eta^{\varepsilon}}=0. \tag{5.53}$$

The corresponding local fractional iteration algorithms become

$$\Phi_{k+1}(\eta,\mu) = L_{\eta}^{(-5\varepsilon)}\left(\frac{\partial^{2\varepsilon}\Phi_k(\eta,\mu)}{\partial\mu^{2\varepsilon}} + \frac{\partial^{\varepsilon}\Phi_k(\eta,\mu)}{\partial\mu^{\varepsilon}} + \Phi_k(\eta,\mu)\right), \quad k \geq 0, \tag{5.54}$$

where

$$\Phi_0(\eta,\mu) = E_{\varepsilon}(\mu^{\varepsilon}). \tag{5.55}$$

The components of the algorithm are given below:

$$\Phi_0(\eta,\mu) = E_{\varepsilon}(\mu^{\varepsilon}), \tag{5.56}$$

$$\begin{aligned}\Phi_1(\eta,\mu) &= L_{\eta}^{(-5\varepsilon)}\left(\frac{\partial^{2\varepsilon}\Phi_0(\eta,\mu)}{\partial\mu^{2\varepsilon}} + \frac{\partial^{\varepsilon}\Phi_0(\eta,\mu)}{\partial\mu^{\varepsilon}} + \Phi_0(\eta,\mu)\right)\\ &= L_{\eta}^{(-5\varepsilon)}\left(3E_{\varepsilon}(\mu^{\varepsilon})\right)\\ &= \frac{3\eta^{5\varepsilon}}{\Gamma(1+5\varepsilon)}E_{\varepsilon}(\mu^{\varepsilon}),\end{aligned} \tag{5.57}$$

$$\begin{aligned}\Phi_2(\eta,\mu) &= L_{\eta}^{(-5\varepsilon)}\left(\frac{\partial^{2\varepsilon}\Phi_1(\eta,\mu)}{\partial\mu^{2\varepsilon}} + \frac{\partial^{\varepsilon}\Phi_1(\eta,\mu)}{\partial\mu^{\varepsilon}} + \Phi_1(\eta,\mu)\right)\\ &= L_{\eta}^{(-5\varepsilon)}\left(\frac{6\eta^{5\varepsilon}}{\Gamma(1+5\varepsilon)}E_{\varepsilon}(\mu^{\varepsilon})\right)\\ &= \frac{6\eta^{10\varepsilon}}{\Gamma(1+10\varepsilon)}E_{\varepsilon}(\mu^{\varepsilon}),\end{aligned} \tag{5.58}$$

$$\begin{aligned}\Phi_3(\eta,\mu) &= L_{\eta}^{(-5\varepsilon)}\left(\frac{\partial^{2\varepsilon}\Phi_2(\eta,\mu)}{\partial\mu^{2\varepsilon}} + \frac{\partial^{\varepsilon}\Phi_2(\eta,\mu)}{\partial\mu^{\varepsilon}} + \Phi_2(\eta,\mu)\right)\\ &= L_{\eta}^{(-5\varepsilon)}\left(\frac{9\eta^{10\varepsilon}}{\Gamma(1+10\varepsilon)}E_{\varepsilon}(\mu^{\varepsilon})\right)\\ &= \frac{9\eta^{15\varepsilon}}{\Gamma(1+15\varepsilon)}E_{\varepsilon}(\mu^{\varepsilon}),\end{aligned} \tag{5.59}$$

$$\vdots$$

$$\Phi_n(\eta,\mu) = \frac{3n\eta^{3n\varepsilon}}{\Gamma(1+3n\varepsilon)}E_{\varepsilon}(\mu^{\varepsilon}), \tag{5.60}$$

and so on.

Thus, the corresponding solution is given by

$$\Phi(\mu,\eta) = \lim_{n\to\infty}\sum_{n=0}^{n}\Phi_n(\mu,\eta) = E_{\varepsilon}\left(\mu^{\varepsilon}\right)\sum_{n=0}^{n}\frac{3n\eta^{3n\varepsilon}}{\Gamma(1+3n\varepsilon)}. \tag{5.61}$$

5.4 Coupling the Laplace transform with variational iteration method of the local fractional operator

In this section, we consider the idea of the local fractional Laplace variational iteration method [113, 114], which is coupled by the variational iteration method and Laplace transform of the local fractional operator.

Using the local fractional Laplace transform, we present the new iteration algorithm in the following form:

$$M\{\Phi_{n+1}(\mu)\} = M\{\Phi_n(\mu)\} + (-1)^k \frac{1}{s^{k\varepsilon}} M\{L_\varepsilon \Phi_n(\mu) + R_\varepsilon \Phi_n(\mu)\}, \tag{5.62}$$

where the initial value condition is indicated as

$$\frac{s^{(k-1)\varepsilon}\Phi(0) + s^{(k-2)\varepsilon}\Phi^{(\varepsilon)}(0) + \cdots + \Phi^{((k-1)\varepsilon)}(0)}{s^{k\varepsilon}} = 0. \tag{5.63}$$

Therefore, we report that

$$M\{\Phi\} = \lim_{n\to\infty} M\{\Phi_n\} \tag{5.64}$$

such that

$$\begin{aligned}\Phi_{n+1}(\mu) &= M^{-1}[M\{\Phi_{n+1}(\mu)\}] \\ &= M^{-1}[M\{\Phi_n(\mu)\}] + (-1)^k M^{-1}\left[\frac{1}{s^{k\varepsilon}} M\left\{\left(L_\varepsilon^{(n)}\Phi_n(\mu) + R_\varepsilon \Phi_n(\mu)\right)\right\}\right].\end{aligned} \tag{5.65}$$

Thus, we finally conclude

$$\Phi(\mu) = \lim_{n\to\infty} M^{-1}\{M\{\Phi_n(\mu)\}\}. \tag{5.66}$$

Next, we analyze the local fractional partial differential equation:

$$\frac{\partial^{3\varepsilon}\Phi(\eta,\mu)}{\partial\mu^{3\varepsilon}} + \Phi(\eta,\mu) = \frac{\partial^{2\varepsilon}\Phi(\eta,\mu)}{\partial\eta^{2\varepsilon}}, \tag{5.67}$$

subjected to the following initial value:

$$\frac{\partial^{\varepsilon}\Phi(0,\mu)}{\partial\eta^{\varepsilon}} = 0, \tag{5.68}$$

$$\Phi(0,\mu) = E_\varepsilon(-\mu^\varepsilon). \tag{5.69}$$

Below, we show the local fractional Laplace variational iteration algorithm, namely,

$$\begin{aligned}\Phi_{n+1}(\eta,\mu) &= M^{-1}[M\{\Phi_{n+1}(\eta,\mu)\}] \\ &= M^{-1}[M\{\Phi_n(\eta,\mu)\}] + M^{-1}\left[\frac{1}{s^{2\varepsilon}} M\left\{\frac{\partial^{2\varepsilon}\Phi_n(\eta,\mu)}{\partial\eta^{2\varepsilon}}\right.\right. \\ &\quad \left.\left. -\Phi_n(\eta,\mu) - \frac{\partial^{3\varepsilon}\Phi_n(\eta,\mu)}{\partial\mu^{3\varepsilon}}\right\}\right],\end{aligned} \tag{5.70}$$

where

$$\Phi_0(\eta,\mu) = E_\varepsilon(\mu^\varepsilon). \tag{5.71}$$

The corresponding approximations are given by

$$\begin{aligned}
\Phi_1(\eta,\mu) &= M^{-1}\left[M\left\{\Phi_1(\eta,\mu)\right\}\right] \\
&= M^{-1}\left[M\left\{\Phi_0(\eta,\mu)\right\}\right] + M^{-1}\left[\frac{1}{s^{2\varepsilon}}M\left\{\frac{\partial^{2\varepsilon}\Phi_0(\eta,\mu)}{\partial\eta^{2\varepsilon}}\right.\right. \\
&\quad \left.\left. -\Phi_0(\eta,\mu) - \frac{\partial^{3\varepsilon}\Phi_0(\eta,\mu)}{\partial\mu^{3\varepsilon}}\right\}\right] \\
&= E_\varepsilon(-\mu^\varepsilon) + M^{-1}\left[\frac{1}{s^{2\varepsilon}}M\left\{\frac{\partial^{3\varepsilon}E_\varepsilon(-\mu^\varepsilon)}{\partial\mu^{3\varepsilon}} + E_\varepsilon(-\mu^\varepsilon) - \frac{\partial^{2\varepsilon}E_\varepsilon(-\mu^\varepsilon)}{\partial\eta^{2\varepsilon}}\right\}\right] \\
&= E_\varepsilon(-\mu^\varepsilon) + M^{-1}\left[\frac{1}{s^{2\varepsilon}}M\left\{-E_\varepsilon(-\mu^\varepsilon) + E_\varepsilon(-\mu^\varepsilon) - 0\right\}\right] \\
&= E_\varepsilon(-\mu^\varepsilon),
\end{aligned} \tag{5.72}$$

$$\begin{aligned}
\Phi_2(\eta,\mu) &= M^{-1}\left[M\left\{\Phi_2(\eta,\mu)\right\}\right] \\
&= M^{-1}\left[M\left\{\Phi_1(\eta,\mu)\right\}\right] + M^{-1}\left[\frac{1}{s^{2\varepsilon}}M\left\{\frac{\partial^{3\varepsilon}\Phi_1(\eta,\mu)}{\partial\mu^{3\varepsilon}}\right.\right. \\
&\quad \left.\left. +\Phi_1(\eta,\mu) - \frac{\partial^{2\varepsilon}\Phi_1(\eta,\mu)}{\partial\eta^{2\varepsilon}}\right\}\right] \\
&= E_\varepsilon(-\mu^\varepsilon) + M^{-1}\left[\frac{1}{s^{2\varepsilon}}M\left\{-E_\varepsilon(-\mu^\varepsilon) + E_\varepsilon(-\mu^\varepsilon) - 0\right\}\right] \\
&= E_\varepsilon(-\mu^\varepsilon),
\end{aligned} \tag{5.73}$$

$$\begin{aligned}
\Phi_3(\eta,\mu) &= M^{-1}\left[M\left\{\Phi_3(\eta,\mu)\right\}\right] \\
&= M^{-1}\left[M\left\{\Phi_2(\eta,\mu)\right\}\right] + M^{-1}\left[\frac{1}{s^{2\varepsilon}}M\left\{\frac{\partial^{3\varepsilon}\Phi_2(\eta,\mu)}{\partial\mu^{3\varepsilon}}\right.\right. \\
&\quad \left.\left. +\Phi_2(\eta,\mu) - \frac{\partial^{2\varepsilon}\Phi_2(\eta,\mu)}{\partial\eta^{2\varepsilon}}\right\}\right] \\
&= E_\varepsilon(-\mu^\varepsilon) + M^{-1}\left[\frac{1}{s^{2\varepsilon}}M\left\{-E_\varepsilon(-\mu^\varepsilon) + E_\varepsilon(-\mu^\varepsilon) - 0\right\}\right] \\
&= E_\varepsilon(-\mu^\varepsilon),
\end{aligned} \tag{5.74}$$

$$\vdots$$

$$\begin{aligned}
\Phi_n(\eta,\mu) &= M^{-1}\left[M\left\{\Phi_n(\eta,\mu)\right\}\right] \\
&= E_\varepsilon(-\mu^\varepsilon).
\end{aligned} \tag{5.75}$$

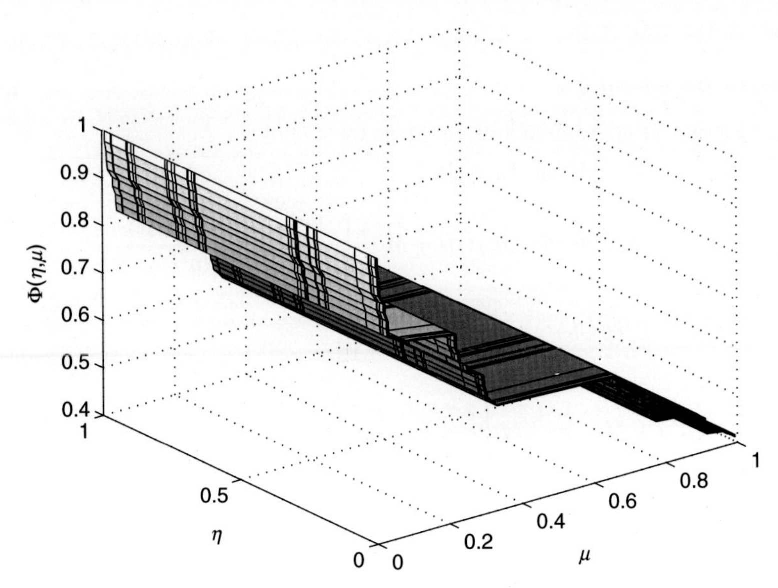

Figure 5.3 The plot of $\Phi(\eta,\mu)$ in fractal dimension $\varepsilon = \ln 2/\ln 3$.

As a result, we conclude that

$$\Phi(\eta,\mu) = \lim_{n\to\infty} M^{-1}\left\{M\left\{\Phi_n(\eta,\mu)\right\}\right\} = E_\varepsilon(-\mu^\varepsilon). \tag{5.76}$$

In Figure 5.3, we show the corresponding graph.

Let us analyze now the local fractional partial differential equation:

$$\frac{\partial^{3\varepsilon}\Phi(\eta,\mu)}{\partial\mu^{3\varepsilon}} - \frac{\partial^{2\varepsilon}\Phi(\eta,\mu)}{\partial\mu^{2\varepsilon}} = \frac{\partial^{3\varepsilon}\Phi(\eta,\mu)}{\partial\eta^{3\varepsilon}}, \tag{5.77}$$

equipped with the initial value

$$\frac{\partial^{2\varepsilon}\Phi(0,\mu)}{\partial\eta^{2\varepsilon}} = 0, \tag{5.78}$$

$$\frac{\partial^{\varepsilon}\Phi(0,\mu)}{\partial\eta^{\varepsilon}} = E_\varepsilon(\mu^\varepsilon), \tag{5.79}$$

$$\Phi(0,\mu) = 0. \tag{5.80}$$

In this case, we calculate the local fractional Laplace variational iteration algorithm, namely,

$$\begin{aligned}\Phi_{n+1}(\eta,\mu) &= M^{-1}\left[M\left\{\Phi_{n+1}(\eta,\mu)\right\}\right]\\ &= M^{-1}\left[M\left\{\Phi_n(\eta,\mu)\right\}\right] + M^{-1}\left[\frac{1}{s^{3\varepsilon}}M\left\{\frac{\partial^{3\varepsilon}\Phi_n(\eta,\mu)}{\partial\mu^{3\varepsilon}}\right.\right.\\ &\qquad \left.\left.-\frac{\partial^{2\varepsilon}\Phi_n(\eta,\mu)}{\partial\mu^{2\varepsilon}} - \frac{\partial^{3\varepsilon}\Phi_n(\eta,\mu)}{\partial\eta^{3\varepsilon}}\right\}\right],\end{aligned} \tag{5.81}$$

where

$$\Phi_0(\eta,\mu) = \frac{\eta^{\varepsilon}}{\Gamma(1+\varepsilon)} E_{\varepsilon}(\mu^{\varepsilon}). \tag{5.82}$$

Therefore, we obtain the following approximations:

$$\begin{aligned}
\Phi_1(\eta,\mu) &= M^{-1}[M\{\Phi_1(\eta,\mu)\}] \\
&= M^{-1}[M\{\Phi_0(\eta,\mu)\}] + M^{-1}\left[\frac{1}{s^{3\varepsilon}} M\left\{\frac{\partial^{3\varepsilon}\Phi_0(\eta,\mu)}{\partial\mu^{3\varepsilon}}\right.\right. \\
&\quad \left.\left. - \frac{\partial^{2\varepsilon}\Phi_0(\eta,\mu)}{\partial\mu^{2\varepsilon}} - \frac{\partial^{3\varepsilon}\Phi_0(\eta,\mu)}{\partial\eta^{3\varepsilon}}\right\}\right] \\
&= \frac{\eta^{\varepsilon}}{\Gamma(1+\varepsilon)} E_{\varepsilon}(\mu^{\varepsilon}),
\end{aligned} \tag{5.83}$$

$$\begin{aligned}
\Phi_2(\eta,\mu) &= M^{-1}[M\{\Phi_2(\eta,\mu)\}] \\
&= M^{-1}[M\{\Phi_1(\eta,\mu)\}] + M^{-1}\left[\frac{1}{s^{3\varepsilon}} M\left\{\frac{\partial^{3\varepsilon}\Phi_1(\eta,\mu)}{\partial\mu^{3\varepsilon}}\right.\right. \\
&\quad \left.\left. - \frac{\partial^{2\varepsilon}\Phi_1(\eta,\mu)}{\partial\mu^{2\varepsilon}} - \frac{\partial^{3\varepsilon}\Phi_1(\eta,\mu)}{\partial\eta^{3\varepsilon}}\right\}\right] \\
&= \frac{\eta^{\varepsilon}}{\Gamma(1+\varepsilon)} E_{\varepsilon}(\mu^{\varepsilon}),
\end{aligned} \tag{5.84}$$

$$\begin{aligned}
\Phi_3(\eta,\mu) &= M^{-1}[M\{\Phi_3(\eta,\mu)\}] \\
&= M^{-1}[M\{\Phi_2(\eta,\mu)\}] + M^{-1}\left[\frac{1}{s^{3\varepsilon}} M\left\{\frac{\partial^{3\varepsilon}\Phi_2(\eta,\mu)}{\partial\mu^{3\varepsilon}}\right.\right. \\
&\quad \left.\left. - \frac{\partial^{2\varepsilon}\Phi_2(\eta,\mu)}{\partial\mu^{2\varepsilon}} - \frac{\partial^{3\varepsilon}\Phi_2(\eta,\mu)}{\partial\eta^{3\varepsilon}}\right\}\right] \\
&= \frac{\eta^{\varepsilon}}{\Gamma(1+\varepsilon)} E_{\varepsilon}(\mu^{\varepsilon}),
\end{aligned} \tag{5.85}$$

$$\vdots$$

$$\begin{aligned}
\Phi_n(\eta,\mu) &= M^{-1}[M\{\Phi_n(\eta,\mu)\}] \\
&= \frac{\eta^{\varepsilon}}{\Gamma(1+\varepsilon)} E_{\varepsilon}(\mu^{\varepsilon}).
\end{aligned} \tag{5.86}$$

At this point, we conclude that

$$\Phi(\eta,\mu) = \lim_{n\to\infty} M^{-1}\{M\{\Phi_n(\eta,\mu)\}\} = \frac{\eta^{\varepsilon}}{\Gamma(1+\varepsilon)} E_{\varepsilon}(\mu^{\varepsilon}), \tag{5.87}$$

and we depict the corresponding graph in Figure 5.4.

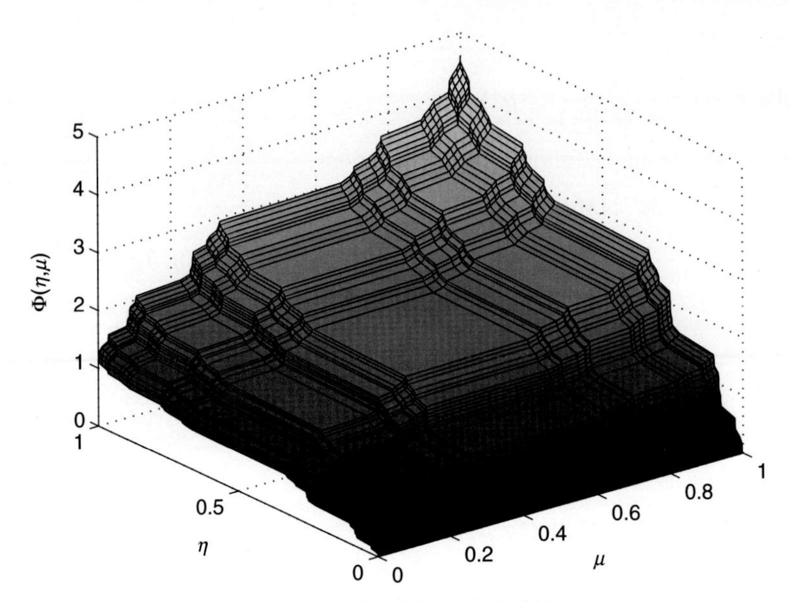

Figure 5.4 The plot of $\Phi(\eta, \mu)$ in fractal dimension $\varepsilon = \ln 2/\ln 3$.

5.5 Coupling the Laplace transform with decomposition method of the local fractional operator

In this section, we show the core of the local fractional Laplace decomposition method [88], which is coupled with the decomposition method and the Laplace transform of the local fractional operator.

Taking the local fractional Laplace transform in (5.1), we have

$$M\left\{L_{\varepsilon}^{(n)}\Phi(\mu,\eta)\right\} + M\left\{R_{\varepsilon}\Phi(\mu,\eta)\right\} = 0, \tag{5.88}$$

such that

$$M\{\Phi(\mu,\eta)\} = \sum_{k=1}^{n}\frac{1}{s^{k\varepsilon}}\Phi^{((k-1)\varepsilon)}(0,\eta) - M\{R_{\varepsilon}\Phi(\mu,\eta)\} - \frac{1}{s^{k\varepsilon}}M\{\Phi(\mu,\eta)\}. \tag{5.89}$$

Taking the inverse of the local fractional Laplace transform in (5.89), it gives

$$\Phi(\mu,\eta) = M^{-1}\left[\sum_{k=1}^{n}\frac{1}{s^{k\varepsilon}}\Phi^{((k-1)\varepsilon)}(0,\eta) - M\{R_{\varepsilon}\Phi(\mu,\eta)\} - \frac{1}{s^{k\varepsilon}}M\{\Phi(\mu,\eta)\}\right], \tag{5.90}$$

which leads to the local fractional recursive relation as

$$\Phi_{n+1}(\mu,\eta) = M^{-1}\left\{-\frac{1}{s^{k\varepsilon}}M\{\Phi(\mu,\eta)\}\right\}, \tag{5.91}$$

where

$$\Phi_0(\mu,\eta) = M^{-1}\left[\sum_{k=1}^{n}\frac{1}{s^{k\varepsilon}}\Phi^{((k-1)\varepsilon)}(0,\eta) - M\{R_\varepsilon\Phi(\mu,\eta)\}\right]. \tag{5.92}$$

Below, we investigate the local fractional partial differential equation

$$\frac{\partial^{2\varepsilon}\Phi(\eta,\mu)}{\partial\mu^{2\varepsilon}} + \frac{\partial^{\varepsilon}\Phi(\eta,\mu)}{\partial\mu^{\varepsilon}} = \frac{\partial^{3\varepsilon}\Phi(\eta,\mu)}{\partial\eta^{3\varepsilon}} \tag{5.93}$$

subjected to the initial values

$$\Phi(0,\mu) = 0, \quad \frac{\partial^{\varepsilon}\Phi(0,\mu)}{\partial\eta^{\varepsilon}} = E_\varepsilon(\mu^\varepsilon). \tag{5.94}$$

In our case, the local fractional iteration algorithms are constructed as

$$\Phi_{k+1}(\eta,\mu) = M^{-1}\left(\frac{1}{s^{3\varepsilon}}M\left\{\frac{\partial^{2\varepsilon}\Phi_k(\eta,\mu)}{\partial\mu^{2\varepsilon}} + \frac{\partial^{\varepsilon}\Phi_k(\eta,\mu)}{\partial\mu^{\varepsilon}}\right\}\right), \quad k \geq 0, \tag{5.95}$$

where

$$\Phi_0(\eta,\mu) = \frac{\eta^\varepsilon}{\Gamma(1+\varepsilon)}E_\varepsilon(\mu^\varepsilon). \tag{5.96}$$

Therefore, from (5.96), we present the components as

$$\begin{aligned}\Phi_1(\eta,\mu) &= M^{-1}\left(\frac{1}{s^{3\varepsilon}}M\left\{\frac{\partial^{2\varepsilon}\Phi_0(\eta,\mu)}{\partial\mu^{2\varepsilon}} + \frac{\partial^{\varepsilon}\Phi_0(\eta,\mu)}{\partial\mu^{\varepsilon}}\right\}\right)\\ &= M^{-1}\left(\frac{2}{s^{5\varepsilon}}E_\varepsilon\left(\mu^\varepsilon\right)\right)\\ &= \frac{2\eta^{4\varepsilon}}{\Gamma(1+4\varepsilon)}E_\varepsilon\left(\mu^\varepsilon\right),\end{aligned} \tag{5.97}$$

$$\begin{aligned}\Phi_2(\eta,\mu) &= M^{-1}\left(\frac{1}{s^{3\varepsilon}}M\left\{\frac{\partial^{2\varepsilon}\Phi_1(\eta,\mu)}{\partial\mu^{2\varepsilon}} + \frac{\partial^{\varepsilon}\Phi_1(\eta,\mu)}{\partial\mu^{\varepsilon}}\right\}\right)\\ &= M^{-1}\left(\frac{4}{s^{8\varepsilon}}E_\varepsilon\left(\mu^\varepsilon\right)\right)\\ &= \frac{4\eta^{7\varepsilon}}{\Gamma(1+7\varepsilon)}E_\varepsilon\left(\mu^\varepsilon\right),\end{aligned} \tag{5.98}$$

$$\begin{aligned}\Phi_3(\eta,\mu) &= M^{-1}\left(\frac{1}{s^{3\varepsilon}}M\left\{\frac{\partial^{2\varepsilon}\Phi_2(\eta,\mu)}{\partial\mu^{2\varepsilon}}+\frac{\partial^{\varepsilon}\Phi_2(\eta,\mu)}{\partial\mu^{\varepsilon}}\right\}\right)\\ &= M^{-1}\left(\frac{8}{s^{11\varepsilon}}E_\varepsilon(\mu^\varepsilon)\right)\\ &= \frac{8\eta^{10\varepsilon}}{\Gamma(1+10\varepsilon)}E_\varepsilon(\mu^\varepsilon),\end{aligned} \tag{5.99}$$

and so on.

As a result, there is

$$\begin{aligned}\Phi(\eta,\mu) &= \lim_{n\to\infty}\sum_{n=0}^{n}M^{-1}\left\{M\left\{\Phi_n(\eta,\mu)\right\}\right\}\\ &= \sum_{n=0}^{\infty}\frac{2^n\eta^{(1+3n)\varepsilon}}{\Gamma(1+(1+3n)\varepsilon)}E_\varepsilon\left(\mu^\varepsilon\right).\end{aligned} \tag{5.100}$$

The next step is to examine the local fractional partial differential equation

$$\frac{\partial^{\varepsilon}\Phi(\eta,\mu)}{\partial\mu^{\varepsilon}}+\Phi(\eta,\mu)=\frac{\partial^{4\varepsilon}\Phi(\eta,\mu)}{\partial\eta^{4\varepsilon}}, \tag{5.101}$$

in the presence of the following initial values:

$$\Phi(0,\mu)=E_\varepsilon(\mu^\varepsilon). \tag{5.102}$$

For this specific case, the local fractional iteration algorithms are written as

$$\Phi_{k+1}(\eta,\mu)=M^{-1}\left(\frac{1}{s^{4\varepsilon}}M\left\{\frac{\partial^{\varepsilon}\Phi_k(\eta,\mu)}{\partial\mu^{\varepsilon}}+\Phi_k(\eta,\mu)\right\}\right),\quad k\geq 0, \tag{5.103}$$

where

$$\Phi_0(\eta,\mu)=E_\varepsilon(\mu^\varepsilon). \tag{5.104}$$

Hence, we evaluate the components, namely,

$$\begin{aligned}\Phi_1(\eta,\mu) &= M^{-1}\left(\frac{1}{s^{4\varepsilon}}M\left\{\frac{\partial^{\varepsilon}\Phi_0(\eta,\mu)}{\partial\mu^{\varepsilon}}+\Phi_0(\eta,\mu)\right\}\right)\\ &= M^{-1}\left(\frac{2}{s^{5\varepsilon}}E_\varepsilon(\mu^\varepsilon)\right)\\ &= \frac{2\eta^{4\varepsilon}}{\Gamma(1+4\varepsilon)}E_\varepsilon(\mu^\varepsilon),\end{aligned} \tag{5.105}$$

$$\begin{aligned}\Phi_2(\eta,\mu) &= M^{-1}\left(\frac{1}{s^{4\varepsilon}}M\left\{\frac{\partial^{\varepsilon}\Phi_1(\eta,\mu)}{\partial\mu^{\varepsilon}}+\Phi_1(\eta,\mu)\right\}\right)\\ &= M^{-1}\left(\frac{4}{s^{9\varepsilon}}E_\varepsilon(\mu^\varepsilon)\right)\\ &= \frac{4\eta^{8\varepsilon}}{\Gamma(1+8\varepsilon)}E_\varepsilon(\mu^\varepsilon),\end{aligned} \tag{5.106}$$

$$\begin{aligned}\Phi_3(\eta,\mu) &= M^{-1}\left(\frac{1}{s^{4\varepsilon}}M\left\{\frac{\partial^{\varepsilon}\Phi_2(\eta,\mu)}{\partial\mu^{\varepsilon}}+\Phi_2(\eta,\mu)\right\}\right)\\ &= M^{-1}\left(\frac{8}{s^{13\varepsilon}}E_{\varepsilon}(\mu^{\varepsilon})\right)\\ &= \frac{8\eta^{12\varepsilon}}{\Gamma(1+12\varepsilon)}E_{\varepsilon}(\mu^{\varepsilon}),\end{aligned} \tag{5.107}$$

and so on.

Therefore, we report

$$\begin{aligned}\Phi(\eta,\mu) &= \lim_{n\to\infty}\sum_{n=0}^{n}M^{-1}\{M\{\Phi_n(\eta,\mu)\}\}\\ &= \sum_{n=0}^{\infty}\frac{2^n\eta^{(1+4n)\varepsilon}}{\Gamma(1+(1+4n)\varepsilon)}E_{\varepsilon}\left(\mu^{\varepsilon}\right).\end{aligned} \tag{5.108}$$

Now, we concentrate on solving the following local fractional partial differential equation:

$$\frac{\partial^{3\varepsilon}\Phi(\eta,\mu)}{\partial\mu^{3\varepsilon}}+\frac{\partial^{2\varepsilon}\Phi(\eta,\mu)}{\partial\mu^{2\varepsilon}}=\frac{\partial^{2\varepsilon}\Phi(\eta,\mu)}{\partial\eta^{2\varepsilon}}, \tag{5.109}$$

subjected to the initial values

$$\Phi(0,\mu)=0, \tag{5.110}$$

$$\frac{\partial^{\varepsilon}\Phi(0,\mu)}{\partial\eta^{\varepsilon}}=0, \tag{5.111}$$

$$\frac{\partial^{2\varepsilon}\Phi(0,\mu)}{\partial\eta^{2\varepsilon}}=E_{\varepsilon}(\mu^{\varepsilon}). \tag{5.112}$$

The local fractional iteration algorithms have the forms

$$\Phi_{k+1}(\eta,\mu)=M^{-1}\left(\frac{1}{s^{2\varepsilon}}M\left\{\frac{\partial^{3\varepsilon}\Phi_k(\eta,\mu)}{\partial\mu^{3\varepsilon}}+\frac{\partial^{2\varepsilon}\Phi_k(\eta,\mu)}{\partial\mu^{2\varepsilon}}\right\}\right),\quad k\geq 0, \tag{5.113}$$

where

$$\Phi_0(\eta,\mu)=\frac{\eta^{2\varepsilon}}{\Gamma(1+2\varepsilon)}E_{\varepsilon}(\mu^{\varepsilon}). \tag{5.114}$$

The components containing nondifferentiable terms are written as

$$\begin{aligned}\Phi_1(\eta,\mu) &= M^{-1}\left(\frac{1}{s^{2\varepsilon}}M\left\{\frac{\partial^{3\varepsilon}\Phi_0(\eta,\mu)}{\partial\mu^{3\varepsilon}}+\frac{\partial^{2\varepsilon}\Phi_0(\eta,\mu)}{\partial\mu^{2\varepsilon}}\right\}\right)\\ &= M^{-1}\left(\frac{2}{s^{5\varepsilon}}E_\varepsilon(\mu^\varepsilon)\right)\\ &= \frac{2\eta^{4\varepsilon}}{\Gamma(1+4\varepsilon)}E_\varepsilon(\mu^\varepsilon),\end{aligned} \tag{5.115}$$

$$\begin{aligned}\Phi_2(\eta,\mu) &= M^{-1}\left(\frac{1}{s^{2\varepsilon}}M\left\{\frac{\partial^{3\varepsilon}\Phi_1(\eta,\mu)}{\partial\mu^{3\varepsilon}}+\frac{\partial^{2\varepsilon}\Phi_1(\eta,\mu)}{\partial\mu^{2\varepsilon}}\right\}\right)\\ &= M^{-1}\left(\frac{4}{s^{7\varepsilon}}E_\varepsilon(\mu^\varepsilon)\right)\\ &= \frac{4\eta^{6\varepsilon}}{\Gamma(1+6\varepsilon)}E_\varepsilon(\mu^\varepsilon),\end{aligned} \tag{5.116}$$

$$\begin{aligned}\Phi_3(\eta,\mu) &= M^{-1}\left(\frac{1}{s^{2\varepsilon}}M\left\{\frac{\partial^{3\varepsilon}\Phi_2(\eta,\mu)}{\partial\mu^{3\varepsilon}}+\frac{\partial^{2\varepsilon}\Phi_2(\eta,\mu)}{\partial\mu^{2\varepsilon}}\right\}\right)\\ &= M^{-1}\left(\frac{6}{s^{9\varepsilon}}E_\varepsilon(\mu^\varepsilon)\right)\\ &= \frac{6\eta^{8\varepsilon}}{\Gamma(1+8\varepsilon)}E_\varepsilon(\mu^\varepsilon),\end{aligned} \tag{5.117}$$

and so on.

Thus, we finally conclude

$$\begin{aligned}\Phi(\eta,\mu) &= \lim_{n\to\infty}\sum_{n=0}^{n}M^{-1}\left\{M\left\{\Phi_n(\eta,\mu)\right\}\right\}\\ &= \sum_{n=0}^{\infty}\frac{2n\eta^{(2+2n)\varepsilon}}{\Gamma(1+(2+2n)\varepsilon)}E_\varepsilon\left(\mu^\varepsilon\right).\end{aligned} \tag{5.118}$$

Appendix A
The analogues of trigonometric functions defined on Cantor sets

In order to understand the first chapter, we present the analogues of the classical trigonometric functions now defined on Cantor sets. Here, we provide the proofs of the analogous trigonometric functions [1, 16, 21], namely,

$$\sin_\varepsilon^2\left(\mu^\varepsilon\right) = \frac{1 - \cos_\varepsilon\left(2\mu\right)^\varepsilon}{2}.$$

Proof.

$$\begin{aligned}
\sin_\varepsilon^2\left(\mu^\varepsilon\right) &= \left(\frac{E_\varepsilon\left(i^\varepsilon\mu^\varepsilon\right) - E_\varepsilon\left(-i^\varepsilon\mu^\varepsilon\right)}{2i^\varepsilon}\right)^2 \\
&= \frac{\left(E_\varepsilon\left(i^\varepsilon\mu^\varepsilon\right) - E_\varepsilon\left(-i^\varepsilon\mu^\varepsilon\right)\right)^2}{4i^{2\varepsilon}} \\
&= \frac{E_\varepsilon^2\left(i^\varepsilon\mu^\varepsilon\right) + E_\varepsilon^2\left(-i^\varepsilon\mu^\varepsilon\right) - 2}{4i^{2\varepsilon}} \\
&= \frac{E_\varepsilon\left(i^\varepsilon\left(2\mu\right)^\varepsilon\right) + E_\varepsilon\left(-i^\varepsilon\left(2\mu\right)^\varepsilon\right) - 2}{4i^{2\varepsilon}} \\
&= \frac{1 - \cos_\varepsilon\left(2\mu\right)^\varepsilon}{2}.
\end{aligned}$$

Thus, the proof is completed. □

$$\cos_\varepsilon^2\left(\mu^\varepsilon\right) = \frac{1 + \cos_\varepsilon\left(2\mu\right)^\varepsilon}{2}.$$

Proof.

$$\begin{aligned}\cos_\varepsilon^2\left(\mu^\varepsilon\right) &= \left(\frac{E_\varepsilon\left(i^\varepsilon\mu^\varepsilon\right)+E_\varepsilon\left(-i^\varepsilon\mu^\varepsilon\right)}{2}\right)^{22} \\ &= \frac{\left(E_\varepsilon\left(i^\varepsilon\mu^\varepsilon\right)+E_\varepsilon\left(-i^\varepsilon\mu^\varepsilon\right)\right)^2}{4i^{2\varepsilon}} \\ &= \frac{E_\varepsilon^2\left(i^\varepsilon\mu^\varepsilon\right)+E_\varepsilon^2\left(-i^\varepsilon\mu^\varepsilon\right)+2}{4} \\ &= \frac{E_\varepsilon\left(i^\varepsilon\left(2\mu\right)^\varepsilon\right)+E_\varepsilon\left(-i^\varepsilon\left(2\mu\right)^\varepsilon\right)+2}{4} \\ &= \frac{1+\cos_\varepsilon\left(2\mu\right)^\varepsilon}{2}.\end{aligned}$$

This completes the proofs. □

$$\cos_\varepsilon\left(2\mu\right)^\varepsilon = \cos_\varepsilon^2\left(\mu^\varepsilon\right)-\sin_\varepsilon^2\left(\mu^\varepsilon\right).$$

Proof.

$$\begin{aligned}\cos_\varepsilon^2\left(\mu^\varepsilon\right)-\sin_\varepsilon^2\left(\mu^\varepsilon\right) &= \frac{1+\cos_\varepsilon\left(2\mu\right)^\varepsilon}{2}-\frac{1-\cos_\varepsilon\left(2\mu\right)^\varepsilon}{2} \\ &= \cos_\varepsilon\left(2\mu\right)^\varepsilon.\end{aligned}$$

Hence, the proof is completed. □

$$\cos_\varepsilon^2\left(\mu^\varepsilon\right)+\sin_\varepsilon^2\left(\mu^\varepsilon\right) = 1.$$

Proof.

$$\begin{aligned}\cos_\varepsilon^2\left(\mu^\varepsilon\right)+\sin_\varepsilon^2\left(\mu^\varepsilon\right) &= \frac{1+\cos_\varepsilon\left(2\mu\right)^\varepsilon}{2}+\frac{1-\cos_\varepsilon\left(2\mu\right)^\varepsilon}{2} \\ &= 1.\end{aligned}$$

The proof is completed. □

$$\cos_\varepsilon\left(\mu^\varepsilon\right)\cos_\varepsilon\left(\eta^\varepsilon\right) = \frac{\cos_\varepsilon\left(\mu^\varepsilon+\eta^\varepsilon\right)+\cos_\varepsilon\left(\mu^\varepsilon-\eta^\varepsilon\right)}{2}.$$

Proof.

$$\begin{aligned}
\cos_{\varepsilon}\left(\mu^{\varepsilon}\right)\cos_{\varepsilon}\left(\eta^{\varepsilon}\right) &= \left(\frac{E_{\varepsilon}\left(i^{\varepsilon}\mu^{\varepsilon}\right)+E_{\varepsilon}\left(-i^{\varepsilon}\mu^{\varepsilon}\right)}{2}\right)\left(\frac{E_{\varepsilon}\left(i^{\varepsilon}\eta^{\varepsilon}\right)+E_{\varepsilon}\left(-i^{\varepsilon}\eta^{\varepsilon}\right)}{2}\right)\\
&= \frac{E_{\varepsilon}\left(i^{\varepsilon}\mu^{\varepsilon}+i^{\varepsilon}\eta^{\varepsilon}\right)+E_{\varepsilon}\left(i^{\varepsilon}\mu^{\varepsilon}-i^{\varepsilon}\eta^{\varepsilon}\right)+E_{\varepsilon}\left(i^{\varepsilon}\eta^{\varepsilon}-i^{\varepsilon}\mu^{\varepsilon}\right)+E_{\varepsilon}\left(-i^{\varepsilon}\mu^{\varepsilon}-i^{\varepsilon}\eta^{\varepsilon}\right)}{4}\\
&= \frac{\cos_{\varepsilon}\left(\mu^{\varepsilon}+\eta^{\varepsilon}\right)+i^{\varepsilon}\sin_{\varepsilon}\left(\mu^{\varepsilon}+\eta^{\varepsilon}\right)}{4}+\frac{\cos_{\varepsilon}\left(\mu^{\varepsilon}-\eta^{\varepsilon}\right)+i^{\varepsilon}\sin_{\varepsilon}\left(\mu^{\varepsilon}-\eta^{\varepsilon}\right)}{4}\\
&\quad+\frac{\cos_{\varepsilon}\left(\mu^{\varepsilon}-\eta^{\varepsilon}\right)-i^{\varepsilon}\sin_{\varepsilon}\left(\mu^{\varepsilon}-\eta^{\varepsilon}\right)}{4}+\frac{\cos_{\varepsilon}\left(\mu^{\varepsilon}+\eta^{\varepsilon}\right)-i^{\varepsilon}\sin_{\varepsilon}\left(\mu^{\varepsilon}+\eta^{\varepsilon}\right)}{4}\\
&= \frac{\cos_{\varepsilon}\left(\mu^{\varepsilon}+\eta^{\varepsilon}\right)+\cos_{\varepsilon}\left(\mu^{\varepsilon}-\eta^{\varepsilon}\right)}{2}
\end{aligned}$$

that completes the proof. □

$$\sin_{\varepsilon}\left(\mu^{\varepsilon}\right)\sin_{\varepsilon}\left(\eta^{\varepsilon}\right) = -\frac{\cos_{\varepsilon}\left(\mu^{\varepsilon}+\eta^{\varepsilon}\right)-\cos_{\varepsilon}\left(\mu^{\varepsilon}-\eta^{\varepsilon}\right)}{2}.$$

Proof.

$$\begin{aligned}
\sin_{\varepsilon}\left(\mu^{\varepsilon}\right)\sin_{\varepsilon}\left(\eta^{\varepsilon}\right) &= \left(\frac{E_{\varepsilon}\left(i^{\varepsilon}\mu^{\varepsilon}\right)-E_{\varepsilon}\left(-i^{\varepsilon}\mu^{\varepsilon}\right)}{2i^{\varepsilon}}\right)\left(\frac{E_{\varepsilon}\left(i^{\varepsilon}\eta^{\varepsilon}\right)-E_{\varepsilon}\left(-i^{\varepsilon}\eta^{\varepsilon}\right)}{2i^{\varepsilon}}\right)\\
&= -\frac{E_{\varepsilon}\left(i^{\varepsilon}\mu^{\varepsilon}+i^{\varepsilon}\eta^{\varepsilon}\right)-E_{\varepsilon}\left(i^{\varepsilon}\eta^{\varepsilon}-i^{\varepsilon}\mu^{\varepsilon}\right)-E_{\varepsilon}\left(i^{\varepsilon}\mu^{\varepsilon}-i^{\varepsilon}\eta^{\varepsilon}\right)+E_{\varepsilon}\left(-i^{\varepsilon}\mu^{\varepsilon}-i^{\varepsilon}\eta^{\varepsilon}\right)}{4}\\
&= -\frac{\cos_{\varepsilon}\left(\mu^{\varepsilon}+\eta^{\varepsilon}\right)+i^{\varepsilon}\sin_{\varepsilon}\left(\mu^{\varepsilon}+\eta^{\varepsilon}\right)}{4}+\frac{\cos_{\varepsilon}\left(\eta^{\varepsilon}-\mu^{\varepsilon}\right)+i^{\varepsilon}\sin_{\varepsilon}\left(\eta^{\varepsilon}-\mu^{\varepsilon}\right)}{4}\\
&\quad+\frac{\cos_{\varepsilon}\left(\mu^{\varepsilon}-\eta^{\varepsilon}\right)+i^{\varepsilon}\sin_{\varepsilon}\left(\mu^{\varepsilon}-\eta^{\varepsilon}\right)}{4}-\frac{\cos_{\varepsilon}\left(\mu^{\varepsilon}+\eta^{\varepsilon}\right)-i^{\varepsilon}\sin_{\varepsilon}\left(\mu^{\varepsilon}+\eta^{\varepsilon}\right)}{4}\\
&= -\frac{\cos_{\varepsilon}\left(\mu^{\varepsilon}+\eta^{\varepsilon}\right)+i^{\varepsilon}\sin_{\varepsilon}\left(\mu^{\varepsilon}+\eta^{\varepsilon}\right)}{4}+\frac{\cos_{\varepsilon}\left(\mu^{\varepsilon}-\eta^{\varepsilon}\right)-i^{\varepsilon}\sin_{\varepsilon}\left(\mu^{\varepsilon}-\eta^{\varepsilon}\right)}{4}\\
&\quad+\frac{\cos_{\varepsilon}\left(\mu^{\varepsilon}-\eta^{\varepsilon}\right)+i^{\varepsilon}\sin_{\varepsilon}\left(\mu^{\varepsilon}-\eta^{\varepsilon}\right)}{4}-\frac{\cos_{\varepsilon}\left(\mu^{\varepsilon}+\eta^{\varepsilon}\right)-i^{\varepsilon}\sin_{\varepsilon}\left(\mu^{\varepsilon}+\eta^{\varepsilon}\right)}{4}\\
&= -\frac{\cos_{\varepsilon}\left(\mu^{\varepsilon}+\eta^{\varepsilon}\right)-\cos_{\varepsilon}\left(\mu^{\varepsilon}-\eta^{\varepsilon}\right)}{2}.
\end{aligned}$$

This completes the proof. □

$$\sin_{\varepsilon}\left(\mu^{\varepsilon}\right)\cos_{\varepsilon}\left(\eta^{\varepsilon}\right) = \frac{\sin_{\varepsilon}\left(\mu^{\varepsilon}+\eta^{\varepsilon}\right)+\sin_{\varepsilon}\left(\mu^{\varepsilon}-\eta^{\varepsilon}\right)}{2}.$$

Proof.

$$\begin{aligned}
\sin_\varepsilon\left(\mu^\varepsilon\right)\cos_\varepsilon\left(\eta^\varepsilon\right) &= \left(\frac{E_\varepsilon\left(i^\varepsilon\mu^\varepsilon\right)-E_\varepsilon\left(-i^\varepsilon\mu^\varepsilon\right)}{2i^\varepsilon}\right)\left(\frac{E_\varepsilon\left(i^\varepsilon\eta^\varepsilon\right)+E_\varepsilon\left(-i^\varepsilon\eta^\varepsilon\right)}{2}\right)\\
&= \frac{E_\varepsilon\left(i^\varepsilon\mu^\varepsilon+i^\varepsilon\eta^\varepsilon\right)+E_\varepsilon\left(i^\varepsilon\mu^\varepsilon-i^\varepsilon\eta^\varepsilon\right)-E_\varepsilon\left(-i^\varepsilon\mu^\varepsilon+i^\varepsilon\eta^\varepsilon\right)-E_\varepsilon\left(-i^\varepsilon\mu^\varepsilon-i^\varepsilon\eta^\varepsilon\right)}{4i^\varepsilon}\\
&= \frac{\cos_\varepsilon\left(\mu^\varepsilon+\eta^\varepsilon\right)+i^\varepsilon\sin_\varepsilon\left(\mu^\varepsilon+\eta^\varepsilon\right)}{4i^\varepsilon}+\frac{\cos_\varepsilon\left(\mu^\varepsilon-\eta^\varepsilon\right)+i^\varepsilon\sin_\varepsilon\left(\mu^\varepsilon-\eta^\varepsilon\right)}{4i^\varepsilon}\\
&\quad -\frac{\cos_\varepsilon\left(\mu^\varepsilon-\eta^\varepsilon\right)-i^\varepsilon\sin_\varepsilon\left(\mu^\varepsilon-\eta^\varepsilon\right)}{4i^\varepsilon}-\frac{\cos_\varepsilon\left(\mu^\varepsilon+\eta^\varepsilon\right)-i^\varepsilon\sin_\varepsilon\left(\mu^\varepsilon+\eta^\varepsilon\right)}{4i^\varepsilon}\\
&= \frac{\sin_\varepsilon\left(\mu^\varepsilon+\eta^\varepsilon\right)+\sin_\varepsilon\left(\mu^\varepsilon-\eta^\varepsilon\right)}{2}.
\end{aligned}$$

Hence, it completes the proof. □

$$\cos_\varepsilon\left(\mu^\varepsilon\right)\sin_\varepsilon\left(\eta^\varepsilon\right)=\frac{\sin_\varepsilon\left(\mu^\varepsilon+\eta^\varepsilon\right)-\sin_\varepsilon\left(\mu^\varepsilon-\eta^\varepsilon\right)}{2}.$$

Proof.

$$\begin{aligned}
\cos_\varepsilon\left(\mu^\varepsilon\right)\sin_\varepsilon\left(\eta^\varepsilon\right) &= \left(\frac{E_\varepsilon\left(i^\varepsilon\mu^\varepsilon\right)+E_\varepsilon\left(-i^\varepsilon\mu^\varepsilon\right)}{2}\right)\left(\frac{E_\varepsilon\left(i^\varepsilon\eta^\varepsilon\right)-E_\varepsilon\left(-i^\varepsilon\eta^\varepsilon\right)}{2i^\varepsilon}\right)\\
&= \frac{E_\varepsilon\left(i^\varepsilon\mu^\varepsilon+i^\varepsilon\eta^\varepsilon\right)+E_\varepsilon\left(i^\varepsilon\eta^\varepsilon-i^\varepsilon\mu^\varepsilon\right)-E_\varepsilon\left(i^\varepsilon\mu^\varepsilon-i^\varepsilon\eta^\varepsilon\right)-E_\varepsilon\left(-i^\varepsilon\mu^\varepsilon-i^\varepsilon\eta^\varepsilon\right)}{4i^\varepsilon}\\
&= \frac{\cos_\varepsilon\left(\mu^\varepsilon+\eta^\varepsilon\right)+i^\varepsilon\sin_\varepsilon\left(\mu^\varepsilon+\eta^\varepsilon\right)}{4i^\varepsilon}+\frac{\cos_\varepsilon\left(\mu^\varepsilon-\eta^\varepsilon\right)-i^\varepsilon\sin_\varepsilon\left(\mu^\varepsilon-\eta^\varepsilon\right)}{4i^\varepsilon}\\
&\quad -\frac{\cos_\varepsilon\left(\mu^\varepsilon-\eta^\varepsilon\right)+i^\varepsilon\sin_\varepsilon\left(\mu^\varepsilon-\eta^\varepsilon\right)}{4i^\varepsilon}-\frac{\cos_\varepsilon\left(\mu^\varepsilon+\eta^\varepsilon\right)-i^\varepsilon\sin_\varepsilon\left(\mu^\varepsilon+\eta^\varepsilon\right)}{4i^\varepsilon}\\
&= \frac{\sin_\varepsilon\left(\mu^\varepsilon+\eta^\varepsilon\right)-\sin_\varepsilon\left(\mu^\varepsilon-\eta^\varepsilon\right)}{2}.
\end{aligned}$$

Hence, it completes the proof. □

$$\sin_\varepsilon\left(\mu^\varepsilon+\eta^\varepsilon\right)=\sin_\varepsilon\left(\mu^\varepsilon\right)\cos_\varepsilon\left(\eta^\varepsilon\right)+\cos_\varepsilon\left(\mu^\varepsilon\right)\sin_\varepsilon\left(\eta^\varepsilon\right).$$

Proof.

$$\begin{aligned}
\sin_\varepsilon\left(\mu^\varepsilon+\eta^\varepsilon\right) &= \frac{\sin_\varepsilon\left(\mu^\varepsilon+\eta^\varepsilon\right)-\sin_\varepsilon\left(\mu^\varepsilon-\eta^\varepsilon\right)}{2}\\
&\quad +\frac{\sin_\varepsilon\left(\mu^\varepsilon+\eta^\varepsilon\right)+\sin_\varepsilon\left(\mu^\varepsilon-\eta^\varepsilon\right)}{2}\\
&= \sin_\varepsilon\left(\mu^\varepsilon\right)\cos_\varepsilon\left(\eta^\varepsilon\right)+\cos_\varepsilon\left(\mu^\varepsilon\right)\sin_\varepsilon\left(\eta^\varepsilon\right).
\end{aligned}$$

Hence, it completes the proof. □

$$\sin_\varepsilon\left(\mu^\varepsilon-\eta^\varepsilon\right)=\sin_\varepsilon\left(\mu^\varepsilon\right)\cos_\varepsilon\left(\eta^\varepsilon\right)+\cos_\varepsilon\left(\mu^\varepsilon\right)\sin_\varepsilon\left(\eta^\varepsilon\right).$$

Proof.

$$\sin_\varepsilon\left(\mu^\varepsilon-\eta^\varepsilon\right)=\frac{\sin_\varepsilon\left(\mu^\varepsilon+\eta^\varepsilon\right)+\sin_\varepsilon\left(\mu^\varepsilon-\eta^\varepsilon\right)}{2}-\frac{\sin_\varepsilon\left(\mu^\varepsilon+\eta^\varepsilon\right)-\sin_\varepsilon\left(\mu^\varepsilon-\eta^\varepsilon\right)}{2}=\sin_\varepsilon\left(\mu^\varepsilon\right)\cos_\varepsilon\left(\eta^\varepsilon\right)-\cos_\varepsilon\left(\mu^\varepsilon\right)\sin_\varepsilon\left(\eta^\varepsilon\right).$$

Hence, completing the proof, we get □

$$\cos_\varepsilon\left(\mu^\varepsilon+\eta^\varepsilon\right)=\cos_\varepsilon\left(\mu^\varepsilon\right)\cos_\varepsilon\left(\eta^\varepsilon\right)-\sin_\varepsilon\left(\mu^\varepsilon\right)\sin_\varepsilon\left(\eta^\varepsilon\right).$$

Proof.

$$\cos_\varepsilon\left(\mu^\varepsilon+\eta^\varepsilon\right)=\frac{\cos_\varepsilon\left(\mu^\varepsilon+\eta^\varepsilon\right)+\cos_\varepsilon\left(\mu^\varepsilon-\eta^\varepsilon\right)}{2}+\frac{\cos_\varepsilon\left(\mu^\varepsilon+\eta^\varepsilon\right)-\cos_\varepsilon\left(\mu^\varepsilon-\eta^\varepsilon\right)}{2}=\cos_\varepsilon\left(\mu^\varepsilon\right)\cos_\varepsilon\left(\eta^\varepsilon\right)-\sin_\varepsilon\left(\mu^\varepsilon\right)\sin_\varepsilon\left(\eta^\varepsilon\right)$$

and this proof is completed. □

$$\cos_\varepsilon\left(\mu^\varepsilon-\eta^\varepsilon\right)=\cos_\varepsilon\left(\mu^\varepsilon\right)\cos_\varepsilon\left(\eta^\varepsilon\right)+\sin_\varepsilon\left(\mu^\varepsilon\right)\sin_\varepsilon\left(\eta^\varepsilon\right).$$

Proof.

$$\cos_\varepsilon\left(\mu^\varepsilon-\eta^\varepsilon\right)=\frac{\cos_\varepsilon\left(\mu^\varepsilon+\eta^\varepsilon\right)+\cos_\varepsilon\left(\mu^\varepsilon-\eta^\varepsilon\right)}{2}-\frac{\cos_\varepsilon\left(\mu^\varepsilon+\eta^\varepsilon\right)-\cos_\varepsilon\left(\mu^\varepsilon-\eta^\varepsilon\right)}{2}=\cos_\varepsilon\left(\mu^\varepsilon\right)\cos_\varepsilon\left(\eta^\varepsilon\right)+\sin_\varepsilon\left(\mu^\varepsilon\right)\sin_\varepsilon\left(\eta^\varepsilon\right).$$

This proof is completed. □

$$\sin_\varepsilon\left(\mu^\varepsilon\right)+\sin_\varepsilon\left(\mu^\varepsilon\right)=2\sin_\varepsilon\left(\left(\frac{\mu+\eta}{2}\right)^\varepsilon\right)\cos_\varepsilon\left(\left(\frac{\mu-\eta}{2}\right)^\varepsilon\right)$$

Proof.

$$2\sin_\varepsilon\left[\left(\frac{\mu+\eta}{2}\right)^\varepsilon\right]\cos_\varepsilon\left[\left(\frac{\mu-\eta}{2}\right)^\varepsilon\right]=\sin_\varepsilon\left[\left(\frac{\mu+\eta}{2}\right)^\varepsilon+\left(\frac{\mu-\eta}{2}\right)^\varepsilon\right]+\sin_\varepsilon\left[\left(\frac{\mu+\eta}{2}\right)^\varepsilon-\left(\frac{\mu-\eta}{2}\right)^\varepsilon\right]=\sin_\varepsilon\left(\mu^\varepsilon\right)+\sin_\varepsilon\left(\eta^\varepsilon\right)$$

This proof is completed. □

$$\sin_\varepsilon\left(\mu^\varepsilon\right)-\sin_\varepsilon\left(\mu^\varepsilon\right)=2\cos_\varepsilon\left[\left(\frac{\mu+\eta}{2}\right)^\varepsilon\right]\sin_\varepsilon\left[\left(\frac{\mu-\eta}{2}\right)^\varepsilon\right].$$

Proof.

$$\begin{aligned}2\cos_\varepsilon\left[\left(\frac{\mu+\eta}{2}\right)^\varepsilon\right]\sin_\varepsilon\left[\left(\frac{\mu-\eta}{2}\right)^\varepsilon\right]&=\sin_\varepsilon\left[\left(\frac{\mu+\eta}{2}\right)^\varepsilon+\left(\frac{\mu-\eta}{2}\right)^\varepsilon\right]\\&\quad-\sin_\varepsilon\left[\left(\frac{\mu+\eta}{2}\right)^\varepsilon-\left(\frac{\mu-\eta}{2}\right)^\varepsilon\right]\\&=\sin_\varepsilon\left(\mu^\varepsilon\right)+\sin_\varepsilon\left(\eta^\varepsilon\right).\end{aligned}$$

This proof is completed. □

$$\cos_\varepsilon\left(\mu^\varepsilon\right)+\cos_\varepsilon\left(\mu^\varepsilon\right)=2\cos_\varepsilon\left(\left(\frac{\mu+\eta}{2}\right)^\varepsilon\right)\cos_\varepsilon\left(\left(\frac{\mu-\eta}{2}\right)^\varepsilon\right).$$

Proof.

$$\begin{aligned}2\cos_\varepsilon\left(\left(\frac{\mu+\eta}{2}\right)^\varepsilon\right)\cos_\varepsilon\left(\left(\frac{\mu-\eta}{2}\right)^\varepsilon\right)&=\cos_\varepsilon\left[\left(\frac{\mu+\eta}{2}\right)^\varepsilon+\left(\frac{\mu-\eta}{2}\right)^\varepsilon\right]\\&\quad+\cos_\varepsilon\left[\left(\frac{\mu+\eta}{2}\right)^\varepsilon-\left(\frac{\mu-\eta}{2}\right)^\varepsilon\right]\\&=\cos_\varepsilon\left(\mu^\varepsilon\right)+\cos_\varepsilon\left(\eta^\varepsilon\right).\end{aligned}$$

This proof is completed. □

$$\cos_\varepsilon\left(\mu^\varepsilon\right)-\cos_\varepsilon\left(\mu^\varepsilon\right)=-2\sin_\varepsilon\left[\left(\frac{\mu+\eta}{2}\right)^\varepsilon\right]\sin_\varepsilon\left[\left(\frac{\mu-\eta}{2}\right)^\varepsilon\right].$$

Proof.

$$\begin{aligned}-2\sin_\varepsilon\left[\left(\frac{\mu+\eta}{2}\right)^\varepsilon\right]\sin_\varepsilon\left[\left(\frac{\mu-\eta}{2}\right)^\varepsilon\right]&=\frac{\cos_\varepsilon\left(\mu^\varepsilon+\eta^\varepsilon\right)-\cos_\varepsilon\left(\mu^\varepsilon-\eta^\varepsilon\right)}{2}\\&=\cos_\varepsilon\left[\left(\frac{\mu+\eta}{2}\right)^\varepsilon+\left(\frac{\mu-\eta}{2}\right)^\varepsilon\right]\\&\quad-\cos_\varepsilon\left[\left(\frac{\mu+\eta}{2}\right)^\varepsilon-\left(\frac{\mu-\eta}{2}\right)^\varepsilon\right]\\&=\cos_\varepsilon\left(\mu^\varepsilon\right)-\cos_\varepsilon\left(\mu^\varepsilon\right).\end{aligned}$$

This completes the proof. □

$$\left[E_\varepsilon\left(i^\varepsilon\mu^\varepsilon\right)\right]^k=\cos_\varepsilon\left[(k\mu)^\varepsilon\right]+i^\varepsilon\sin_\varepsilon\left[(k\mu)^\varepsilon\right].$$

Proof. Let $k = 0$; it is right.
Let $k = 1$; it is right.
When $k = n$, we suppose that

$$\left[E_\varepsilon\left(i^\varepsilon \mu^\varepsilon\right)\right]^k = \left(\cos_\varepsilon\left[(k\mu)^\varepsilon\right] + i^\varepsilon \sin_\varepsilon\left[(k\mu)^\varepsilon\right]\right).$$

When $k = n + 1$, we present that

$$\begin{aligned}
\left[E_\varepsilon\left(i^\varepsilon \mu^\varepsilon\right)\right]^{k+1} &= \left(\cos_\varepsilon\left[(k\mu)^\varepsilon\right] + i^\varepsilon \sin_\varepsilon\left[(k\mu)^\varepsilon\right]\right)\left(\cos_\varepsilon\left(\mu^\varepsilon\right) + i^\varepsilon \sin_\varepsilon\left(\mu^\varepsilon\right)\right) \\
&= \cos_\varepsilon\left[(k\mu)^\varepsilon\right]\cos_\varepsilon\left(\mu^\varepsilon\right) + i^\varepsilon \cos_\varepsilon\left[(k\mu)^\varepsilon\right]\sin_\varepsilon\left(\mu^\varepsilon\right) \\
&\quad + i^\varepsilon \sin_\varepsilon\left[(k\mu)^\varepsilon\right]\cos_\varepsilon\left(\mu^\varepsilon\right) - \sin_\varepsilon\left[(k\mu)^\varepsilon\right]\sin_\varepsilon\left(\mu^\varepsilon\right) \\
&= \frac{\cos_\varepsilon\left[(k\mu)^\varepsilon + \mu^\varepsilon\right] + \cos_\varepsilon\left[(k\mu)^\varepsilon - \mu^\varepsilon\right]}{2} \\
&\quad + i^\varepsilon \frac{\sin_\varepsilon\left[(k\mu)^\varepsilon + \mu^\varepsilon\right] - \sin_\varepsilon\left[(k\mu)^\varepsilon - \mu^\varepsilon\right]}{2} \\
&\quad + i^\varepsilon \frac{\sin_\varepsilon\left[(k\mu)^\varepsilon + \mu^\varepsilon\right] + \sin_\varepsilon\left[(k\mu)^\varepsilon - \mu^\varepsilon\right]}{2} \\
&\quad + \frac{\cos_\varepsilon\left[(k\mu)^\varepsilon + \mu^\varepsilon\right] - \cos_\varepsilon\left[(k\mu)^\varepsilon - \mu^\varepsilon\right]}{2} \\
&= \cos_\varepsilon\left[(k\mu)^\varepsilon + \mu^\varepsilon\right] + i^\varepsilon \sin_\varepsilon\left[(k\mu)^\varepsilon + \mu^\varepsilon\right] \\
&= \cos_\varepsilon\left[(k+1)^\varepsilon \mu^\varepsilon\right] + i^\varepsilon \sin_\varepsilon\left[(k+1)^\varepsilon \mu^\varepsilon\right].
\end{aligned}$$

Hence, we completed the proof. □

In this case, we use the following formula:

$$\left[E_\varepsilon\left(i^\varepsilon \mu^\varepsilon\right)\right]^k = \left(\cos_\varepsilon\left[(k\mu)^\varepsilon\right] + i^\varepsilon \sin_\varepsilon\left[(k\mu)^\varepsilon\right]\right) = E_\varepsilon\left[i^\varepsilon (k\mu)^\varepsilon\right].$$

Let us define

$$\theta^\varepsilon = \sum_{n=1}^{k} \cos_\varepsilon (n\mu)^\varepsilon$$

and

$$\vartheta^\varepsilon = \sum_{n=1}^{k} \sin_\varepsilon (n\mu)^\varepsilon.$$

Then, we may structure a function

$$\begin{aligned}
\Theta &= \theta^\varepsilon + i^\varepsilon \vartheta^\varepsilon \\
&= \sum_{n=1}^{k} E_\varepsilon\left[i^\varepsilon (n\mu)^\varepsilon\right].
\end{aligned}$$

This allows to obtain

$$
\begin{aligned}
E_\varepsilon\left(i^\varepsilon \mu^\varepsilon\right)\Theta &= E_\varepsilon\left(i^\varepsilon \mu^\varepsilon\right)\left(\theta^\varepsilon + i^\varepsilon \vartheta^\varepsilon\right) \\
&= E_\varepsilon\left(i^\varepsilon \mu^\varepsilon\right)\left\{\sum_{n=1}^{k} E_\varepsilon\left[i^\varepsilon (n\mu)^\varepsilon\right]\right\} \\
&= \sum_{n=2}^{k+1} E_\varepsilon\left[i^\varepsilon (n\mu)^\varepsilon\right].
\end{aligned}
$$

Therefore, we have

$$
\Theta\left(E_\varepsilon\left(i^\varepsilon \mu^\varepsilon\right) - 1\right) = E_\varepsilon\left[i^\varepsilon\left((k+1)\mu\right)^\varepsilon\right] - E_\varepsilon\left(i^\varepsilon \mu^\varepsilon\right),
$$

which leads to

$$
\begin{aligned}
\Theta &= \theta^\varepsilon + i^\varepsilon \vartheta^\varepsilon \\
&= \frac{E_\varepsilon\left[i^\varepsilon\left((k+1)\mu\right)^\varepsilon\right] - E_\varepsilon\left(i^\varepsilon \mu^\varepsilon\right)}{\left(E_\varepsilon\left(i^\varepsilon \mu^\varepsilon\right) - 1\right)} \\
&= \frac{E_\varepsilon\left(-i^\varepsilon\left(\frac{\mu}{2}\right)^\varepsilon\right) E_\varepsilon\left[i^\varepsilon\left((k+1)\mu\right)^\varepsilon\right] - E_\varepsilon\left(i^\varepsilon \mu^\varepsilon\right)}{E_\varepsilon\left(-i^\varepsilon\left(\frac{\mu}{2}\right)^\varepsilon\right)\left(E_\varepsilon\left(i^\varepsilon \mu^\varepsilon\right) - 1\right)} \\
&= \frac{E_\varepsilon\left[i^\varepsilon\left(\left(k+\frac{1}{2}\right)\mu\right)^\varepsilon\right] - E_\varepsilon\left(i^\varepsilon\left(\frac{\mu}{2}\right)^\varepsilon\right)}{E_\varepsilon\left[i^\varepsilon\left(\frac{\mu}{2}\right)^\varepsilon\right] - E_\varepsilon\left[-i^\varepsilon\left(\frac{\mu}{2}\right)^\varepsilon\right]} \\
&= E_\varepsilon\left[i^\varepsilon\left(\left(\frac{k+1}{2}\right)\mu\right)^\varepsilon\right]\left\{\frac{E_\varepsilon\left[i^\varepsilon\left(\frac{k\mu}{2}\right)^\varepsilon\right] - E_\varepsilon\left(-i^\varepsilon\left(\frac{k\mu}{2}\right)^\varepsilon\right)}{E_\varepsilon\left[i^\varepsilon\left(\frac{\mu}{2}\right)^\varepsilon\right] - E_\varepsilon\left[-i^\varepsilon\left(\frac{\mu}{2}\right)^\varepsilon\right]}\right\} \\
&= E_\varepsilon\left[i^\varepsilon\left(\left(\frac{k+1}{2}\right)\mu\right)^\varepsilon\right]\left\{\frac{\frac{E_\varepsilon\left[i^\varepsilon\left(\frac{k\mu}{2}\right)^\varepsilon\right] - E_\varepsilon\left(-i^\varepsilon\left(\frac{k\mu}{2}\right)^\varepsilon\right)}{2i^\varepsilon}}{\frac{E_\varepsilon\left[i^\varepsilon\left(\frac{\mu}{2}\right)^\varepsilon\right] - E_\varepsilon\left[-i^\varepsilon\left(\frac{\mu}{2}\right)^\varepsilon\right]}{2i^\varepsilon}}\right\} \\
&= E_\varepsilon\left[i^\varepsilon\left(\left(\frac{k+1}{2}\right)\mu\right)^\varepsilon\right]\left\{\frac{\sin_\varepsilon\left[\left(\frac{k\mu}{2}\right)^\varepsilon\right]}{\sin_\varepsilon\left[\left(\frac{\mu}{2}\right)^\varepsilon\right]}\right\} \\
&= \left\{\frac{\sin_\varepsilon\left[\left(\frac{k\mu}{2}\right)^\varepsilon\right]}{\sin_\varepsilon\left[\left(\frac{\mu}{2}\right)^\varepsilon\right]}\right\}\left\{\cos_\varepsilon\left[\left(\left(\frac{k+1}{2}\right)\mu\right)^\varepsilon\right] + i^\varepsilon \sin_\varepsilon\left[\left(\left(\frac{k+1}{2}\right)\mu\right)^\varepsilon\right]\right\}.
\end{aligned}
$$

Consequently, we obtain

$$
\theta^\varepsilon = \sum_{n=1}^{k} \cos_\varepsilon (n\mu)^\varepsilon = \left\{\frac{\sin_\varepsilon\left[\left(\frac{k\mu}{2}\right)^\varepsilon\right]}{\sin_\varepsilon\left[\left(\frac{\mu}{2}\right)^\varepsilon\right]}\right\} \cos_\varepsilon\left[\left(\left(\frac{k+1}{2}\right)\mu\right)^\varepsilon\right]
$$

and

$$\vartheta^{\varepsilon} = \sum_{n=1}^{k} \sin_{\varepsilon} (n\mu)^{\varepsilon} = \left\{ \frac{\sin_{\varepsilon}\left[\left(\frac{k\mu}{2}\right)^{\varepsilon}\right]}{\sin_{\varepsilon}\left[\left(\frac{\mu}{2}\right)^{\varepsilon}\right]} \right\} \sin_{\varepsilon}\left[\left(\left(\frac{k+1}{2}\right)\mu\right)^{\varepsilon}\right],$$

where

$$\sin_{\varepsilon}\left[\left(\frac{\mu}{2}\right)^{\varepsilon}\right] \neq 0.$$

Now, we can present the following formula as

$$\begin{aligned}\frac{1}{2} + \sum_{n=1}^{k} \cos_{\varepsilon} (n\mu)^{\varepsilon} &= \frac{1}{2} + \left\{ \frac{\sin_{\varepsilon}\left[\left(\frac{k\mu}{2}\right)^{\varepsilon}\right]}{\sin_{\varepsilon}\left[\left(\frac{\mu}{2}\right)^{\varepsilon}\right]} \right\} \cos_{\varepsilon}\left[\left(\left(\frac{k+1}{2}\right)\mu\right)^{\varepsilon}\right] \\ &= \frac{\sin_{\varepsilon}\left[\left(\frac{\mu}{2}\right)^{\varepsilon}\right] + 2\sin_{\varepsilon}\left[\left(\frac{k\mu}{2}\right)^{\varepsilon}\right]\cos_{\varepsilon}\left[\left(\left(\frac{k+1}{2}\right)\mu\right)^{\varepsilon}\right]}{2\sin_{\varepsilon}\left[\left(\frac{\mu}{2}\right)^{\varepsilon}\right]} \\ &= \frac{\sin_{\varepsilon}\left[\left(\frac{\mu}{2}\right)^{\varepsilon}\right] + \sin_{\varepsilon}\left[\left(\frac{k\mu}{2}\right)^{\varepsilon}\right]\cos_{\varepsilon}\left[\left(\left(\frac{k+1}{2}\right)\mu\right)^{\varepsilon}\right]}{2\sin_{\varepsilon}\left[\left(\frac{\mu}{2}\right)^{\varepsilon}\right]} \\ &= \frac{\sin_{\varepsilon}\left[\left(\left(k+\frac{1}{2}\right)\mu\right)^{\varepsilon}\right]}{2\sin_{\varepsilon}\left[\left(\frac{\mu}{2}\right)^{\varepsilon}\right]}\end{aligned}$$

$$D_{k,\varepsilon}(t) = \frac{1}{2} + \sum_{n=1}^{k} \cos_{\varepsilon} (n\mu)^{\varepsilon} = \frac{\sin_{\varepsilon}\left[\left(\left(k+\frac{1}{2}\right)\mu\right)^{\varepsilon}\right]}{2\sin_{\varepsilon}\left[\left(\frac{\mu}{2}\right)^{\varepsilon}\right]}.$$

Let us define the tangent function defined on Cantor sets, namely,

$$\tan_{\varepsilon}\left(\mu^{\varepsilon}\right) = \frac{\sin_{\varepsilon}\left(\mu^{\varepsilon}\right)}{\cos_{\varepsilon}\left(\mu^{\varepsilon}\right)}.$$

Similarly, the cotangent function defined on Cantor sets is

$$\cot_{\varepsilon}\left(\mu^{\varepsilon}\right) = \frac{\cos_{\varepsilon}\left(\mu^{\varepsilon}\right)}{\sin_{\varepsilon}\left(\mu^{\varepsilon}\right)},$$

where $\mu \in R$ and $0 < \varepsilon \leq 1$

$$\tan_{\varepsilon}\left(\mu^{\varepsilon}\right) = \frac{\sin_{\varepsilon}\left[(2\mu)^{\varepsilon}\right]}{1+\cos_{\varepsilon}\left[(2\mu)^{\varepsilon}\right]} = \frac{1-\cos_{\varepsilon}\left[(2\mu)^{\varepsilon}\right]}{\sin_{\varepsilon}\left[(2\mu)^{\varepsilon}\right]}.$$

Proof.

$$\tan_{\varepsilon}\left(\mu^{\varepsilon}\right) = \frac{2\sin_{\varepsilon}(\mu^{\varepsilon})\cos_{\varepsilon}(\mu^{\varepsilon})}{2\cos_{\varepsilon}(\mu^{\varepsilon})\cos_{\varepsilon}(\mu^{\varepsilon})} = \frac{\sin_{\varepsilon}\left[(2\mu)^{\varepsilon}\right]}{1+\cos_{\varepsilon}\left[(2\mu)^{\varepsilon}\right]}$$

and consequently, one obtains

$$\tan_{\varepsilon}\left(\mu^{\varepsilon}\right) = \frac{2\sin_{\varepsilon}(\mu^{\varepsilon})\sin_{\varepsilon}(\mu^{\varepsilon})}{2\sin_{\varepsilon}(\mu^{\varepsilon})\cos_{\varepsilon}(\mu^{\varepsilon})} = \frac{1-\cos_{\varepsilon}\left[(2\mu)^{\varepsilon}\right]}{\sin_{\varepsilon}\left[(2\mu)^{\varepsilon}\right]}.$$

□

Appendix B Local fractional derivatives of elementary functions

Consider the function

$$E_\varepsilon\left(C\mu^\varepsilon\right)=\sum_{k=0}^{\infty}\frac{C^k\mu^{k\varepsilon}}{\Gamma\left(1+k\varepsilon\right)}.$$

Then, we have

$$\begin{aligned}\frac{\mathrm{d}^\varepsilon}{\mathrm{d}\mu^\varepsilon}E_\varepsilon\left(C\mu^\varepsilon\right)&=\frac{\mathrm{d}^\varepsilon}{\mathrm{d}\mu^\varepsilon}\left(\sum_{k=0}^{\infty}\frac{C^k\mu^{k\varepsilon}}{\Gamma\left(1+k\varepsilon\right)}\right)\\&=\sum_{k=1}^{\infty}\frac{C^k\mu^{(k-1)\varepsilon}}{\Gamma\left(1+(k-1)\varepsilon\right)}\\&=C\sum_{k=1}^{\infty}\frac{C^{(k-1)}\mu^{(k-1)\varepsilon}}{\Gamma\left(1+(k-1)\varepsilon\right)}.\end{aligned}$$

Hence, we get

$$\frac{\mathrm{d}^\varepsilon}{\mathrm{d}\mu^\varepsilon}E_\varepsilon\left(C\mu^\varepsilon\right)=CE_\varepsilon\left(C\mu^\varepsilon\right).$$

Further, when $C=-1$, we have

$$\frac{\mathrm{d}^\varepsilon}{\mathrm{d}\mu^\varepsilon}E_\varepsilon\left(-\mu^\varepsilon\right)=-E_\varepsilon\left(-\mu^\varepsilon\right).$$

Using the chain rule, one obtains

$$\begin{aligned}\frac{\mathrm{d}^\varepsilon}{\mathrm{d}\mu^\varepsilon}\left[E_\varepsilon\left(\mu^{2\varepsilon}\right)\right]&=\frac{\mathrm{d}^\varepsilon E_\varepsilon\left(\mu^{2\varepsilon}\right)}{\mathrm{d}\left(\mu^2\right)^\varepsilon}\left[\frac{\mathrm{d}\mu^2}{\mathrm{d}\mu}\right]^\varepsilon\\&=\left(2\mu\right)^\varepsilon E_\varepsilon\left(\mu^{2\varepsilon}\right).\end{aligned}$$

In a similar manner, we have

$$\frac{\mathrm{d}^\varepsilon}{\mathrm{d}\mu^\varepsilon}E_\varepsilon\left(C\mu^{2\varepsilon}\right)=\left(2\mu\right)^\varepsilon CE_\varepsilon\left(C\mu^{2\varepsilon}\right)$$

and

$$\frac{d^{\varepsilon}}{d\mu^{\varepsilon}} E_{\varepsilon}\left(-\mu^{2\varepsilon}\right) = -(2\mu)^{\varepsilon} E_{\varepsilon}\left(-\mu^{2\varepsilon}\right)$$

$$\begin{aligned}
\frac{d^{\varepsilon}}{d\mu^{\varepsilon}} \sin_{\varepsilon}\left(\mu^{\varepsilon}\right) &= \frac{d^{\varepsilon}}{d\mu^{\varepsilon}} \left[\frac{E_{\varepsilon}\left(i^{\varepsilon}\mu^{\varepsilon}\right) - E_{\varepsilon}\left(-i^{\varepsilon}\mu^{\varepsilon}\right)}{2i^{\varepsilon}}\right] \\
&= \left[\frac{i^{\varepsilon}E_{\varepsilon}\left(i^{\varepsilon}\mu^{\varepsilon}\right) + i^{\varepsilon}E_{\varepsilon}\left(-i^{\varepsilon}\mu^{\varepsilon}\right)}{2i^{\varepsilon}}\right] \\
&= \frac{E_{\varepsilon}\left(i^{\varepsilon}\mu^{\varepsilon}\right) + E_{\varepsilon}\left(-i^{\varepsilon}\mu^{\varepsilon}\right)}{2} \\
&= \cos_{\varepsilon}\left(\mu^{\varepsilon}\right).
\end{aligned}$$

Therefore, applying the same style of calculations, we get the following formulas:

$$\frac{d^{\varepsilon}}{d\mu^{\varepsilon}} \sin_{\varepsilon}\left(C\mu^{\varepsilon}\right) = C\cos_{\varepsilon}\left(\mu^{\varepsilon}\right);$$

$$\frac{d^{\varepsilon}}{d\mu^{\varepsilon}} \cos_{\varepsilon}\left(\mu^{\varepsilon}\right) = -\sin_{\varepsilon}\left(\mu^{\varepsilon}\right);$$

$$\frac{d^{\varepsilon}}{d\mu^{\varepsilon}} \cos_{\varepsilon}\left(C\mu^{\varepsilon}\right) = -C\sin_{\varepsilon}\left(C\mu^{\varepsilon}\right);$$

$$\begin{aligned}
\frac{d^{\varepsilon}}{d\mu^{\varepsilon}} \sinh_{\varepsilon}\left(\mu^{\varepsilon}\right) &= \frac{d^{\varepsilon}}{d\mu^{\varepsilon}} \left[\frac{E_{\varepsilon}\left(\mu^{\varepsilon}\right) - E_{\varepsilon}\left(-\mu^{\varepsilon}\right)}{2}\right] \\
&= \frac{E_{\varepsilon}\left(\mu^{\varepsilon}\right) + E_{\varepsilon}\left(-\mu^{\varepsilon}\right)}{2} \\
&= \cosh_{\varepsilon}\left(\mu^{\varepsilon}\right);
\end{aligned}$$

$$\begin{aligned}
\frac{d^{\varepsilon}}{d\mu^{\varepsilon}} \cosh_{\varepsilon}\left(\mu^{\varepsilon}\right) &= \frac{d^{\varepsilon}}{d\mu^{\varepsilon}} \left[\frac{E_{\varepsilon}\left(\mu^{\varepsilon}\right) + E_{\varepsilon}\left(-\mu^{\varepsilon}\right)}{2}\right] \\
&= \frac{E_{\varepsilon}\left(\mu^{\varepsilon}\right) - E_{\varepsilon}\left(-\mu^{\varepsilon}\right)}{2} \\
&= \sinh_{\varepsilon}\left(\mu^{\varepsilon}\right).
\end{aligned}$$

Similarly, we may obtain that

$$\begin{aligned}
\frac{d^{\varepsilon}}{d\mu^{\varepsilon}} \sinh_{\varepsilon}\left(C\mu^{\varepsilon}\right) &= \frac{d^{\varepsilon}}{d\mu^{\varepsilon}} \left[\frac{E_{\varepsilon}\left(C\mu^{\varepsilon}\right) - E_{\varepsilon}\left(-C\mu^{\varepsilon}\right)}{2}\right] \\
&= \frac{CE_{\varepsilon}\left(C\mu^{\varepsilon}\right) + CE_{\varepsilon}\left(-C\mu^{\varepsilon}\right)}{2} \\
&= C\cosh_{\varepsilon}\left(C\mu^{\varepsilon}\right)
\end{aligned}$$

$$
\begin{aligned}
\frac{d^{\varepsilon}}{d\mu^{\varepsilon}} \cosh_{\varepsilon}\left(C\mu^{\varepsilon}\right) &= \frac{d^{\varepsilon}}{d\mu^{\varepsilon}}\left[\frac{E_{\varepsilon}\left(C\mu^{\varepsilon}\right)+E_{\varepsilon}\left(-C\mu^{\varepsilon}\right)}{2}\right] \\
&= \frac{E_{\varepsilon}\left(C\mu^{\varepsilon}\right)-CE_{\varepsilon}\left(-C\mu^{\varepsilon}\right)}{2} \\
&= C\sinh_{\varepsilon}\left(C\mu^{\varepsilon}\right).
\end{aligned}
$$

Appendix C
Local fractional Maclaurin's series of elementary functions

In this appendix, we will present the local fractional Maclaurin's series of nondifferentiable elementary functions.

For a given nondifferentiable function $\varphi(\mu)$, the local fractional Maclaurin's series takes the form

$$\varphi(\mu) = \sum_{k=0}^{\infty} \frac{D^{(k\varepsilon)}\varphi(0)}{\Gamma(1+k\varepsilon)}\mu^{k\varepsilon}.$$

Moreover, the local fractional Maclaurin polynomial can be presented as

$$\mathrm{T}_n[\varphi(\mu)] = \sum_{k=0}^{n} \frac{D^{(k\varepsilon)}\varphi(0)}{\Gamma(1+k\varepsilon)}\mu^{k\varepsilon}.$$

Further, using the relation $D^{(k\varepsilon)}\varphi(0) = 1$, where $\varphi(\mu) = E_\varepsilon(\mu^\varepsilon)$, one obtains

$$E_\varepsilon\left(\mu^\varepsilon\right) = \sum_{k=0}^{\infty} \frac{\mu^{k\varepsilon}}{\Gamma(1+k\varepsilon)}.$$

Similarly, we may obtain that

$$D^{(k\varepsilon)}\varphi(0) = (-1)^k,$$

with $\varphi(\mu) = E_\varepsilon(-\mu^\varepsilon)$ and $k \in \mathrm{R}_0$, such that

$$E_\varepsilon\left(-\mu^\varepsilon\right) = \sum_{k=0}^{\infty} \frac{(-1)^k\mu^{k\varepsilon}}{\Gamma(1+k\varepsilon)}.$$

Next, we have

$$D^{(k\varepsilon)}\varphi(0) = \begin{cases} (-1)^k, & 2k+1 \\ 0, & 2k, \end{cases}$$

where $\varphi(\mu) = \sin_\varepsilon(\mu^\varepsilon)$ and $k \in \mathrm{R}_0$, such that

$$\sin_\varepsilon\left(\mu^\varepsilon\right) = \sum_{k=0}^{\infty} \frac{(-1)^k\mu^{(2k+1)\varepsilon}}{\Gamma(1+(2k+1)\varepsilon)}.$$

In the same way, we have

$$D^{(k\varepsilon)}\varphi(0) = \begin{cases} (-1)^k, & 2k \\ 0, & 2k+1, \end{cases}$$

where $\varphi(\mu) = \cos_\varepsilon(\mu^\varepsilon)$ and $k \in R_0$, such that

$$\cos_\varepsilon\left(\mu^\varepsilon\right) = \sum_{k=0}^{\infty} \frac{(-1)^k \mu^{2k\varepsilon}}{\Gamma(1+2k\varepsilon)}.$$

In addition,

$$D^{(k\varepsilon)}\varphi(0) = \begin{cases} 1, & 2k+1 \\ 0, & 2k, \end{cases}$$

where $\varphi(\mu) = \sinh_\varepsilon(\mu^\varepsilon)$ and $k \in R_0$, such that

$$\sinh_\varepsilon\left(\mu^\varepsilon\right) = \sum_{k=0}^{\infty} \frac{\mu^{(2k+1)\varepsilon}}{\Gamma(1+(2k+1)\varepsilon)}.$$

Similarly,

$$D^{(k\varepsilon)}\varphi(0) = \begin{cases} 1, & 2k \\ 0, & 2k+1, \end{cases}$$

where $\varphi(\mu) = \cosh_\varepsilon(\mu^\varepsilon)$ and $k \in R_0$, such that

$$\cosh_\varepsilon\left(\mu^\varepsilon\right) = \sum_{k=0}^{\infty} \frac{\mu^{2k\varepsilon}}{\Gamma(1+2k\varepsilon)}.$$

In the above formulas, we notice that $R_0 = R \cup 0$.

Appendix D
Coordinate systems of Cantor-type cylindrical and Cantor-type spherical coordinates

Let us consider the coordinate system of the Cantor-type cylindrical coordinates

$$\begin{aligned} r &= R^{\varepsilon} \cos_{\varepsilon} \left(\theta^{\varepsilon}\right) e_1^{\varepsilon} + \mathrm{R}^{\varepsilon} \sin_{\varepsilon} \left(\theta^{\varepsilon}\right) e_2^{\varepsilon} + \sigma^{\varepsilon} e_3^{\varepsilon} \\ &= r_R e_R^{\varepsilon} + \mathrm{r}_{\theta} e_{\theta}^{\varepsilon} + \mathrm{r}_{\sigma} e_{\sigma}^{\varepsilon}, \end{aligned}$$

where

$$\begin{cases} \mu^{\varepsilon} = R^{\varepsilon} \cos_{\varepsilon}(\theta^{\varepsilon}), \\ \eta^{\varepsilon} = R^{\varepsilon} \sin_{\varepsilon}(\theta^{\varepsilon}), \\ \sigma^{\varepsilon} = \sigma^{\varepsilon}, \end{cases}$$

with $R \in (0, +\infty)$, $z \in (-\infty, +\infty)$, $\theta \in (0, \pi]$, and $\mu^{2\varepsilon} + \eta^{2\varepsilon} = R^{2\varepsilon}$.

We have

$$\begin{cases} \mathrm{H}_R^{\varepsilon} = \dfrac{1}{\Gamma(1+\varepsilon)} \dfrac{\partial^{\varepsilon} \mathbf{r}}{\partial R^{\varepsilon}} = \cos_{\varepsilon}(\theta^{\varepsilon})\, \mathbf{e}_1^{\varepsilon} + \sin_{\varepsilon}(\theta^{\varepsilon})\, \mathbf{e}_2^{\varepsilon}, \\ \mathrm{H}_{\theta}^{\varepsilon} = \dfrac{1}{\Gamma(1+\varepsilon)} \dfrac{\partial^{\varepsilon} \mathbf{r}}{\partial \theta^{\varepsilon}} = -\frac{R^{\varepsilon}}{\Gamma(1+\varepsilon)} \sin_{\varepsilon}(\theta^{\varepsilon})\, \mathbf{e}_1^{\varepsilon} + \frac{R^{\varepsilon}}{\Gamma(1+\varepsilon)} \cos_{\varepsilon}(\theta^{\varepsilon})\, \mathbf{e}_2^{\varepsilon}, \\ \mathrm{H}_3^{\varepsilon} = \dfrac{1}{\Gamma(1+\varepsilon)} \dfrac{\partial^{\varepsilon} \mathbf{r}}{\partial \sigma^{\varepsilon}} = \mathbf{e}_3^{\varepsilon}, \end{cases}$$

such that

$$\begin{cases} \mathbf{e}_R^{\varepsilon} = \cos_{\varepsilon}(\theta^{\varepsilon})\, \mathbf{e}_1^{\varepsilon} + \sin_{\varepsilon}(\theta^{\varepsilon})\, \mathbf{e}_2^{\varepsilon}, \\ \mathbf{e}_{\theta}^{\varepsilon} = -\sin_{\varepsilon}(\theta^{\varepsilon})\, \mathbf{e}_1^{\varepsilon} + \cos_{\varepsilon}(\theta^{\varepsilon})\, \mathbf{e}_2^{\varepsilon}, \\ \mathbf{e}_{\sigma}^{\varepsilon} = \mathbf{e}_3^{\varepsilon}, \end{cases}$$

where

$$\begin{cases} \mathrm{H}_R^{\varepsilon} = \mathbf{e}_R^{\varepsilon}, \\ \mathrm{H}_{\theta}^{\varepsilon} = \frac{R^{\varepsilon}}{\Gamma(1+\alpha)} \mathbf{e}_{\theta}^{\varepsilon}, \\ \mathrm{H}_3^{\varepsilon} = \mathbf{e}_{\sigma}^{\varepsilon}. \end{cases}$$

We have

$$\begin{pmatrix} \mathbf{e}_R^{\varepsilon} \\ \mathbf{e}_{\theta}^{\varepsilon} \\ \mathbf{e}_{\varepsilon}^{\varepsilon} \end{pmatrix} = \begin{pmatrix} \cos_{\varepsilon}(\theta^{\varepsilon}) & \sin_{\varepsilon}(\theta^{\varepsilon}) & 0 \\ -\sin_{\varepsilon}(\theta^{\varepsilon}) & \cos_{\varepsilon}(\theta^{\varepsilon}) & 0 \\ 0 & 0 & 1 \end{pmatrix} \begin{pmatrix} \mathbf{e}_1^{\varepsilon} \\ \mathbf{e}_2^{\varepsilon} \\ \mathbf{e}_3^{\varepsilon} \end{pmatrix}$$

or

$$W_i^\varepsilon = G_{ij}^\varepsilon W_j^\varepsilon,$$

where

$$W_i^\varepsilon = \begin{pmatrix} \mathbf{e}_R^\varepsilon \\ \mathbf{e}_\theta^\varepsilon \\ \mathbf{e}_\sigma^\varepsilon \end{pmatrix},$$

$$G_{ij}^\varepsilon = \begin{pmatrix} \cos_\varepsilon(\theta^\varepsilon) & \sin_\varepsilon(\theta^\varepsilon) & 0 \\ -\sin_\varepsilon(\theta^\varepsilon) & \cos_\varepsilon(\theta^\varepsilon) & 0 \\ 0 & 0 & 1 \end{pmatrix},$$

$$W_j^\varepsilon = \begin{pmatrix} \mathbf{e}_1^\varepsilon \\ \mathbf{e}_2^\varepsilon \\ \mathbf{e}_3^\varepsilon. \end{pmatrix}.$$

In this case, we have that

$$\frac{\partial^\varepsilon}{\partial R^\varepsilon} = \left(\frac{\partial \mu}{\partial R}\right)^\varepsilon \frac{\partial^\varepsilon}{\partial \mu^\varepsilon} + \left(\frac{\partial \eta}{\partial R}\right)^\varepsilon \frac{\partial^\varepsilon}{\partial \eta^\varepsilon} + \left(\frac{\partial \sigma}{\partial R}\right)^\varepsilon \frac{\partial^\varepsilon}{\partial \sigma^\varepsilon} = \mathbf{e}_R^\varepsilon \cdot \nabla^\varepsilon = \nabla_R^\varepsilon,$$
$$\frac{\partial^\varepsilon}{\partial \theta^\varepsilon} = \left(\frac{\partial \mu}{\partial \theta}\right)^\varepsilon \frac{\partial^\varepsilon}{\partial \mu^\varepsilon} + \left(\frac{\partial \eta}{\partial \theta}\right)^\varepsilon \frac{\partial^\varepsilon}{\partial \eta^\varepsilon} + \left(\frac{\partial \sigma}{\partial \theta}\right)^\varepsilon \frac{\partial^\varepsilon}{\partial \sigma^\varepsilon} = R^\varepsilon \mathbf{e}_\theta^\varepsilon \cdot \nabla^\varepsilon = R^\varepsilon \nabla_\theta^\varepsilon,$$
$$\frac{\partial^\varepsilon}{\partial \sigma^\varepsilon} = \left(\frac{\partial \mu}{\partial \sigma}\right)^\varepsilon \frac{\partial^\varepsilon}{\partial \mu^\varepsilon} + \left(\frac{\partial \eta}{\partial \sigma}\right)^\varepsilon \frac{\partial^\varepsilon}{\partial \eta^\varepsilon} + \left(\frac{\partial \sigma}{\partial \sigma}\right)^\varepsilon \frac{\partial^\varepsilon}{\partial \sigma^\varepsilon} = \mathbf{e}_\sigma^\varepsilon \cdot \nabla^\varepsilon = \nabla_\sigma^\varepsilon,$$

where

$$\begin{cases} \nabla_R^\varepsilon = \mathbf{e}_R^\varepsilon \cdot \nabla^\varepsilon = \frac{\partial^\varepsilon}{\partial R^\varepsilon}, \\ \nabla_\theta^\varepsilon = R^\varepsilon \mathbf{e}_\theta^\varepsilon \cdot \nabla^\varepsilon = \frac{1}{R^\varepsilon}\frac{\partial^\varepsilon}{\partial \theta^\varepsilon}, \\ \nabla_\sigma^\varepsilon = \mathbf{e}_\sigma^\varepsilon \cdot \nabla^\varepsilon = \frac{\partial^\varepsilon}{\partial \sigma^\varepsilon}. \end{cases}$$

The local fractional gradient operator in the Cantor-type cylindrical coordinates reads as

$$\nabla^\varepsilon = \mathbf{e}_R^\varepsilon \nabla_R^\varepsilon + \mathbf{e}_\theta^\varepsilon \nabla_\theta^\varepsilon + \mathbf{e}_\sigma^\varepsilon \nabla_\sigma^\varepsilon = \mathbf{e}_R^\varepsilon \frac{\partial^\varepsilon}{\partial R^\varepsilon} + \mathbf{e}_\theta^\varepsilon \frac{1}{R^\varepsilon}\frac{\partial^\varepsilon}{\partial \theta^\varepsilon} + \mathbf{e}_\sigma^\varepsilon \frac{\partial^\varepsilon}{\partial \sigma^\varepsilon}.$$

There is

$$\nabla^\varepsilon \cdot \mathbf{r} = \mathbf{e}_R^\varepsilon \cdot \frac{\partial^\varepsilon \mathbf{r}}{\partial R^\varepsilon} + \mathbf{e}_\theta^\varepsilon \cdot \frac{1}{R^\varepsilon}\frac{\partial^\varepsilon \mathbf{r}}{\partial \theta^\varepsilon} + \mathbf{e}_\sigma^\varepsilon \cdot \frac{\partial^\varepsilon \mathbf{r}}{\partial \sigma^\varepsilon},$$

where

$$\begin{aligned} \mathbf{e}_R^\varepsilon \cdot \frac{\partial^\varepsilon \mathbf{r}}{\partial R^\varepsilon} &= \mathbf{e}_R^\varepsilon \cdot \left(\frac{\partial^\varepsilon \mathbf{r}_R}{\partial R^\varepsilon}\mathbf{e}_R^\varepsilon + \frac{\partial^\varepsilon \mathbf{r}_\theta}{\partial R^\varepsilon}\mathbf{e}_\theta^\varepsilon + \frac{\partial^\varepsilon \mathbf{r}_\sigma}{\partial R^\varepsilon}\mathbf{e}_\sigma^\varepsilon + \mathbf{r}_R \frac{\partial^\varepsilon \mathbf{e}_R^\varepsilon}{\partial R^\varepsilon} + \mathbf{r}_\theta \frac{\partial^\varepsilon \mathbf{e}_\theta^\varepsilon}{\partial R^\varepsilon} + \mathbf{r}_\sigma \frac{\partial^\varepsilon \mathbf{e}_\sigma^\varepsilon}{\partial R^\varepsilon}\right) \\ &= \mathbf{e}_R^\varepsilon \cdot \left(\frac{\partial^\varepsilon \mathbf{r}_R}{\partial R^\varepsilon}\mathbf{e}_R^\varepsilon + \frac{\partial^\varepsilon \mathbf{r}_\theta}{\partial R^\varepsilon}\mathbf{e}_\theta^\varepsilon + \frac{\partial^\varepsilon \mathbf{r}_\sigma}{\partial R^\varepsilon}\mathbf{e}_\sigma^\varepsilon + 0 + 0 + 0\right) \end{aligned}$$

$$
\begin{aligned}
&= \mathbf{e}_R^{\varepsilon} \cdot \left(\frac{\partial^{\varepsilon} \mathbf{r}_R}{\partial R^{\varepsilon}} \mathbf{e}_R^{\varepsilon} + \frac{\partial^{\varepsilon} \mathbf{r}_{\theta}}{\partial R^{\varepsilon}} \mathbf{e}_{\theta}^{\varepsilon} + \frac{\partial^{\varepsilon} \mathbf{r}_{\sigma}}{\partial R^{\varepsilon}} \mathbf{e}_{\sigma}^{\varepsilon} \right) \\
&= \frac{\partial^{\varepsilon} \mathbf{r}_R}{\partial R^{\varepsilon}},
\end{aligned}
$$

$$
\begin{aligned}
\mathbf{e}_{\theta}^{\varepsilon} \cdot \frac{1}{R^{\varepsilon}} \frac{\partial^{\varepsilon} \mathbf{r}}{\partial \theta^{\varepsilon}} &= \mathbf{e}_{\theta}^{\varepsilon} \cdot \frac{1}{R^{\varepsilon}} \left(\frac{\partial^{\varepsilon} \mathbf{r}_R}{\partial \theta^{\varepsilon}} \mathbf{e}_R^{\varepsilon} + \frac{\partial^{\varepsilon} \mathbf{r}_{\theta}}{\partial \theta^{\varepsilon}} \mathbf{e}_{\theta}^{\varepsilon} + \frac{\partial^{\varepsilon} \mathbf{r}_{\sigma}}{\partial \theta^{\varepsilon}} \mathbf{e}_{\sigma}^{\varepsilon} + \mathbf{r}_R \frac{\partial^{\varepsilon} \mathbf{e}_R^{\varepsilon}}{\partial \theta^{\varepsilon}} + \mathbf{r}_{\theta} \frac{\partial^{\varepsilon} \mathbf{e}_{\theta}^{\varepsilon}}{\partial \theta^{\varepsilon}} + \mathbf{r}_{\sigma} \frac{\partial^{\varepsilon} \mathbf{e}_{\sigma}^{\varepsilon}}{\partial \theta^{\varepsilon}} \right) \\
&= \mathbf{e}_{\theta}^{\varepsilon} \cdot \frac{1}{R^{\varepsilon}} \left(\frac{\partial^{\varepsilon} \mathbf{r}_R}{\partial \theta^{\varepsilon}} \mathbf{e}_R^{\varepsilon} + \frac{\partial^{\varepsilon} \mathbf{r}_{\theta}}{\partial \theta^{\varepsilon}} \mathbf{e}_{\theta}^{\varepsilon} + \frac{\partial^{\varepsilon} \mathbf{r}_{\sigma}}{\partial \theta^{\varepsilon}} \mathbf{e}_{\sigma}^{\varepsilon} + \mathbf{r}_R \mathbf{e}_{\theta}^{\varepsilon} - \mathbf{r}_{\theta} \mathbf{e}_R^{\varepsilon} + 0 \right) \\
&= \mathbf{e}_{\theta}^{\varepsilon} \cdot \frac{1}{R^{\varepsilon}} \left(\frac{\partial^{\varepsilon} \mathbf{r}_R}{\partial \theta^{\varepsilon}} \mathbf{e}_R^{\varepsilon} + \frac{\partial^{\varepsilon} \mathbf{r}_{\theta}}{\partial \theta^{\varepsilon}} \mathbf{e}_{\theta}^{\varepsilon} + \frac{\partial^{\varepsilon} \mathbf{r}_{\sigma}}{\partial \theta^{\varepsilon}} \mathbf{e}_{\sigma}^{\varepsilon} + \mathbf{r}_R \mathbf{e}_{\theta}^{\varepsilon} - \mathbf{r}_{\theta} \mathbf{e}_R^{\varepsilon} \right) \\
&= \frac{1}{R^{\varepsilon}} \frac{\partial^{\varepsilon} \mathbf{r}_{\theta}}{\partial \theta^{\varepsilon}} + \frac{\mathbf{r}_R}{R^{\varepsilon}},
\end{aligned}
$$

$$
\begin{aligned}
\mathbf{e}_{\sigma}^{\varepsilon} \cdot \frac{\partial^{\varepsilon} \mathbf{r}}{\partial \sigma^{\varepsilon}} &= \mathbf{e}_{\sigma}^{\varepsilon} \cdot \left(\frac{\partial^{\varepsilon} \mathbf{r}_R}{\partial \sigma^{\varepsilon}} \mathbf{e}_R^{\varepsilon} + \frac{\partial^{\varepsilon} \mathbf{r}_{\theta}}{\partial \sigma^{\varepsilon}} \mathbf{e}_{\theta}^{\varepsilon} + \frac{\partial^{\varepsilon} \mathbf{r}_z}{\partial \sigma^{\varepsilon}} \mathbf{e}_{\sigma}^{\varepsilon} + \mathbf{r}_R \frac{\partial^{\varepsilon} \mathbf{e}_R^{\varepsilon}}{\partial \sigma^{\varepsilon}} + \frac{\partial^{\varepsilon} \mathbf{e}_{\theta}^{\varepsilon}}{\partial \sigma^{\varepsilon}} \mathbf{r}_{\theta} + \frac{\partial^{\varepsilon} \mathbf{e}_{\sigma}^{\varepsilon}}{\partial \sigma^{\varepsilon}} \mathbf{r}_{\sigma} \right) \\
&= \mathbf{e}_{\sigma}^{\varepsilon} \cdot \left(\frac{\partial^{\varepsilon} \mathbf{r}_R}{\partial \sigma^{\varepsilon}} \mathbf{e}_R^{\varepsilon} + \frac{\partial^{\varepsilon} \mathbf{r}_{\theta}}{\partial \sigma^{\varepsilon}} \mathbf{e}_{\theta}^{\varepsilon} + \frac{\partial^{\varepsilon} \mathbf{r}_z}{\partial \sigma^{\varepsilon}} \mathbf{e}_{\sigma}^{\varepsilon} + 0 + 0 + 0 \right) \\
&= \mathbf{e}_{\sigma}^{\varepsilon} \cdot \left(\frac{\partial^{\varepsilon} \mathbf{r}_R}{\partial \sigma^{\varepsilon}} \mathbf{e}_R^{\varepsilon} + \frac{\partial^{\varepsilon} \mathbf{r}_{\theta}}{\partial \sigma^{\varepsilon}} \mathbf{e}_{\theta}^{\varepsilon} + \frac{\partial^{\varepsilon} \mathbf{r}_z}{\partial \sigma^{\varepsilon}} \mathbf{e}_{\sigma}^{\varepsilon} \right) \\
&= \frac{\partial^{\varepsilon} \mathbf{r}_z}{\partial \sigma^{\varepsilon}}.
\end{aligned}
$$

Hence, we get the local fractional divergence operator in the Cantor-type cylindrical coordinates

$$
\nabla^{\varepsilon} \cdot \mathbf{r} = \frac{\partial^{\varepsilon} \mathbf{r}_R}{\partial R^{\varepsilon}} + \frac{1}{R^{\varepsilon}} \frac{\partial^{\varepsilon} \mathbf{r}_{\theta}}{\partial \theta^{\varepsilon}} + \frac{\mathbf{r}_R}{R^{\varepsilon}} + \frac{\partial^{\varepsilon} \mathbf{r}_z}{\partial \sigma^{\varepsilon}}.
$$

The local fractional curl operator in the Cantor-type cylindrical coordinates is presented as follows:

$$
\begin{aligned}
\nabla^{\varepsilon} \times \mathbf{r} &= \left(\mathbf{e}_R^{\varepsilon} \frac{\partial^{\varepsilon}}{\partial R^{\varepsilon}} + \mathbf{e}_{\theta}^{\varepsilon} \frac{1}{R^{\varepsilon}} \frac{\partial^{\varepsilon}}{\partial \theta^{\varepsilon}} + \mathbf{e}_{\sigma}^{\varepsilon} \frac{\partial^{\varepsilon}}{\partial \sigma^{\varepsilon}} \right) \times \left(\mathbf{e}_R^{\varepsilon} \mathbf{r}_R + \mathbf{e}_{\theta}^{\varepsilon} \mathbf{r}_{\theta} + \mathbf{e}_{\sigma}^{\varepsilon} \mathbf{r}_{\sigma} \right) \\
&= \mathbf{e}_R^{\varepsilon} \times \frac{\partial^{\varepsilon} \mathbf{r}}{\partial R^{\varepsilon}} + \mathbf{e}_{\theta}^{\varepsilon} \times \frac{1}{R^{\varepsilon}} \frac{\partial^{\varepsilon} \mathbf{r}}{\partial \theta^{\varepsilon}} + \mathbf{e}_{\sigma}^{\varepsilon} \times \frac{\partial^{\varepsilon} \mathbf{r}}{\partial \sigma^{\varepsilon}},
\end{aligned}
$$

where

$$
\begin{aligned}
\mathbf{e}_R^{\varepsilon} \times \frac{\partial^{\varepsilon} \mathbf{r}}{\partial R^{\varepsilon}} &= \mathbf{e}_R^{\varepsilon} \times \left(\frac{\partial^{\varepsilon} \mathbf{r}_R}{\partial R^{\varepsilon}} \mathbf{e}_R^{\varepsilon} + \frac{\partial^{\varepsilon} \mathbf{r}_{\theta}}{\partial R^{\varepsilon}} \mathbf{e}_{\theta}^{\varepsilon} + \frac{\partial^{\varepsilon} \mathbf{r}_{\sigma}}{\partial R^{\varepsilon}} \mathbf{e}_{\sigma}^{\varepsilon} + \mathbf{r}_R \frac{\partial^{\varepsilon} \mathbf{e}_R^{\varepsilon}}{\partial R^{\varepsilon}} + \mathbf{r}_{\theta} \frac{\partial^{\varepsilon} \mathbf{e}_{\theta}^{\varepsilon}}{\partial R^{\varepsilon}} + \mathbf{r}_{\sigma} \frac{\partial^{\varepsilon} \mathbf{e}_{\sigma}^{\varepsilon}}{\partial R^{\varepsilon}} \right) \\
&= \mathbf{e}_R^{\varepsilon} \times \left(\frac{\partial^{\varepsilon} \mathbf{r}_R}{\partial R^{\varepsilon}} \mathbf{e}_R^{\varepsilon} + \frac{\partial^{\varepsilon} \mathbf{r}_{\theta}}{\partial R^{\varepsilon}} \mathbf{e}_{\theta}^{\varepsilon} + \frac{\partial^{\varepsilon} \mathbf{r}_{\sigma}}{\partial R^{\varepsilon}} \mathbf{e}_{\sigma}^{\varepsilon} + 0 + 0 + 0 \right)
\end{aligned}
$$

$$= \mathbf{e}_R^\varepsilon \times \left(\frac{\partial^\varepsilon \mathbf{r}_R}{\partial R^\varepsilon} \mathbf{e}_R^\varepsilon + \frac{\partial^\varepsilon \mathbf{r}_\theta}{\partial R^\varepsilon} \mathbf{e}_\theta^\varepsilon + \frac{\partial^\varepsilon \mathbf{r}_\sigma}{\partial R^\varepsilon} \mathbf{e}_\sigma^\varepsilon \right)$$
$$= \frac{\partial^\varepsilon \mathbf{r}_\theta}{\partial R^\varepsilon} \mathbf{e}_\sigma^\varepsilon - \frac{\partial^\varepsilon \mathbf{r}_\sigma}{\partial R^\varepsilon} \mathbf{e}_\theta^\varepsilon,$$

$$\mathbf{e}_\theta^\varepsilon \times \frac{1}{R^\varepsilon} \frac{\partial^\varepsilon \mathbf{r}}{\partial \theta^\varepsilon} = \mathbf{e}_\theta^\varepsilon \times \frac{1}{R^\varepsilon} \left(\frac{\partial^\varepsilon \mathbf{r}_R}{\partial \theta^\varepsilon} \mathbf{e}_R^\varepsilon + \frac{\partial^\varepsilon \mathbf{r}_\theta}{\partial \theta^\varepsilon} \mathbf{e}_\theta^\varepsilon + \frac{\partial^\varepsilon \mathbf{r}_\sigma}{\partial \theta^\varepsilon} \mathbf{e}_\sigma^\varepsilon + \mathbf{r}_R \frac{\partial^\varepsilon \mathbf{e}_R^\varepsilon}{\partial \theta^\varepsilon} + \mathbf{r}_\theta \frac{\partial^\varepsilon \mathbf{e}_\theta^\varepsilon}{\partial \theta^\varepsilon} + \mathbf{r}_\sigma \frac{\partial^\varepsilon \mathbf{e}_\sigma^\varepsilon}{\partial \theta^\varepsilon} \right)$$
$$= \mathbf{e}_\theta^\varepsilon \times \frac{1}{R^\varepsilon} \left(\frac{\partial^\varepsilon \mathbf{r}_R}{\partial \theta^\varepsilon} \mathbf{e}_R^\varepsilon + \frac{\partial^\varepsilon \mathbf{r}_\theta}{\partial \theta^\varepsilon} \mathbf{e}_\theta^\varepsilon + \frac{\partial^\varepsilon \mathbf{r}_\sigma}{\partial \theta^\varepsilon} \mathbf{e}_\sigma^\varepsilon + \mathbf{r}_R \mathbf{e}_\theta^\varepsilon - \mathbf{r}_\theta \mathbf{e}_R^\varepsilon + 0 \right)$$
$$= -\mathbf{e}_\sigma^\varepsilon \frac{1}{R^\varepsilon} \frac{\partial^\varepsilon \mathbf{r}_R}{\partial \theta^\varepsilon} + \mathbf{e}_R^\varepsilon \frac{1}{R^\varepsilon} \frac{\partial^\varepsilon \mathbf{r}_\sigma}{\partial \theta^\varepsilon} + \mathbf{e}_\sigma^\varepsilon \frac{\mathbf{r}_\theta}{R^\varepsilon},$$

$$\mathbf{e}_\sigma^\varepsilon \times \frac{\partial^\varepsilon \mathbf{r}}{\partial \sigma^\varepsilon} = \mathbf{e}_\sigma^\varepsilon \times \left(\frac{\partial^\varepsilon \mathbf{r}_R}{\partial \sigma^\varepsilon} \mathbf{e}_R^\varepsilon + \frac{\partial^\varepsilon \mathbf{r}_\theta}{\partial \sigma^\varepsilon} \mathbf{e}_\theta^\varepsilon + \frac{\partial^\varepsilon \mathbf{r}_\sigma}{\partial \sigma^\varepsilon} \mathbf{e}_\sigma^\varepsilon + \mathbf{r}_R \frac{\partial^\varepsilon \mathbf{e}_R^\varepsilon}{\partial \sigma^\varepsilon} + \mathbf{r}_\theta \frac{\partial^\varepsilon \mathbf{e}_\theta^\varepsilon}{\partial \sigma^\varepsilon} + \mathbf{r}_\sigma \frac{\partial^\varepsilon \mathbf{e}_\sigma^\varepsilon}{\partial \sigma^\varepsilon} \right)$$
$$= \mathbf{e}_\sigma^\varepsilon \times \left(\frac{\partial^\varepsilon \mathbf{r}_R}{\partial \sigma^\varepsilon} \mathbf{e}_R^\varepsilon + \frac{\partial^\varepsilon \mathbf{r}_\theta}{\partial \sigma^\varepsilon} \mathbf{e}_\theta^\varepsilon + \frac{\partial^\varepsilon \mathbf{r}_\sigma}{\partial \sigma^\varepsilon} \mathbf{e}_\sigma^\varepsilon + 0 + 0 + 0 \right)$$
$$= \mathbf{e}_\theta^\varepsilon \frac{\partial^\varepsilon \mathbf{r}_R}{\partial \sigma^\varepsilon} - \mathbf{e}_R^\varepsilon \frac{\partial^\varepsilon \mathbf{r}_\theta}{\partial \sigma^\varepsilon}.$$

Thus, we obtain the local fractional curl operator in the Cantor-type cylindrical coordinates:

$$\nabla^\varepsilon \times \mathbf{r} = \left(\frac{1}{R^\varepsilon} \frac{\partial^\varepsilon \mathbf{r}_\sigma}{\partial \theta^\varepsilon} - \frac{\partial^\varepsilon \mathbf{r}_\theta}{\partial \sigma^\varepsilon} \right) \mathbf{e}_R^\varepsilon + \left(\frac{\partial^\varepsilon \mathbf{r}_R}{\partial \sigma^\varepsilon} - \frac{\partial^\varepsilon \mathbf{r}_\sigma}{\partial R^\varepsilon} \right) \mathbf{e}_\theta^\varepsilon + \left(\frac{\partial^\varepsilon \mathbf{r}_\theta}{\partial R^\varepsilon} + \frac{\mathbf{r}_\theta}{R^\varepsilon} - \frac{1}{R^\varepsilon} \frac{\partial^\varepsilon \mathbf{r}_R}{\partial \theta^\varepsilon} \right) \mathbf{e}_\sigma^\varepsilon.$$

We get the local fractional Laplace operator in the Cantor-type cylindrical coordinates:

$$\nabla^{2\varepsilon} \psi(R,\theta,\sigma) = \left(\mathbf{e}_R^\varepsilon \frac{\partial^\varepsilon}{\partial R^\varepsilon} + \mathbf{e}_\theta^\varepsilon \frac{1}{R^\varepsilon} \frac{\partial^\varepsilon}{\partial \theta^\varepsilon} + \mathbf{e}_\sigma^\varepsilon \frac{\partial^\varepsilon}{\partial \sigma^\varepsilon} \right) \cdot \left(\mathbf{e}_R^\varepsilon \frac{\partial^\varepsilon \psi}{\partial R^\varepsilon} + \mathbf{e}_\theta^\varepsilon \frac{1}{R^\varepsilon} \frac{\partial^\varepsilon \psi}{\partial \theta^\varepsilon} + \mathbf{e}_\sigma^\varepsilon \frac{\partial^\varepsilon \psi}{\partial \sigma^\varepsilon} \right)$$
$$= \mathbf{e}_R^\varepsilon \cdot \frac{\partial^\varepsilon}{\partial R^\varepsilon} \left(\mathbf{e}_R^\varepsilon \frac{\partial^\varepsilon \psi}{\partial R^\varepsilon} + \mathbf{e}_\theta^\varepsilon \frac{1}{R^\varepsilon} \frac{\partial^\varepsilon \psi}{\partial \theta^\varepsilon} + \mathbf{e}_\sigma^\varepsilon \frac{\partial^\varepsilon \psi}{\partial \sigma^\varepsilon} \right)$$
$$+ \mathbf{e}_\theta^\varepsilon \cdot \frac{1}{R^\varepsilon} \frac{\partial^\varepsilon}{\partial \theta^\varepsilon} \left(\mathbf{e}_R^\varepsilon \frac{\partial^\varepsilon \psi}{\partial R^\varepsilon} + \mathbf{e}_\theta^\varepsilon \frac{1}{R^\varepsilon} \frac{\partial^\varepsilon \psi}{\partial \theta^\varepsilon} + \mathbf{e}_\sigma^\varepsilon \frac{\partial^\varepsilon \psi}{\partial \sigma^\varepsilon} \right)$$
$$+ \mathbf{e}_\sigma^\varepsilon \cdot \frac{\partial^\varepsilon}{\partial \sigma^\varepsilon} \left(\mathbf{e}_R^\varepsilon \frac{\partial^\varepsilon \psi}{\partial R^\varepsilon} + \mathbf{e}_\theta^\varepsilon \frac{1}{R^\varepsilon} \frac{\partial^\varepsilon \psi}{\partial \theta^\varepsilon} + \mathbf{e}_\sigma^\varepsilon \frac{\partial^\varepsilon \psi}{\partial \sigma^\varepsilon} \right),$$

where

$$\mathbf{e}_R^\varepsilon \cdot \frac{\partial^\varepsilon}{\partial R^\varepsilon} \left(\mathbf{e}_R^\varepsilon \frac{\partial^\varepsilon \psi}{\partial R^\varepsilon} + \mathbf{e}_\theta^\varepsilon \frac{1}{R^\varepsilon} \frac{\partial^\varepsilon \psi}{\partial \theta^\varepsilon} + \mathbf{e}_\sigma^\varepsilon \frac{\partial^\varepsilon \psi}{\partial \sigma^\varepsilon} \right)$$
$$= \mathbf{e}_R^\varepsilon \cdot \left(\mathbf{e}_R^\varepsilon \frac{\partial^{2\varepsilon} \psi}{\partial R^{2\varepsilon}} + \mathbf{e}_\theta^\varepsilon \frac{\partial^\varepsilon \phi}{\partial R^\varepsilon} \left(\frac{1}{R^\varepsilon} \frac{\partial^\varepsilon \psi}{\partial \theta^\varepsilon} \right) + \mathbf{e}_\sigma^\varepsilon \frac{\partial^{2\varepsilon} \psi}{\partial R^\varepsilon \partial \sigma^\varepsilon} \right)$$
$$= \frac{\partial^{2\varepsilon} \psi}{\partial R^{2\varepsilon}},$$

$$\mathbf{e}_\theta^\varepsilon \cdot \frac{1}{R^\varepsilon}\frac{\partial^\varepsilon}{\partial\theta^\varepsilon}\left(\mathbf{e}_R^\varepsilon \frac{\partial^\varepsilon\psi}{\partial R^\varepsilon} + \mathbf{e}_\theta^\varepsilon \frac{1}{R^\varepsilon}\frac{\partial^\varepsilon\psi}{\partial\theta^\varepsilon} + \mathbf{e}_\sigma^\varepsilon \frac{\partial^\varepsilon\psi}{\partial\sigma^\varepsilon}\right) = \frac{1}{R^\varepsilon}\frac{\partial^\varepsilon}{\partial\theta^\varepsilon}\left(\frac{1}{R^\varepsilon}\frac{\partial^\varepsilon\psi}{\partial\theta^\varepsilon}\right) + \frac{1}{R^\varepsilon}\frac{\partial^\varepsilon\psi}{\partial R^\varepsilon}$$

$$= \frac{1}{R^{2\varepsilon}}\frac{\partial^{2\varepsilon}\psi}{\partial\theta^{2\varepsilon}} + \frac{1}{R^\varepsilon}\frac{\partial^\varepsilon\psi}{\partial R^\varepsilon},$$

$$\mathbf{e}_\sigma^\varepsilon \cdot \frac{\partial^\varepsilon}{\partial\sigma^\varepsilon}\left(\mathbf{e}_R^\varepsilon \frac{\partial^\varepsilon\psi}{\partial R^\varepsilon} + \mathbf{e}_\theta^\varepsilon \frac{1}{R^\varepsilon}\frac{\partial^\varepsilon\psi}{\partial\theta^\varepsilon} + \mathbf{e}_\sigma^\varepsilon \frac{\partial^\varepsilon\psi}{\partial\sigma^\varepsilon}\right)$$

$$= \mathbf{e}_\sigma^\varepsilon \cdot \left(\mathbf{e}_R^\varepsilon \frac{\partial^{2\varepsilon}\psi}{\partial\sigma^\varepsilon\partial R^\varepsilon} + \mathbf{e}_\theta^\varepsilon \frac{\partial^\varepsilon}{\partial\sigma^\varepsilon}\left(\frac{1}{R^\varepsilon}\frac{\partial^\varepsilon\psi}{\partial\theta^\varepsilon}\right) + \mathbf{e}_\sigma^\varepsilon \frac{\partial^{2\varepsilon}\psi}{\partial\sigma^{2\varepsilon}}\right)$$

$$= \frac{\partial^{2\varepsilon}\psi}{\partial\sigma^{2\varepsilon}}.$$

Hence, the local fractional Laplace operator in the Cantor-type cylindrical coordinates can be written in the form

$$\nabla^{2\varepsilon}\psi(R,\theta,\sigma) = \frac{\partial^{2\varepsilon}\psi}{\partial R^{2\varepsilon}} + \frac{1}{R^{2\varepsilon}}\frac{\partial^{2\varepsilon}\psi}{\partial\theta^{2\varepsilon}} + \frac{1}{R^\varepsilon}\frac{\partial^\varepsilon\psi}{\partial R^\varepsilon} + \frac{\partial^{2\varepsilon}\psi}{\partial\sigma^{2\varepsilon}}.$$

We consider the coordinate system of the Cantor-type spherical coordinates

$$r = R^\varepsilon \cos_\varepsilon\left(\theta^\varepsilon\right)\sin_\varepsilon\left(\vartheta^\varepsilon\right)e_1^\varepsilon + R^\varepsilon \sin_\varepsilon\left(\theta^\varepsilon\right)\sin_\varepsilon\left(\vartheta^\varepsilon\right)e_2^\varepsilon + R^\varepsilon \cos_\varepsilon\left(\vartheta^\varepsilon\right)e_3^\varepsilon$$
$$= \mathrm{r}_R e_R^\varepsilon + \mathrm{r}_\vartheta e_\vartheta^\varepsilon + \mathrm{r}_\theta e_\theta^\varepsilon,$$

where

$$\begin{cases} \mu^\varepsilon = R^\varepsilon \cos_\varepsilon(\theta^\varepsilon)\sin_\varepsilon(\vartheta^\varepsilon), \\ \eta^\varepsilon = R^\varepsilon \sin_\varepsilon(\theta^\varepsilon)\sin_\varepsilon(\vartheta^\varepsilon), \\ \sigma^\varepsilon = R^\varepsilon \cos_\varepsilon(\vartheta^\varepsilon), \end{cases}$$

with $R \in (0,+\infty)$, $\vartheta \in (0,\pi)$, $\theta \in (0,2\pi)$, and $\mu^{2\varepsilon} + \eta^{2\varepsilon} + \sigma^{2\varepsilon} = R^{2\varepsilon}$.

We have

$$\begin{cases} e_R^\varepsilon = \sin_\varepsilon(\vartheta^\varepsilon)\cos_\varepsilon(\theta^\varepsilon)e_1^\varepsilon + \sin_\varepsilon(\vartheta^\varepsilon)\sin_\varepsilon(\theta^\varepsilon)e_2^\varepsilon + \cos_\varepsilon(\vartheta^\varepsilon)e_3^\varepsilon, \\ e_\vartheta^\varepsilon = \cos_\varepsilon(\vartheta^\varepsilon)\cos_\varepsilon(\theta^\varepsilon)e_1^\varepsilon + \cos_\varepsilon(\vartheta^\varepsilon)\sin_\varepsilon(\theta^\varepsilon)e_2^\varepsilon - \sin_\varepsilon(\vartheta^\varepsilon)e_3^\varepsilon, \\ e_\theta^\varepsilon = -\sin_\varepsilon(\theta^\varepsilon)e_1^\varepsilon + \cos_\varepsilon(\theta^\varepsilon)e_2^\varepsilon. \end{cases}$$

We rewrite (5.5) as

$$\begin{pmatrix} \mathbf{e}_R^\varepsilon \\ \mathbf{e}_\vartheta^\varepsilon \\ \mathbf{e}_\theta^\varepsilon \end{pmatrix} = \begin{pmatrix} \sin_\varepsilon(\vartheta^\varepsilon)\cos_\varepsilon(\theta^\varepsilon) & \sin_\varepsilon(\vartheta^\varepsilon)\sin_\varepsilon(\theta^\varepsilon) & \cos_\varepsilon(\vartheta^\varepsilon) \\ \cos_\varepsilon(\vartheta^\varepsilon)\cos_\varepsilon(\theta^\varepsilon) & \cos_\varepsilon(\vartheta^\varepsilon)\sin_\varepsilon(\theta^\varepsilon) & -\sin_\varepsilon(\vartheta^\varepsilon) \\ -\sin_\varepsilon(\theta^\varepsilon) & \cos_\varepsilon(\theta^\varepsilon) & 0 \end{pmatrix} \begin{pmatrix} \mathbf{e}_1^\varepsilon \\ \mathbf{e}_2^\varepsilon \\ \mathbf{e}_3^\varepsilon \end{pmatrix}$$

or

$$S_i^\varepsilon = D_{ij}^\varepsilon S_j^\varepsilon,$$

where

$$S_i^\varepsilon = \begin{pmatrix} \mathbf{e}_R^\varepsilon \\ \mathbf{e}_\theta^\varepsilon \\ \mathbf{e}_\sigma^\varepsilon \end{pmatrix},$$

$$D_{ij}^\varepsilon = \begin{pmatrix} \sin_\varepsilon(\vartheta^\varepsilon)\cos_\varepsilon(\theta^\varepsilon) & \sin_\varepsilon(\vartheta^\varepsilon)\sin_\varepsilon(\theta^\varepsilon) & \cos_\varepsilon(\vartheta^\varepsilon) \\ \cos_\varepsilon(\vartheta^\varepsilon)\cos_\varepsilon(\theta^\varepsilon) & \cos_\varepsilon(\vartheta^\varepsilon)\sin_\varepsilon(\theta^\varepsilon) & -\sin_\varepsilon(\vartheta^\varepsilon) \\ -\sin_\varepsilon(\theta^\varepsilon) & \cos_\varepsilon(\theta^\varepsilon) & 0 \end{pmatrix},$$

$$S_j^\varepsilon = \begin{pmatrix} \mathbf{e}_1^\varepsilon \\ \mathbf{e}_2^\varepsilon \\ \mathbf{e}_3^\varepsilon. \end{pmatrix}.$$

In this case, we present

$$\begin{aligned} \frac{\partial^\varepsilon}{\partial R^\varepsilon} &= \left(\frac{\partial \mu}{\partial R}\right)^\varepsilon \frac{\partial^\varepsilon}{\partial \mu^\varepsilon} + \left(\frac{\partial \eta}{\partial R}\right)^\varepsilon \frac{\partial^\varepsilon}{\partial \eta^\varepsilon} + \left(\frac{\partial \sigma}{\partial R}\right)^\varepsilon \frac{\partial^\varepsilon}{\partial \sigma^\varepsilon} \\ &= \mathbf{e}_R^\varepsilon \cdot \nabla^\varepsilon \\ &= \nabla_R^\varepsilon, \end{aligned}$$

$$\begin{aligned} \frac{\partial^\varepsilon}{\partial \vartheta^\varepsilon} &= \left(\frac{\partial \mu}{\partial \vartheta}\right)^\varepsilon \frac{\partial^\varepsilon}{\partial \mu^\varepsilon} + \left(\frac{\partial \eta}{\partial \vartheta}\right)^\varepsilon \frac{\partial^\varepsilon}{\partial \eta^\varepsilon} + \left(\frac{\partial \sigma}{\partial \vartheta}\right)^\varepsilon \frac{\partial^\varepsilon}{\partial \sigma^\varepsilon} \\ &= R^\varepsilon \mathbf{e}_\vartheta^\varepsilon \cdot \nabla^\varepsilon \\ &= R^\varepsilon \nabla_\vartheta^\varepsilon, \end{aligned}$$

$$\begin{aligned} \frac{\partial^\varepsilon}{\partial \theta^\varepsilon} &= \left(\frac{\partial \mu}{\partial \theta}\right)^\varepsilon \frac{\partial^\varepsilon}{\partial \mu^\varepsilon} + \left(\frac{\partial \eta}{\partial \theta}\right)^\varepsilon \frac{\partial^\varepsilon}{\partial \eta^\varepsilon} + \left(\frac{\partial \sigma}{\partial \theta}\right)^\varepsilon \frac{\partial^\varepsilon}{\partial \sigma^\varepsilon} \\ &= \frac{1}{R^\varepsilon \sin_\varepsilon(\vartheta^\varepsilon)} \mathbf{e}_\theta^\varepsilon \cdot \nabla^\varepsilon \\ &= \frac{1}{R^\varepsilon \sin_\varepsilon(\vartheta^\varepsilon)} \nabla_\theta^\varepsilon, \end{aligned}$$

where

$$\begin{cases} \nabla_R^\varepsilon = \mathbf{e}_R^\varepsilon \cdot \nabla^\varepsilon = \frac{\partial^\varepsilon}{\partial R^\varepsilon}, \\ \nabla_\vartheta^\varepsilon = R^\varepsilon \mathbf{e}_\vartheta^\varepsilon \cdot \nabla^\varepsilon = \frac{1}{R^\varepsilon} \frac{\partial^\varepsilon}{\partial \vartheta^\varepsilon}, \\ \nabla_\theta^\varepsilon = R^\varepsilon \sin_\varepsilon(\vartheta^\varepsilon)\, \mathbf{e}_\theta^\varepsilon \cdot \nabla^\varepsilon = \frac{1}{R^\varepsilon \sin_\varepsilon(\vartheta^\varepsilon)} \frac{\partial^\varepsilon}{\partial \theta^\varepsilon}. \end{cases}$$

The local fractional gradient operator in the Cantor-type spherical coordinates is written as follows:

$$\begin{aligned} \nabla^\varepsilon \psi &= \mathbf{e}_R^\varepsilon \nabla_R^\varepsilon \psi + \mathbf{e}_\vartheta^\varepsilon \nabla_\vartheta^\varepsilon \psi + \mathbf{e}_\theta^\varepsilon \nabla_\theta^\varepsilon \psi \\ &= \mathbf{e}_R^\varepsilon \frac{\partial^\varepsilon \psi}{\partial R^\varepsilon} + \mathbf{e}_\vartheta^\varepsilon \frac{1}{R^\varepsilon} \frac{\partial^\varepsilon \psi}{\partial \vartheta^\varepsilon} + \mathbf{e}_\theta^\varepsilon \frac{1}{R^\varepsilon \sin_\varepsilon(\vartheta^\varepsilon)} \frac{\partial^\varepsilon \psi}{\partial \theta^\varepsilon}. \end{aligned}$$

We obtain the local fractional divergence operator in the Cantor-type spherical coordinates

$$\nabla^{\varepsilon} \cdot \mathbf{r} = \mathbf{e}_R^{\varepsilon} \cdot \frac{\partial^{\varepsilon} \mathbf{r}}{\partial R^{\varepsilon}} + \mathbf{e}_{\vartheta}^{\varepsilon} \cdot \frac{1}{R^{\varepsilon}} \frac{\partial^{\varepsilon} \mathbf{r}}{\partial \vartheta^{\varepsilon}} + \mathbf{e}_{\theta}^{\varepsilon} \cdot \frac{1}{R^{\varepsilon} \sin_{\varepsilon}(\vartheta^{\varepsilon})} \frac{\partial^{\varepsilon} \mathbf{r}}{\partial \theta^{\varepsilon}},$$

where

$$\begin{aligned}
\mathbf{e}_R^{\varepsilon} \cdot \frac{\partial^{\varepsilon} \mathbf{r}}{\partial R^{\varepsilon}} &= \mathbf{e}_R^{\varepsilon} \cdot \frac{\partial^{\varepsilon}}{\partial R^{\varepsilon}} \left(\mathrm{r}_R \mathbf{e}_R^{\varepsilon} + \mathrm{r}_{\vartheta} \mathbf{e}_{\vartheta}^{\varepsilon} + \mathrm{r}_{\theta} \mathbf{e}_{\theta}^{\varepsilon} \right) \\
&= \mathbf{e}_R^{\varepsilon} \cdot \left(\mathbf{e}_R^{\varepsilon} \frac{\partial^{\varepsilon} \mathrm{r}_R}{\partial R^{\varepsilon}} + \mathbf{e}_{\vartheta}^{\varepsilon} \frac{\partial^{\varepsilon} \mathrm{r}_{\vartheta}}{\partial R^{\varepsilon}} + \mathbf{e}_{\theta}^{\varepsilon} \frac{\partial^{\varepsilon} \mathrm{r}_{\theta}}{\partial R^{\varepsilon}} + \mathrm{r}_R \frac{\partial^{\varepsilon} \mathbf{e}_R^{\varepsilon}}{\partial R^{\varepsilon}} + \mathrm{r}_{\vartheta} \frac{\partial^{\varepsilon} \mathbf{e}_{\vartheta}^{\varepsilon}}{\partial R^{\varepsilon}} + \mathrm{r}_{\theta} \frac{\partial^{\varepsilon} \mathbf{e}_{\theta}^{\varepsilon}}{\partial R^{\varepsilon}} \right) \\
&= \frac{\partial^{\varepsilon} \mathrm{r}_R}{\partial R^{\varepsilon}},
\end{aligned}$$

$$\begin{aligned}
\mathbf{e}_{\vartheta}^{\varepsilon} \cdot \frac{1}{R^{\varepsilon}} \frac{\partial^{\varepsilon} \mathbf{r}}{\partial \vartheta^{\varepsilon}} &= \mathbf{e}_{\vartheta}^{\varepsilon} \cdot \frac{1}{R^{\varepsilon}} \frac{\partial^{\varepsilon}}{\partial \vartheta^{\varepsilon}} \left(\mathrm{r}_R \mathbf{e}_R^{\varepsilon} + \mathrm{r}_{\vartheta} \mathbf{e}_{\vartheta}^{\varepsilon} + \mathrm{r}_{\theta} \mathbf{e}_{\theta}^{\varepsilon} \right) \\
&= \mathbf{e}_{\vartheta}^{\varepsilon} \cdot \frac{1}{R^{\varepsilon}} \left(\mathbf{e}_R^{\varepsilon} \frac{\partial^{\varepsilon} \mathrm{r}_R}{\partial \vartheta^{\varepsilon}} + \mathbf{e}_{\vartheta}^{\varepsilon} \frac{\partial^{\varepsilon} \mathrm{r}_{\vartheta}}{\partial \vartheta^{\varepsilon}} + \mathbf{e}_{\theta}^{\varepsilon} \frac{\partial^{\varepsilon} \mathrm{r}_{\theta}}{\partial \vartheta^{\varepsilon}} + \mathrm{r}_R \frac{\partial^{\varepsilon} \mathbf{e}_R^{\varepsilon}}{\partial \vartheta^{\varepsilon}} + \mathrm{r}_{\vartheta} \frac{\partial^{\varepsilon} \mathbf{e}_{\vartheta}^{\varepsilon}}{\partial \vartheta^{\varepsilon}} + \mathrm{r}_{\theta} \frac{\partial^{\varepsilon} \mathbf{e}_{\theta}^{\varepsilon}}{\partial \vartheta^{\varepsilon}} \right) \\
&= \frac{1}{R^{\varepsilon}} \frac{\partial^{\varepsilon} \mathrm{r}_{\vartheta}}{\partial \vartheta^{\varepsilon}} + \frac{\mathrm{r}_R}{R^{\varepsilon}},
\end{aligned}$$

$$\begin{aligned}
\mathbf{e}_{\theta}^{\varepsilon} \cdot \frac{1}{R^{\varepsilon} \sin_{\varepsilon}(\vartheta^{\varepsilon})} \frac{\partial^{\varepsilon} \mathbf{r}}{\partial \theta^{\varepsilon}} &= \mathbf{e}_{\theta}^{\varepsilon} \cdot \frac{1}{R^{\varepsilon} \sin_{\varepsilon}(\vartheta^{\varepsilon})} \frac{\partial^{\varepsilon}}{\partial \theta^{\varepsilon}} \left(\mathrm{r}_R \mathbf{e}_R^{\varepsilon} + \mathrm{r}_{\vartheta} \mathbf{e}_{\vartheta}^{\varepsilon} + \mathrm{r}_{\theta} \mathbf{e}_{\theta}^{\varepsilon} \right) \\
&= \mathbf{e}_{\theta}^{\varepsilon} \cdot \frac{1}{R^{\varepsilon} \sin_{\varepsilon}(\vartheta^{\varepsilon})} \left(\mathbf{e}_R^{\varepsilon} \frac{\partial^{\varepsilon} \mathrm{r}_R}{\partial \theta^{\varepsilon}} + \mathbf{e}_{\vartheta}^{\varepsilon} \frac{\partial^{\varepsilon} \mathrm{r}_{\vartheta}}{\partial \theta^{\varepsilon}} + \mathbf{e}_{\theta}^{\varepsilon} \frac{\partial^{\varepsilon} \mathrm{r}_{\theta}}{\partial \theta^{\varepsilon}} \right. \\
&\quad \left. + \mathrm{r}_R \frac{\partial^{\varepsilon} \mathbf{e}_R^{\varepsilon}}{\partial \theta^{\varepsilon}} + \mathrm{r}_{\vartheta} \frac{\partial^{\varepsilon} \mathbf{e}_{\vartheta}^{\varepsilon}}{\partial \theta^{\varepsilon}} + \mathrm{r}_{\theta} \frac{\partial^{\varepsilon} \mathbf{e}_{\theta}^{\varepsilon}}{\partial \theta^{\varepsilon}} \right) \\
&= \frac{1}{R^{\varepsilon} \sin_{\varepsilon}(\vartheta^{\varepsilon})} \left(\frac{\partial^{\varepsilon} \mathrm{r}_{\theta}}{\partial \theta^{\varepsilon}} + \mathrm{r}_{\vartheta} \cos_{\varepsilon}(\vartheta^{\varepsilon}) \right) + \frac{\mathrm{r}_{\theta}}{R^{\varepsilon}}.
\end{aligned}$$

Thus, we rewrite the local fractional gradient operator in the Cantor-type spherical coordinates

$$\nabla^{\varepsilon} \cdot \mathbf{r} = \frac{\partial^{\varepsilon} \mathrm{r}_R}{\partial R^{\varepsilon}} + \frac{2\mathrm{r}_R}{R^{\varepsilon}} + \frac{1}{R^{\varepsilon}} \frac{\partial^{\varepsilon} \mathrm{r}_{\vartheta}}{\partial \vartheta^{\varepsilon}} + \frac{1}{R^{\varepsilon} \sin_{\varepsilon}(\vartheta^{\varepsilon})} \left(\frac{\partial^{\varepsilon} \mathrm{r}_{\theta}}{\partial \theta^{\varepsilon}} + \mathrm{r}_{\vartheta} \cos_{\varepsilon}(\vartheta^{\varepsilon}) \right).$$

The local fractional curl operator in the Cantor-type spherical coordinates takes the form

$$\nabla^{\varepsilon} \times \mathbf{r} = \mathbf{e}_R^{\varepsilon} \times \frac{\partial^{\varepsilon} \mathbf{r}}{\partial R^{\varepsilon}} + \mathbf{e}_{\vartheta}^{\varepsilon} \times \frac{1}{R^{\varepsilon}} \frac{\partial^{\varepsilon} \mathbf{r}}{\partial \vartheta^{\varepsilon}} + \mathbf{e}_{\theta}^{\varepsilon} \times \frac{1}{R^{\varepsilon} \sin_{\varepsilon}(\vartheta^{\varepsilon})} \frac{\partial^{\varepsilon} \mathbf{r}}{\partial \theta^{\varepsilon}},$$

where

$$\mathbf{e}_R^{\varepsilon} \times \frac{\partial^{\varepsilon} \mathbf{r}}{\partial R^{\varepsilon}} = \mathbf{e}_R^{\varepsilon} \times \frac{\partial^{\varepsilon}}{\partial R^{\varepsilon}} \left(\mathrm{r}_R \mathbf{e}_R^{\varepsilon} + \mathrm{r}_{\vartheta} \mathbf{e}_{\vartheta}^{\varepsilon} + \mathrm{r}_{\theta} \mathbf{e}_{\theta}^{\varepsilon} \right)$$

$$= \mathbf{e}_R^\varepsilon \times \left(\mathbf{e}_R^\varepsilon \frac{\partial^\varepsilon \mathrm{r}_R}{\partial R^\varepsilon} + \mathbf{e}_\vartheta^\varepsilon \frac{\partial^\varepsilon \mathrm{r}_\vartheta}{\partial R^\varepsilon} + \mathbf{e}_\theta^\varepsilon \frac{\partial^\varepsilon \mathrm{r}_\theta}{\partial R^\varepsilon} + \mathrm{r}_R \frac{\partial^\varepsilon \mathbf{e}_R^\varepsilon}{\partial R^\varepsilon} + \mathrm{r}_\vartheta \frac{\partial^\varepsilon \mathbf{e}_\vartheta^\varepsilon}{\partial R^\varepsilon} + \mathrm{r}_\theta \frac{\partial^\varepsilon \mathbf{e}_\theta^\varepsilon}{\partial R^\varepsilon} \right)$$
$$= \mathbf{e}_\theta^\varepsilon \frac{\partial^\varepsilon \mathrm{r}_R}{\partial R^\varepsilon} - \mathbf{e}_\vartheta^\varepsilon \frac{\partial^\varepsilon \mathrm{r}_\theta}{\partial R^\varepsilon},$$

$$\begin{aligned} \mathbf{e}_\vartheta^\varepsilon \times \frac{1}{R^\varepsilon} \frac{\partial^\varepsilon \mathbf{r}}{\partial \vartheta^\varepsilon} &= \mathbf{e}_\vartheta^\varepsilon \times \frac{1}{R^\varepsilon} \frac{\partial^\varepsilon}{\partial \vartheta^\varepsilon} \left(\mathrm{r}_R \mathbf{e}_R^\varepsilon + \mathrm{r}_\vartheta \mathbf{e}_\vartheta^\varepsilon + \mathrm{r}_\theta \mathbf{e}_\theta^\varepsilon \right) \\ &= \mathbf{e}_\vartheta^\varepsilon \times \frac{1}{R^\varepsilon} \left(\mathbf{e}_R^\varepsilon \frac{\partial^\varepsilon \mathrm{r}_R}{\partial \vartheta^\varepsilon} + \mathbf{e}_\vartheta^\varepsilon \frac{\partial^\varepsilon \mathrm{r}_\vartheta}{\partial \vartheta^\varepsilon} + \mathbf{e}_\theta^\varepsilon \frac{\partial^\varepsilon \mathrm{r}_\theta}{\partial \vartheta^\varepsilon} + \mathrm{r}_R \frac{\partial^\varepsilon \mathbf{e}_R^\varepsilon}{\partial \vartheta^\varepsilon} \right. \\ &\quad \left. + \mathrm{r}_\vartheta \frac{\partial^\varepsilon \mathbf{e}_\vartheta^\varepsilon}{\partial \vartheta^\varepsilon} + \mathrm{r}_\theta \frac{\partial^\varepsilon \mathbf{e}_\theta^\varepsilon}{\partial \vartheta^\varepsilon} \right) \\ &= -\frac{1}{R^\varepsilon} \frac{\partial^\varepsilon \mathrm{r}_R}{\partial \vartheta^\varepsilon} \mathbf{e}_\theta^\varepsilon + \frac{1}{R^\varepsilon} \frac{\partial^\varepsilon \mathrm{r}_\theta}{\partial \vartheta^\varepsilon} \mathbf{e}_R^\varepsilon - \frac{\mathrm{r}_\vartheta}{R^\varepsilon} \mathbf{e}_\theta^\varepsilon, \end{aligned}$$

$$\begin{aligned} \mathbf{e}_\theta^\varepsilon \times \frac{1}{R^\varepsilon \sin_\varepsilon (\vartheta^\varepsilon)} \frac{\partial^\varepsilon \mathbf{r}}{\partial \theta^\varepsilon} &= \mathbf{e}_\theta^\varepsilon \times \frac{1}{R^\varepsilon \sin_\varepsilon (\vartheta^\varepsilon)} \frac{\partial^\varepsilon}{\partial \theta^\varepsilon} \left(\mathrm{r}_R \mathbf{e}_R^\varepsilon + \mathrm{r}_\vartheta \mathbf{e}_\vartheta^\varepsilon + \mathrm{r}_\theta \mathbf{e}_\theta^\varepsilon \right) \\ &= \mathbf{e}_\theta^\varepsilon \times \frac{1}{R^\varepsilon \sin_\varepsilon (\vartheta^\varepsilon)} \left(\mathbf{e}_R^\varepsilon \frac{\partial^\varepsilon \mathrm{r}_R}{\partial \theta^\varepsilon} + \mathbf{e}_\vartheta^\varepsilon \frac{\partial^\varepsilon \mathrm{r}_\vartheta}{\partial \theta^\varepsilon} + \mathbf{e}_\theta^\varepsilon \frac{\partial^\varepsilon \mathrm{r}_\theta}{\partial \theta^\varepsilon} \right. \\ &\quad \left. + \mathrm{r}_R \frac{\partial^\varepsilon \mathbf{e}_R^\varepsilon}{\partial \theta^\varepsilon} + \mathrm{r}_\vartheta \frac{\partial^\varepsilon \mathbf{e}_\vartheta^\varepsilon}{\partial \theta^\varepsilon} + \mathrm{r}_\theta \frac{\partial^\varepsilon \mathbf{e}_\theta^\varepsilon}{\partial \theta^\varepsilon} \right) \\ &= \frac{1}{R^\varepsilon \sin_\varepsilon (\vartheta^\varepsilon)} \left(\mathbf{e}_\vartheta^\varepsilon \frac{\partial^\varepsilon \mathrm{r}_R}{\partial \theta^\varepsilon} - \mathbf{e}_R^\varepsilon \frac{\partial^\varepsilon \mathrm{r}_\vartheta}{\partial \theta^\varepsilon} - \sin_\varepsilon \left(\vartheta^\varepsilon \right) \mathrm{r}_\theta \mathbf{e}_\vartheta^\varepsilon \right. \\ &\quad \left. + \mathrm{r}_\theta \cos_\varepsilon \left(\vartheta^\varepsilon \right) \mathbf{e}_R^\varepsilon \right). \end{aligned}$$

Therefore, we obtain the local fractional curl operator in the Cantor-type spherical coordinates

$$\begin{aligned} \nabla^\varepsilon \times \mathbf{r} &= \mathbf{e}_\theta^\varepsilon \frac{\partial^\varepsilon \mathrm{r}_R}{\partial R^\varepsilon} - \mathbf{e}_\vartheta^\varepsilon \frac{\partial^\varepsilon \mathrm{r}_\theta}{\partial R^\varepsilon} - \frac{1}{R^\varepsilon} \frac{\partial^\varepsilon \mathrm{r}_R}{\partial \vartheta^\varepsilon} \mathbf{e}_\theta^\varepsilon + \frac{1}{R^\varepsilon} \frac{\partial^\varepsilon \mathrm{r}_\theta}{\partial \vartheta^\varepsilon} \mathbf{e}_R^\varepsilon - \frac{\mathrm{r}_\vartheta}{R^\varepsilon} \mathbf{e}_\theta^\varepsilon \\ &\quad + \frac{1}{R^\varepsilon \sin_\varepsilon (\vartheta^\varepsilon)} \left(\mathbf{e}_\vartheta^\varepsilon \frac{\partial^\varepsilon \mathrm{r}_R}{\partial \theta^\varepsilon} - \mathbf{e}_R^\varepsilon \frac{\partial^\varepsilon \mathrm{r}_\vartheta}{\partial \theta^\varepsilon} - \sin_\varepsilon \left(\vartheta^\varepsilon \right) \mathrm{r}_\theta \mathbf{e}_\vartheta^\varepsilon + \mathrm{r}_\theta \cos_\varepsilon \left(\vartheta^\varepsilon \right) \mathbf{e}_R^\varepsilon \right) \\ &= \mathbf{e}_R^\varepsilon \left(\frac{1}{R^\varepsilon} \frac{\partial^\varepsilon \mathrm{r}_\theta}{\partial \vartheta^\varepsilon} + \frac{1}{R^\varepsilon \sin_\varepsilon (\vartheta^\varepsilon)} \frac{\partial^\varepsilon \mathrm{r}_\vartheta}{\partial \theta^\varepsilon} + \frac{\mathrm{r}_\theta \cos_\varepsilon (\vartheta^\varepsilon)}{R^\varepsilon \sin_\varepsilon (\vartheta^\varepsilon)} \right) \\ &\quad + \mathbf{e}_\vartheta^\varepsilon \left(\frac{1}{R^\varepsilon \sin_\varepsilon (\vartheta^\varepsilon)} \frac{\partial^\varepsilon \mathrm{r}_R}{\partial \theta^\varepsilon} - \frac{\partial^\varepsilon \mathrm{r}_\theta}{\partial R^\varepsilon} - \frac{\mathrm{r}_\theta}{R^\varepsilon} \right) + \mathbf{e}_\theta^\varepsilon \left(\frac{\partial^\varepsilon \mathrm{r}_R}{\partial R^\varepsilon} - \frac{1}{R^\varepsilon} \frac{\partial^\varepsilon \mathrm{r}_R}{\partial \vartheta^\varepsilon} - \frac{\mathrm{r}_\vartheta}{R^\varepsilon} \right) \\ &= \mathbf{e}_R^\varepsilon \frac{1}{R^\varepsilon \sin_\varepsilon (\vartheta^\varepsilon)} \left(\frac{\partial^\varepsilon \left(\mathrm{r}_\theta \sin_\varepsilon (\vartheta^\varepsilon) \right)}{\partial \vartheta^\varepsilon} - \frac{\partial^\varepsilon \mathrm{r}_\theta}{\partial \vartheta^\varepsilon} \right) \\ &\quad + \mathbf{e}_\vartheta^\varepsilon \left(\frac{1}{R^\varepsilon \sin_\varepsilon (\vartheta^\varepsilon)} \frac{\partial^\varepsilon \mathrm{r}_R}{\partial \theta^\varepsilon} - \frac{\partial^\varepsilon \mathrm{r}_\theta}{\partial R^\varepsilon} - \frac{\mathrm{r}_\theta}{R^\varepsilon} \right) + \mathbf{e}_\theta^\varepsilon \left(\frac{\partial^\varepsilon \mathrm{r}_R}{\partial R^\varepsilon} - \frac{1}{R^\varepsilon} \frac{\partial^\varepsilon \mathrm{r}_R}{\partial \vartheta^\varepsilon} - \frac{\mathrm{r}_\vartheta}{R^\varepsilon} \right). \end{aligned}$$

The local fractional Laplace operator in Cantor-type spherical coordinates is presented as follows:

$$\begin{aligned}\nabla^{2\varepsilon}\psi(R,\theta,\sigma) &= \left(\mathbf{e}_R^{\varepsilon}\frac{\partial^{\varepsilon}}{\partial R^{\varepsilon}}+\frac{\mathbf{e}_{\vartheta}^{\varepsilon}}{R^{\varepsilon}}\frac{\partial^{\varepsilon}}{\partial\vartheta^{\varepsilon}}+\frac{\mathbf{e}_{\theta}^{\varepsilon}}{R^{\varepsilon}\sin_{\varepsilon}(\vartheta^{\varepsilon})}\frac{\partial^{\varepsilon}}{\partial\theta^{\varepsilon}}\right)\cdot\left(\mathbf{e}_R^{\varepsilon}\frac{\partial^{\varepsilon}\psi}{\partial R^{\varepsilon}}+\frac{\mathbf{e}_{\vartheta}^{\varepsilon}}{R^{\varepsilon}}\frac{\partial^{\varepsilon}\psi}{\partial\vartheta^{\varepsilon}}\right.\\
&\quad\left.+\frac{\mathbf{e}_{\theta}^{\varepsilon}}{R^{\varepsilon}\sin_{\varepsilon}(\vartheta^{\varepsilon})}\frac{\partial^{\varepsilon}\psi}{\partial\theta^{\varepsilon}}\right)\\
&= \mathbf{e}_R^{\varepsilon}\cdot\frac{\partial^{\varepsilon}}{\partial R^{\varepsilon}}\left(\mathbf{e}_R^{\varepsilon}\frac{\partial^{\varepsilon}\psi}{\partial R^{\varepsilon}}+\mathbf{e}_{\vartheta}^{\varepsilon}\frac{1}{R^{\varepsilon}}\frac{\partial^{\varepsilon}\psi}{\partial\vartheta^{\varepsilon}}+\mathbf{e}_{\theta}^{\varepsilon}\frac{1}{R^{\varepsilon}\sin_{\varepsilon}(\vartheta^{\varepsilon})}\frac{\partial^{\varepsilon}\psi}{\partial\theta^{\varepsilon}}\right)\\
&\quad+\mathbf{e}_{\vartheta}^{\varepsilon}\cdot\frac{1}{R^{\varepsilon}}\frac{\partial^{\varepsilon}}{\partial\vartheta^{\varepsilon}}\left(\mathbf{e}_R^{\varepsilon}\frac{\partial^{\varepsilon}\psi}{\partial R^{\varepsilon}}+\mathbf{e}_{\vartheta}^{\varepsilon}\frac{1}{R^{\varepsilon}}\frac{\partial^{\varepsilon}\psi}{\partial\vartheta^{\varepsilon}}+\mathbf{e}_{\theta}^{\varepsilon}\frac{1}{R^{\varepsilon}\sin_{\varepsilon}(\vartheta^{\varepsilon})}\frac{\partial^{\varepsilon}\psi}{\partial\theta^{\varepsilon}}\right)\\
&\quad+\mathbf{e}_{\theta}^{\varepsilon}\cdot\frac{1}{R^{\varepsilon}\sin_{\varepsilon}(\vartheta^{\varepsilon})}\frac{\partial^{\varepsilon}}{\partial\theta^{\varepsilon}}\left(\mathbf{e}_R^{\varepsilon}\frac{\partial^{\varepsilon}\psi}{\partial R^{\varepsilon}}+\mathbf{e}_{\vartheta}^{\varepsilon}\frac{1}{R^{\varepsilon}}\frac{\partial^{\varepsilon}\psi}{\partial\vartheta^{\varepsilon}}\right.\\
&\quad\left.+\mathbf{e}_{\theta}^{\varepsilon}\frac{1}{R^{\varepsilon}\sin_{\varepsilon}(\vartheta^{\varepsilon})}\frac{\partial^{\varepsilon}\psi}{\partial\theta^{\varepsilon}}\right),\end{aligned}$$

where

$$\mathbf{e}_R^{\varepsilon}\cdot\frac{\partial^{\varepsilon}}{\partial R^{\varepsilon}}\left(\mathbf{e}_R^{\varepsilon}\frac{\partial^{\varepsilon}\psi}{\partial R^{\varepsilon}}+\mathbf{e}_{\vartheta}^{\varepsilon}\frac{1}{R^{\varepsilon}}\frac{\partial^{\varepsilon}\psi}{\partial\vartheta^{\varepsilon}}+\mathbf{e}_{\theta}^{\varepsilon}\frac{1}{R^{\varepsilon}\sin_{\varepsilon}(\vartheta^{\varepsilon})}\frac{\partial^{\varepsilon}\psi}{\partial\theta^{\varepsilon}}\right)=\frac{\partial^{2\varepsilon}\psi}{\partial R^{2\varepsilon}},$$

$$\begin{aligned}&\mathbf{e}_{\vartheta}^{\varepsilon}\cdot\frac{1}{R^{\varepsilon}}\frac{\partial^{\varepsilon}}{\partial\vartheta^{\varepsilon}}\left(\mathbf{e}_R^{\varepsilon}\frac{\partial^{\varepsilon}\psi}{\partial R^{\varepsilon}}+\mathbf{e}_{\vartheta}^{\varepsilon}\frac{1}{R^{\varepsilon}}\frac{\partial^{\varepsilon}\psi}{\partial\vartheta^{\varepsilon}}+\mathbf{e}_{\theta}^{\varepsilon}\frac{1}{R^{\varepsilon}\sin_{\varepsilon}(\vartheta^{\varepsilon})}\frac{\partial^{\varepsilon}\psi}{\partial\theta^{\varepsilon}}\right)\\
&=\frac{1}{R^{\varepsilon}}\frac{\partial^{\varepsilon}\psi}{\partial R^{\varepsilon}}+\frac{1}{R^{\varepsilon}}\frac{\partial^{\varepsilon}}{\partial\vartheta^{\varepsilon}}\left(\frac{1}{R^{\varepsilon}}\frac{\partial^{\varepsilon}\psi}{\partial\vartheta^{\varepsilon}}\right)\\
&=\frac{1}{R^{2\varepsilon}}\frac{\partial^{2\varepsilon}\psi}{\partial\vartheta^{2\varepsilon}}+\frac{1}{R^{\varepsilon}}\frac{\partial^{\varepsilon}\psi}{\partial R^{\varepsilon}},\end{aligned}$$

$$\begin{aligned}&\mathbf{e}_{\theta}^{\varepsilon}\cdot\frac{1}{R^{\varepsilon}\sin_{\varepsilon}(\vartheta^{\varepsilon})}\frac{\partial^{\varepsilon}}{\partial\theta^{\varepsilon}}\left(\mathbf{e}_R^{\varepsilon}\frac{\partial^{\varepsilon}\psi}{\partial R^{\varepsilon}}+\mathbf{e}_{\vartheta}^{\varepsilon}\frac{1}{R^{\varepsilon}}\frac{\partial^{\varepsilon}\psi}{\partial\vartheta^{\varepsilon}}+\mathbf{e}_{\theta}^{\varepsilon}\frac{1}{R^{\varepsilon}\sin_{\varepsilon}(\vartheta^{\varepsilon})}\frac{\partial^{\varepsilon}\psi}{\partial\theta^{\varepsilon}}\right)\\
&=\frac{1}{R^{\varepsilon}}\frac{\partial^{\varepsilon}\psi}{\partial R^{\varepsilon}}+\frac{\cos_{\varepsilon}(\vartheta^{\varepsilon})}{R^{2\varepsilon}\sin_{\varepsilon}(\vartheta^{\varepsilon})}\frac{\partial^{\varepsilon}\psi}{\partial\vartheta^{\varepsilon}}+\frac{1}{R^{2\varepsilon}\sin_{\varepsilon}^{2\varepsilon}(\vartheta^{\varepsilon})}\frac{\partial^{2\varepsilon}\psi}{\partial\theta^{2\varepsilon}}.\end{aligned}$$

Thus, the local fractional Laplace operator in Cantor-type spherical coordinates takes the form

$$\begin{aligned}\nabla^{2\varepsilon}\psi(R,\theta,\sigma) &= \frac{1}{R^{2\varepsilon}}\frac{\partial^{2\varepsilon}\psi}{\partial\vartheta^{2\varepsilon}}+\frac{2}{R^{\varepsilon}}\frac{\partial^{\varepsilon}\psi}{\partial R^{\varepsilon}}+\frac{1}{R^{2\varepsilon}}\frac{\partial^{2\varepsilon}\psi}{\partial\vartheta^{2\varepsilon}}+\frac{\cos_{\varepsilon}(\vartheta^{\varepsilon})}{R^{2\varepsilon}\sin_{\varepsilon}(\vartheta^{\varepsilon})}\frac{\partial^{\varepsilon}\psi}{\partial\vartheta^{\varepsilon}}\\
&\quad+\frac{1}{R^{2\varepsilon}\sin_{\varepsilon}^{2\varepsilon}(\vartheta^{\varepsilon})}\frac{\partial^{2\varepsilon}\psi}{\partial\theta^{2\varepsilon}}.\end{aligned}$$

Appendix E
Tables of local fractional Fourier transform operators

We present the list of the local fractional Fourier transforms (Table E.1):

$$\Im\left[\theta\left(\tau\right)\right]=\Theta\left(\omega\right)=\frac{1}{\Gamma\left(1+\varepsilon\right)}\int_{-\infty}^{\infty}\theta\left(\tau\right)E_{\varepsilon}\left(-i^{\varepsilon}\tau^{\varepsilon}\omega^{\varepsilon}\right)\left(\mathrm{d}\tau\right)^{\varepsilon}.$$

Table E.1 Tables for local fractional Fourier transform operators

Transforms	Functions
$\frac{(2\pi)^{\varepsilon}}{\Gamma(1+\varepsilon)}\delta_{\varepsilon}\left(\omega\right)$	1
1	$\delta_{\varepsilon}\left(\tau\right)$
$E_{\varepsilon}\left(-i^{\varepsilon}\tau_0^{\varepsilon}\omega^{\varepsilon}\right)$	$\delta_{\varepsilon}\left(\tau-\tau_0\right)$
$\theta\left(\tau_0\right)E_{\varepsilon}\left(-i^{\varepsilon}\tau_0^{\varepsilon}\omega^{\varepsilon}\right)$	$\delta_{\varepsilon}\left(\tau-\tau_0\right)\theta\left(\tau\right)$
$i^{\varepsilon}\omega^{\varepsilon}$	$\delta_{\varepsilon}^{(\varepsilon)}\left(\tau\right)$
$\frac{(2\pi)^{\varepsilon}}{\Gamma(1+\varepsilon)}\delta_{\varepsilon}\left(\omega-\omega_0\right)$	$E_{\varepsilon}\left(i^{\varepsilon}\tau^{\varepsilon}\omega_0^{\varepsilon}\right)$
$\frac{1}{2}\frac{(2\pi)^{\varepsilon}}{\Gamma(1+\varepsilon)}\delta_{\varepsilon}\left(\omega\right)+\frac{1}{i^{\varepsilon}\omega^{\varepsilon}}$	$H_{\varepsilon}\left(\tau\right)$
$\frac{2\sin_{\varepsilon}\left(\frac{\omega}{2}\right)^{\varepsilon}}{\omega^{\varepsilon}}$	$\mathrm{rect}_{\varepsilon}\left(\tau\right)$
$\frac{4\sin_{\varepsilon}^{2}\left(\frac{\omega}{2}\right)^{\varepsilon}}{\omega^{2\varepsilon}}$	$\mathrm{trig}_{\varepsilon}\left(\tau\right)$
$\frac{2a^{\varepsilon}}{a^{2\varepsilon}+\omega^{2\varepsilon}}$	$\Xi_{a.\varepsilon}\left(\tau\right)$
$\frac{2}{i^{\varepsilon}\omega^{\varepsilon}}$	$\mathrm{sgn}_{\varepsilon}\left(\tau\right)$
$\frac{\pi^{\frac{\varepsilon}{2}}\sqrt{\frac{1}{a}}}{\Gamma(1+\varepsilon)}E_{\varepsilon}\left[-\frac{1}{a}\left(\frac{\omega}{2}\right)^{2\varepsilon}\right]$	$E_{\varepsilon}\left(-a\tau^{2\varepsilon}\right)$
$\frac{\pi^{\frac{\varepsilon}{2}}}{\Gamma(1+\varepsilon)}E_{\varepsilon}\left[-\left(\frac{\omega}{2}\right)^{2\varepsilon}\right]$	$E_{\varepsilon}\left(-\tau^{2\varepsilon}\right)$
$\frac{(2\pi)^{\varepsilon}}{\Gamma(1+\varepsilon)}\frac{\left[\delta_{\varepsilon}(\omega+a)-\delta_{\varepsilon}(\omega-a)\right]}{2i^{\varepsilon}}$	$\sin_{\varepsilon}\left(a^{\varepsilon}\tau^{\varepsilon}\right)$
$\frac{(2\pi)^{\varepsilon}}{\Gamma(1+\varepsilon)}\frac{\left[\delta_{\varepsilon}(\omega+a)+\delta_{\varepsilon}(\omega-a)\right]}{2}$	$\cos_{\varepsilon}\left(a^{\varepsilon}\tau^{\varepsilon}\right)$
$\frac{(2\pi i)^{\varepsilon}}{\Gamma(1+\varepsilon)}\delta_{\varepsilon}^{(\varepsilon)}\left(\omega\right)$	τ^{ε}
$\frac{1}{\omega^{\varepsilon}i^{\varepsilon}+a^{\varepsilon}}$	$H_{\varepsilon}\left(\tau\right)E_{\varepsilon}\left(-\tau^{\varepsilon}a^{\varepsilon}\right)$

Appendix F
Tables of local fractional Laplace transform operators

We start with the local fractional Laplace transform of some elementary functions.

Find the local fractional Laplace transform of the alternative definition of analogous rectangular pulse, denoted by $\mathrm{rect}_\varepsilon(\tau, \tau_0, \tau_1) = H_\varepsilon(\tau - \tau_0) - H_\varepsilon(\tau - \tau_1)$, namely,

$$\begin{aligned} \mathrm{M}\left[\mathrm{rect}_\varepsilon(\tau, \tau_0, \tau_1)\right] &= \frac{1}{\Gamma(1+\varepsilon)} \int_0^\infty \left(H_\varepsilon(\tau - \tau_0)\right. \\ &\quad \left. - H_\varepsilon(\tau - \tau_1)\right) E_\varepsilon\left(-\tau^\varepsilon s^\varepsilon\right) (\mathrm{d}\tau)^\varepsilon \\ &= \frac{1}{\Gamma(1+\varepsilon)} \int_{\tau_0}^{\tau_1} E_\varepsilon\left(-\tau^\varepsilon s^\varepsilon\right) (\mathrm{d}\tau)^\varepsilon \\ &= \frac{1}{s^\varepsilon}\left[E_\varepsilon\left(-\tau_0^\varepsilon s^\varepsilon\right) - E_\varepsilon\left(-\tau_1^\varepsilon s^\varepsilon\right)\right], \end{aligned}$$

where $0 < \tau_0 < \tau_1$.

Find the local fractional Laplace transform of the analogous Heaviside function, defined by

$$H_\varepsilon(\tau - \tau_0) = \begin{cases} 0, & \tau \le \tau_0, \\ 1, & \tau > \tau_0, \end{cases}$$

that is,

$$\begin{aligned} \mathrm{M}\left[H_\varepsilon(\tau)\right] &= \frac{1}{\Gamma(1+\varepsilon)} \int_0^\infty H_\varepsilon(\tau - \tau_0) E_\varepsilon\left(-\tau^\varepsilon s^\varepsilon\right) (\mathrm{d}\tau)^\varepsilon \\ &= \frac{1}{\Gamma(1+\varepsilon)} \int_{\tau_0}^\infty E_\varepsilon\left(-\tau^\varepsilon s^\varepsilon\right) (\mathrm{d}\tau)^\varepsilon \\ &= \frac{1}{s^\varepsilon} E_\varepsilon\left(-\tau_0^\varepsilon s^\varepsilon\right), \end{aligned}$$

where $\tau_0 > 0$.

$$\mathrm{M}\left[\frac{1}{a^2 - b^2}\left(\cos_\varepsilon\left(b\tau^\varepsilon\right) - \cos_\varepsilon\left(a\tau^\varepsilon\right)\right)\right] = \frac{s^\varepsilon}{\left(s^{2\varepsilon} + a^2\right)\left(s^{2\varepsilon} + b^2\right)}.$$

Proof. It follows that

$$\mathrm{M}\left[\frac{1}{a^2-b^2}\left(\cos_\varepsilon\left(b\tau^\varepsilon\right)-\cos_\varepsilon\left(a\tau^\varepsilon\right)\right)\right]=\frac{1}{\Gamma(1+\varepsilon)}\int_0^\infty\left[\frac{1}{a^2-b^2}(\cos_\varepsilon(b\tau^\varepsilon)\right.$$
$$\left.-\cos_\varepsilon(a\tau^\varepsilon))\right]E_\varepsilon\left(-\tau^\varepsilon s^\varepsilon\right)(\mathrm{d}\tau)^\varepsilon$$
$$=\frac{1}{a^2-b^2}\left[\frac{s^\varepsilon}{s^{2\varepsilon}+b^2}-\frac{s^\varepsilon}{s^{2\varepsilon}+a^2}\right]$$
$$=\frac{s^\varepsilon}{\left(s^{2\varepsilon}+a^2\right)\left(s^{2\varepsilon}+b^2\right)}.$$

□

$$\mathrm{M}\left[\frac{1}{a^2-b^2}\left(\frac{\sinh_\varepsilon\left(a\tau^\varepsilon\right)}{a}-\frac{\sinh_\varepsilon\left(b\tau^\varepsilon\right)}{b}\right)\right]=\frac{1}{\left(s^{2\varepsilon}-a^2\right)\left(s^{2\varepsilon}-b^2\right)}.$$

Proof. We put

$$\mathrm{M}\left[\frac{1}{a^2-b^2}\left(\frac{\sinh_\varepsilon\left(a\tau^\varepsilon\right)}{a}-\frac{\sinh_\varepsilon\left(b\tau^\varepsilon\right)}{b}\right)\right]$$
$$=\frac{1}{\Gamma(1+\varepsilon)}\int_0^\infty\left\{\frac{1}{a^2-b^2}\left[\frac{\sinh_\varepsilon\left(a\tau^\varepsilon\right)}{a}-\frac{\sinh_\varepsilon\left(\eta\tau^\varepsilon\right)}{b}\right]\right\}E_\varepsilon\left(-\tau^\varepsilon s^\varepsilon\right)(\mathrm{d}\tau)^\varepsilon$$
$$=\frac{1}{a^2-b^2}\left[\frac{1}{s^{2\varepsilon}-a^2}-\frac{1}{s^{2\varepsilon}-b^2}\right]$$
$$=\frac{1}{\left(s^{2\varepsilon}-a^2\right)\left(s^{2\varepsilon}-b^2\right)}.$$

□

$$\mathrm{M}\left[\frac{1}{a^2-b^2}\left(\cosh_\varepsilon\left(a\tau^\varepsilon\right)-\cosh_\varepsilon\left(b\tau^\varepsilon\right)\right)\right]=\frac{s^\varepsilon}{\left(s^{2\varepsilon}-a^2\right)\left(s^{2\varepsilon}-b^2\right)}.$$

Proof. This gives

$$\mathrm{M}\left[\frac{1}{a^2-b^2}\left(\cosh_\varepsilon\left(a\tau^\varepsilon\right)-\cosh_\varepsilon\left(b\tau^\varepsilon\right)\right)\right]=\frac{1}{\Gamma(1+\varepsilon)}\int_0^\infty\left[\frac{1}{a^2-b^2}\left(\cosh_\varepsilon\left(a\tau^\varepsilon\right)\right.\right.$$
$$\left.\left.-\cosh_\varepsilon\left(b\tau^\varepsilon\right)\right)\right]E_\varepsilon\left(-\tau^\varepsilon s^\varepsilon\right)(\mathrm{d}\tau)^\varepsilon$$
$$=\frac{1}{a^2-b^2}\left(\frac{s^\varepsilon}{s^{2\varepsilon}-a^2}-\frac{s^\varepsilon}{s^{2\varepsilon}-b^2}\right)$$
$$=\frac{s^\varepsilon}{\left(s^{2\varepsilon}-a^2\right)\left(s^{2\varepsilon}-b^2\right)}.$$

□

$$\mathrm{M}\left[\frac{\tau^{\varepsilon}}{\Gamma(1+\varepsilon)}\sin_{\varepsilon}\left(\zeta\tau^{\varepsilon}\right)\right]=\frac{\zeta s^{\varepsilon}}{\left(s^{2\varepsilon}+\zeta^{2\varepsilon}\right)^{2}}.$$

Proof. We observe that

$$\begin{aligned}\mathrm{M}\left[\frac{\tau^{\varepsilon}}{\Gamma(1+\varepsilon)}\sin_{\varepsilon}\left(\zeta\tau^{\varepsilon}\right)\right]&=\frac{1}{\Gamma(1+\varepsilon)}\int_{0}^{\infty}\left[\frac{\tau^{\varepsilon}}{\Gamma(1+\varepsilon)}\right.\\&\quad\times\left.\frac{E_{\varepsilon}\left(i^{\varepsilon}\zeta\tau^{\varepsilon}\right)-E_{\varepsilon}\left(-i^{\varepsilon}\zeta\tau^{\varepsilon}\right)}{2i^{\varepsilon}}\right]E_{\varepsilon}\left(-\tau^{\varepsilon}s^{\varepsilon}\right)(\mathrm{d}\tau)^{\varepsilon}\\&=\frac{1}{2i^{\varepsilon}}\left[\frac{1}{(s^{\varepsilon}-\zeta i^{\varepsilon})^{2}}-\frac{1}{(s^{\varepsilon}+\zeta i^{\varepsilon})^{2}}\right]\\&=\frac{\zeta s^{\varepsilon}}{\left(s^{2\varepsilon}+\zeta^{2\varepsilon}\right)^{2}}.\end{aligned}$$

□

$$\mathrm{M}\left[\frac{\tau^{\varepsilon}}{\Gamma(1+\varepsilon)}\cos_{\varepsilon}\left(\zeta\tau^{\varepsilon}\right)\right]=\frac{s^{2\varepsilon}-\zeta^{2}}{\left(s^{2\varepsilon}+\zeta^{2}\right)^{2}}.$$

Proof. We have that

$$\begin{aligned}\mathrm{M}\left[\frac{\tau^{\varepsilon}}{\Gamma(1+\varepsilon)}\cos_{\varepsilon}\left(\zeta\tau^{\varepsilon}\right)\right]&=\frac{1}{\Gamma(1+\varepsilon)}\int_{0}^{\infty}\left[\frac{\tau^{\varepsilon}}{\Gamma(1+\varepsilon)}\right.\\&\quad\times\left.\frac{E_{\varepsilon}\left(i^{\varepsilon}\zeta\tau^{\varepsilon}\right)+E_{\varepsilon}\left(-i^{\varepsilon}\zeta\tau^{\varepsilon}\right)}{2}\right]E_{\varepsilon}\left(-\tau^{\varepsilon}s^{\varepsilon}\right)(\mathrm{d}\tau)^{\varepsilon}\\&=\frac{1}{2}\left[\frac{1}{(s^{\varepsilon}-\zeta i^{\varepsilon})^{2}}+\frac{1}{(s^{\varepsilon}+\zeta i^{\varepsilon})^{2}}\right]\\&=\frac{s^{2\varepsilon}-\zeta^{2}}{\left(s^{2\varepsilon}+\zeta^{2}\right)^{2}}.\end{aligned}$$

□

$$\mathrm{M}\left[\frac{\tau^{\varepsilon}}{\Gamma(1+\varepsilon)}\sinh_{\varepsilon}\left(\zeta\tau^{\varepsilon}\right)\right]=\frac{2\zeta s^{\varepsilon}}{\left(s^{2\varepsilon}-\zeta^{2\varepsilon}\right)^{2}}.$$

Proof. It creates that

$$\begin{aligned}\mathrm{M}\left[\frac{\tau^{\varepsilon}}{\Gamma(1+\varepsilon)}\sinh_{\varepsilon}\left(\zeta\tau^{\varepsilon}\right)\right]&=\frac{1}{\Gamma(1+\varepsilon)}\int_{0}^{\infty}\left[\frac{\tau^{\varepsilon}}{\Gamma(1+\varepsilon)}\right.\\&\quad\times\left.\frac{E_{\varepsilon}\left(\zeta\tau^{\varepsilon}\right)-E_{\varepsilon}\left(-\zeta\tau^{\varepsilon}\right)}{2}\right]E_{\varepsilon}\left(-\tau^{\varepsilon}s^{\varepsilon}\right)(\mathrm{d}\tau)^{\varepsilon}\end{aligned}$$

$$= \frac{1}{2}\left[\frac{1}{(s^{\varepsilon}-\zeta)^2} - \frac{1}{(s^{\varepsilon}+\zeta)^2}\right]$$
$$= \frac{2\zeta s^{\varepsilon}}{\left(s^{2\varepsilon}-\zeta^{2\varepsilon}\right)^2}.$$

□

$$\mathrm{M}\left[\frac{\tau^{\varepsilon}}{\Gamma(1+\varepsilon)}\cosh_{\varepsilon}\left(\zeta\tau^{\varepsilon}\right)\right] = \frac{s^{2\varepsilon}+\zeta^2}{\left(s^{2\varepsilon}-\zeta^2\right)^2}.$$

Proof. It results to

$$\mathrm{M}\left[\frac{\tau^{\varepsilon}}{\Gamma(1+\varepsilon)}\cosh_{\varepsilon}\left(\zeta\tau^{\varepsilon}\right)\right] = \frac{1}{\Gamma(1+\varepsilon)}\int_0^{\infty}\left[\frac{\tau^{\varepsilon}}{\Gamma(1+\varepsilon)} \times \frac{E_{\varepsilon}\left(\zeta\tau^{\varepsilon}\right)+E_{\varepsilon}\left(-\zeta\tau^{\varepsilon}\right)}{2}\right]E_{\varepsilon}\left(-\tau^{\varepsilon}s^{\varepsilon}\right)(\mathrm{d}\tau)^{\varepsilon}$$
$$= \frac{1}{2}\left[\frac{1}{(s^{\varepsilon}-\zeta)^2} + \frac{1}{(s^{\varepsilon}+\zeta)^2}\right]$$
$$= \frac{s^{2\varepsilon}+\zeta^2}{\left(s^{2\varepsilon}-\zeta^2\right)^2}.$$

□

$$\mathrm{M}\left[\frac{1}{2}\left(\sin_{\varepsilon}\left(\zeta\tau^{\varepsilon}\right) - \frac{\zeta\tau^{\varepsilon}}{\Gamma(1+\varepsilon)}\cos_{\varepsilon}\left(\zeta\tau^{\varepsilon}\right)\right)\right] = \frac{\zeta^3}{\left(s^{2\varepsilon}+\zeta^2\right)^2}.$$

Proof. It is found that

$$\mathrm{M}\left[\frac{1}{2}\left(\sin_{\varepsilon}\left(\zeta\tau^{\varepsilon}\right) - \frac{\zeta\tau^{\varepsilon}}{\Gamma(1+\varepsilon)}\cos_{\varepsilon}\left(\zeta\tau^{\varepsilon}\right)\right)\right]$$
$$= \frac{1}{\Gamma(1+\varepsilon)}\int_0^{\infty}\left[\frac{1}{2}\left(\sin_{\varepsilon}\left(\zeta\tau^{\varepsilon}\right) - \frac{\zeta\tau^{\varepsilon}}{\Gamma(1+\varepsilon)}\cos_{\varepsilon}\left(\zeta\tau^{\varepsilon}\right)\right)\right]E_{\varepsilon}\left(-\tau^{\varepsilon}s^{\varepsilon}\right)(\mathrm{d}\tau)^{\varepsilon}$$
$$= \frac{1}{2}\left[\frac{\zeta}{s^{2\varepsilon}+\zeta^2} - \frac{\zeta\left(s^{2\varepsilon}-\zeta^2\right)}{\left(s^{2\varepsilon}+\zeta^2\right)^2}\right],$$
$$= \frac{\zeta^3}{\left(s^{2\varepsilon}+\zeta^2\right)^2}.$$

□

$$\mathrm{M}\left[\frac{1}{2}\left(\sin_{\varepsilon}\left(\zeta\tau^{\varepsilon}\right) + \frac{\zeta\tau^{\varepsilon}}{\Gamma(1+\varepsilon)}\cos_{\varepsilon}\left(\zeta\tau^{\varepsilon}\right)\right)\right] = \frac{\zeta s^{2\varepsilon}}{\left(s^{2\varepsilon}+\zeta^2\right)^2}.$$

Proof. We can observe that

$$\begin{aligned}
&\mathrm{M}\left[\frac{1}{2}\left(\sin_{\varepsilon}\left(\zeta\tau^{\varepsilon}\right)+\frac{\zeta\tau^{\varepsilon}}{\Gamma(1+\varepsilon)}\cos_{\varepsilon}\left(\zeta\tau^{\varepsilon}\right)\right)\right] \\
&=\frac{1}{\Gamma(1+\varepsilon)}\int_{0}^{\infty}\left[\frac{1}{2}\left(\sin_{\varepsilon}\left(\zeta\tau^{\varepsilon}\right)\right.\right. \\
&\quad\left.\left.+\frac{\zeta\tau^{\varepsilon}}{\Gamma(1+\varepsilon)}\cos_{\varepsilon}\left(\zeta\tau^{\varepsilon}\right)\right)\right]E_{\varepsilon}\left(-\tau^{\varepsilon}s^{\varepsilon}\right)(\mathrm{d}\tau)^{\varepsilon} \\
&=\frac{1}{2}\left[\frac{\zeta}{s^{2\varepsilon}+\zeta^{2}}+\frac{\zeta\left(s^{2\varepsilon}-\zeta^{2}\right)}{\left(s^{2\varepsilon}+\zeta^{2}\right)^{2}}\right], \\
&=\frac{\zeta s^{2\varepsilon}}{\left(s^{2\varepsilon}+\zeta^{2}\right)^{2}}.
\end{aligned}$$

□

$$\mathrm{M}\left[\cosh_{\varepsilon}\left(\zeta\tau^{\varepsilon}\right)+\frac{\zeta}{2}\frac{\tau^{\varepsilon}}{\Gamma(1+\varepsilon)}\sinh_{\varepsilon}\left(\zeta\tau^{\varepsilon}\right)\right]=\frac{s^{3\varepsilon}}{\left(s^{2\varepsilon}-\zeta^{2\varepsilon}\right)^{2}}.$$

Proof. We can see that

$$\begin{aligned}
&\mathrm{M}\left[\cosh_{\varepsilon}\left(\zeta\tau^{\varepsilon}\right)+\frac{\zeta}{2}\frac{\tau^{\varepsilon}}{\Gamma(1+\varepsilon)}\sinh_{\varepsilon}\left(\zeta\tau^{\varepsilon}\right)\right] \\
&=\frac{1}{\Gamma(1+\varepsilon)}\int_{0}^{\infty}\left[\cosh_{\varepsilon}\left(\zeta\tau^{\varepsilon}\right)\right. \\
&\quad\left.+\frac{\zeta}{2}\frac{\tau^{\varepsilon}}{\Gamma(1+\varepsilon)}\sinh_{\varepsilon}\left(\zeta\tau^{\varepsilon}\right)\right]E_{\varepsilon}\left(-\tau^{\varepsilon}s^{\varepsilon}\right)(\mathrm{d}\tau)^{\varepsilon} \\
&=\frac{s^{\varepsilon}}{s^{2\varepsilon}-\zeta^{2}}+\frac{\zeta^{2}s^{\varepsilon}}{\left(s^{2\varepsilon}-\zeta^{2}\right)^{2}} \\
&=\frac{s^{3\varepsilon}}{\left(s^{2\varepsilon}-\zeta^{2\varepsilon}\right)^{2}}.
\end{aligned}$$

□

$$\mathrm{M}\left[\frac{1}{2}\left(\frac{\zeta\tau^{\varepsilon}}{\Gamma(1+\varepsilon)}\cosh_{\varepsilon}\left(\zeta\tau^{\varepsilon}\right)-\sinh_{\varepsilon}\left(\zeta\tau^{\varepsilon}\right)\right)\right]=\frac{\zeta^{3}}{\left(s^{2\varepsilon}-\zeta^{2}\right)^{2}}.$$

Proof. It results in

$$\begin{aligned}
&\mathrm{M}\left[\frac{1}{2}\left(\frac{\zeta\tau^{\varepsilon}}{\Gamma(1+\varepsilon)}\cosh_{\varepsilon}\left(\zeta\tau^{\varepsilon}\right)-\sinh_{\varepsilon}\left(\zeta\tau^{\varepsilon}\right)\right)\right] \\
&=\frac{1}{\Gamma(1+\varepsilon)}\int_{0}^{\infty}\left[\frac{1}{2}\left(\frac{\zeta\tau^{\varepsilon}}{\Gamma(1+\varepsilon)}\cosh_{\varepsilon}\left(\zeta\tau^{\varepsilon}\right)-\sinh_{\varepsilon}\left(\zeta\tau^{\varepsilon}\right)\right)\right]E_{\varepsilon}\left(-\tau^{\varepsilon}s^{\varepsilon}\right)(\mathrm{d}\tau)^{\varepsilon}
\end{aligned}$$

$$= \frac{1}{2}\left[\frac{\zeta\left(s^{2\varepsilon}+\zeta^{2}\right)}{\left(s^{2\varepsilon}-\zeta^{2}\right)^{2}}-\frac{\zeta}{s^{2\varepsilon}-\zeta^{2}}\right]$$

$$= \frac{\zeta^{3}}{\left(s^{2\varepsilon}-\zeta^{2}\right)^{2}}.$$

□

$$M\left[\sigma_{\varepsilon}\left(\tau-\tau_{0}\right)\right]=E_{\varepsilon}\left(-\tau_{0}^{\varepsilon} s^{\varepsilon}\right).$$

Proof.

$$\begin{aligned} M\left[\sigma_{\varepsilon}\left(\tau-\tau_{0}\right)\right] &= \frac{1}{\Gamma(1+\varepsilon)} \int_{0}^{\infty} \sigma_{\varepsilon}\left(\tau-\tau_{0}\right) E_{\varepsilon}\left(-\tau^{\varepsilon} s^{\varepsilon}\right)(\mathrm{d}\tau)^{\varepsilon} \\ &= E_{\varepsilon}\left(-\tau_{0}^{\varepsilon} s^{\varepsilon}\right). \end{aligned}$$

We present the list of the local fractional Laplace transforms (Table F.1):

$$M[\theta(\tau)]=\Theta(s)=\frac{1}{\Gamma(1+\varepsilon)} \int_{0}^{\infty} \theta(\tau) E_{\varepsilon}\left(-\tau^{\varepsilon} s^{\varepsilon}\right)(\mathrm{d}\tau)^{\varepsilon}.$$

□

Table F.1 **Tables for local fractional Laplace transform operators**

Transforms	Functions
$\frac{1}{s^{\varepsilon}}$	1
1	$\delta_{\varepsilon}(\tau)$
$E_{\varepsilon}\left(-\tau_{0}^{\varepsilon} s^{\varepsilon}\right)$	$\sigma_{\varepsilon}\left(\tau-\tau_{0}\right)$
$\frac{1}{s^{\varepsilon}}\left[E_{\varepsilon}\left(-\tau_{0}^{\varepsilon} s^{\varepsilon}\right)-E_{\varepsilon}\left(-\tau_{1}^{\varepsilon} s^{\varepsilon}\right)\right]$	$\operatorname{rect}_{\varepsilon}\left(\tau, \tau_{0}, \tau_{1}\right)$
$\frac{\Gamma(1+2\varepsilon)}{2\Gamma^{3}(1+\varepsilon)}\left(\frac{\pi}{s}\right)^{\frac{\varepsilon}{2}}$	$\tau^{-\frac{\varepsilon}{2}}$
$\frac{1}{s^{\varepsilon}-a^{\varepsilon}}$	$E_{\varepsilon}\left(a^{\varepsilon} \tau^{\varepsilon}\right)$
$\frac{s^{\varepsilon}}{s^{2\varepsilon}+\eta^{2\varepsilon}}$	$\cos_{\varepsilon}\left(\eta^{\varepsilon} \tau^{\varepsilon}\right)$
$\frac{\eta^{\varepsilon}}{s^{2\varepsilon}+\eta^{2\varepsilon}}$	$\sin_{\varepsilon}\left(\eta^{\varepsilon} \tau^{\varepsilon}\right)$
$\frac{s^{\varepsilon}}{s^{2\varepsilon}-\eta^{2\varepsilon}}$	$\cosh_{\varepsilon}\left(\eta^{\varepsilon} \tau^{\varepsilon}\right)$
$\frac{\eta^{\varepsilon}}{s^{2\varepsilon}-\eta^{2\varepsilon}}$	$\sinh_{\varepsilon}\left(\eta^{\varepsilon} \tau^{\varepsilon}\right)$
$\frac{1}{s^{\varepsilon(k+1)}}$	$\frac{\tau^{k\varepsilon}}{\Gamma(1+k\varepsilon)}$
$\frac{1}{(s-a)^{(k+1)\varepsilon}}$	$\frac{\tau^{k\varepsilon}}{\Gamma(1+k\varepsilon)} E_{\varepsilon}\left(a^{\varepsilon} \tau^{\varepsilon}\right)$

Table F.1 Continued

Transforms	Functions
$\frac{(s-a)^{2\varepsilon}}{(s-a)^{2\varepsilon}+\eta^{2\varepsilon}}$	$E_\varepsilon(a^\varepsilon\tau^\varepsilon)\cos_\varepsilon(\eta^\varepsilon\tau^\varepsilon)$
$\frac{\eta^\varepsilon}{(s-a)^{2\varepsilon}+\eta^{2\varepsilon}}$	$E_\varepsilon(a^\varepsilon\tau^\varepsilon)\sin_\varepsilon(\eta^\varepsilon\tau^\varepsilon)$
$\frac{(s-a)^\varepsilon}{(s-a)^{2\varepsilon}-\eta^{2\varepsilon}}$	$E_\varepsilon(a^\varepsilon\tau^\varepsilon)\cosh_\varepsilon(\eta^\varepsilon\tau^\varepsilon)$
$\frac{\eta^\varepsilon}{(s-a)^{2\varepsilon}-\eta^{2\varepsilon}}$	$E_\varepsilon(a^\varepsilon\tau^\varepsilon)\sinh_\varepsilon(\eta^\varepsilon\tau^\varepsilon)$
$\frac{b-a}{(s^\varepsilon+a)(s^\varepsilon+b)}$	$E_\varepsilon(-a\tau^\varepsilon)-E_\varepsilon(-b\tau^\varepsilon)$
$\frac{a^2}{(s^\varepsilon+a)s^{2\varepsilon}}$	$E_\varepsilon(-a\tau^\varepsilon)+\frac{a\tau^\varepsilon}{\Gamma(1+\varepsilon)}-1$
$\frac{1}{s^{2\varepsilon}(s^{2\varepsilon}+\eta^2)}$	$\frac{1}{\eta^2}\frac{\tau^\varepsilon}{\Gamma(1+\varepsilon)}-\frac{1}{\eta^3}\sin_\varepsilon(\eta\tau^\varepsilon)$
$\frac{1}{(s^{2\varepsilon}+\mu^2)(s^{2\varepsilon}+\eta^2)}$	$\frac{1}{\eta^2-\mu^2}\left(\frac{\sin_\varepsilon(\mu\tau^\varepsilon)}{\mu}-\frac{\sin_\varepsilon(\eta\tau^\varepsilon)}{\eta}\right)$
$\frac{s^\varepsilon}{(s^{2\varepsilon}+a^2)(s^{2\varepsilon}+b^2)}$	$\frac{1}{a^2-b^2}(\cos_\varepsilon(b\tau^\varepsilon)-\cos_\varepsilon(a\tau^\varepsilon))$
$\frac{1}{(s^{2\varepsilon}-a^2)(s^{2\varepsilon}-b^2)}$	$\frac{1}{a^2-b^2}\left(\frac{\sinh_\varepsilon(a\tau^\varepsilon)}{a}-\frac{\sinh_\varepsilon(b\tau^\varepsilon)}{b}\right)$
$\frac{s^\varepsilon}{(s^{2\varepsilon}-a^2)(s^{2\varepsilon}-b^2)}$	$\frac{1}{a^2-b^2}(\cosh_\varepsilon(a\tau^\varepsilon)-\cosh_\varepsilon(b\tau^\varepsilon))$
$\frac{1}{s^\varepsilon}E_\varepsilon(-\tau_0^\varepsilon s^\varepsilon)$	$H_\varepsilon(\tau-\tau_0)$
$\frac{\zeta s^\varepsilon}{(s^{2\varepsilon}+\zeta^{2\varepsilon})^2}$	$\frac{\tau^\varepsilon}{\Gamma(1+\varepsilon)}\sin_\varepsilon(\zeta\tau^\varepsilon)$
$\frac{s^{2\varepsilon}-\zeta^2}{(s^{2\varepsilon}+\zeta^2)^2}$	$\frac{\tau^\varepsilon}{\Gamma(1+\varepsilon)}\cos_\varepsilon(\zeta\tau^\varepsilon)$
$\frac{2\zeta s^\varepsilon}{(s^{2\varepsilon}-\zeta^{2\varepsilon})^2}$	$\frac{\tau^\varepsilon}{\Gamma(1+\varepsilon)}\sinh_\varepsilon(\zeta\tau^\varepsilon)$
$\frac{s^{2\varepsilon}+\zeta^2}{(s^{2\varepsilon}-\zeta^2)^2}$	$\frac{\tau^\varepsilon}{\Gamma(1+\varepsilon)}\cosh_\varepsilon(\zeta\tau^\varepsilon)$
$\frac{\zeta^3}{(s^{2\varepsilon}+\zeta^2)^2}$	$\frac{1}{2}\left(\sin_\varepsilon(\zeta\tau^\varepsilon)-\frac{\zeta\tau^\varepsilon}{\Gamma(1+\varepsilon)}\cos_\varepsilon(\zeta\tau^\varepsilon)\right)$
$\frac{\zeta s^{2\varepsilon}}{(s^{2\varepsilon}+\zeta^2)^2}$	$\frac{1}{2}\left(\sin_\varepsilon(\zeta\tau^\varepsilon)+\frac{\zeta\tau^\varepsilon}{\Gamma(1+\varepsilon)}\cos_\varepsilon(\zeta\tau^\varepsilon)\right)$
$\frac{s^{3\varepsilon}}{(s^{2\varepsilon}-\zeta^{2\varepsilon})^2}$	$\cosh_\varepsilon(\zeta\tau^\varepsilon)+\frac{\zeta}{2}\frac{\tau^\varepsilon}{\Gamma(1+\varepsilon)}\sinh_\varepsilon(\zeta\tau^\varepsilon)$
$\frac{\zeta^3}{(s^{2\varepsilon}-\zeta^2)^2}$	$\frac{1}{2}\left(\frac{\zeta\tau^\varepsilon}{\Gamma(1+\varepsilon)}\cosh_\varepsilon(\zeta\tau^\varepsilon)-\sinh_\varepsilon(\zeta\tau^\varepsilon)\right)$

Bibliography

[1] X.-J. Yang, Advanced Local Fractional Calculus and Its Applications, World Science, New York, 2012.

[2] K.M. Kolwankar, A.D. Gangal, Fractional differentiability of nowhere differentiable functions and dimensions, Chaos 6 (4) (1996) 505–513.

[3] A.A. Kilbas, H.M. Srivastava, J.J. Trujillo, Theory and Applications of Fractional Differential Equations North-Holland Mathematical Studies, vol. 204, Elsevier (North-Holland) Science Publishers, Amsterdam, London, New York, 2006.

[4] J. Sabatier, O.P. Agrawal, J.A.T. Machado, Advances in Fractional Calculus, Springer, Berlin, Heidelberg, New York, 2007.

[5] A. Carpinteri, F. Mainardi, Fractals and Fractional Calculus in Continuum Mechanics, Springer, New York, 1997.

[6] K.B. Oldham, J. Spanier, The Fractional Calculus, Academic Press, New York, 1974.

[7] K.M. Kolwankar, A.D. Gangal, Hölder exponents of irregular signals and local fractional derivatives, Pramana 48 (1) (1997) 49–68.

[8] K.M. Kolwankar, A.D. Gangal, Local fractional Fokker-Planck equation, Phys. Rev. Lett. 80 (2) (1998) 214.

[9] A. Carpinteri, P. Cornetti, A fractional calculus approach to the description of stress and strain localization in fractal media, Chaos Soliton. Fract. 13 (1) (2002) 85–94.

[10] A. Carpinteri, B. Chiaia, P. Cornetti, Static-kinematic duality and the principle of virtual work in the mechanics of fractal media, Comput. Methods Appl. Mech. Eng. 191 (1) (2001) 3–19.

[11] A. Carpinteri, B. Chiaia, P. Cornetti, The elastic problem for fractal media: basic theory and finite element formulation, Comput. Struct. 82 (6) (2004) 499–508.

[12] A. Carpinteri, P. Cornetti, K.M. Kolwankar, Calculation of the tensile and flexural strength of disordered materials using fractional calculus, Chaos Soliton. Fract. 21 (3) (2004) 623–632.

[13] Y. Chen, Y. Yan, K. Zhang, On the local fractional derivative, J. Math. Anal. Appl. 362 (1) (2010) 17–33.

[14] F.B. Adda, J. Cresson, About non-differentiable functions, J. Math. Anal. Appl. 263 (2) (2001) 721–737.

[15] A. Babakhani, V. Daftardar-Gejji, On calculus of local fractional derivatives, J. Math. Anal. Appl. 270 (1) (2002) 66–79.

[16] X.-J. Yang, Local fractional integral transforms, Prog. Nonlinear Sci. 4 (1) (2011) 1–225.

[17] W. Chen, Time-space fabric underlying anomalous diffusion, Chaos Soliton. Fract. 28 (4) (2006) 923–929.

[18] W. Chen, H. Sun, X. Zhang, D. Korovsak, Anomalous diffusion modeling by fractal and fractional derivatives, Comput. Math. Appl. 59 (5) (2010) 1754–1758.

[19] J.-H. He, A new fractal derivation, Therm. Sci. 15 (Suppl. 1) (2011) 145–147.

[20] J.-H. He, S.K. Elagan, Z.-B. Li, Geometrical explanation of the fractional complex transform and derivative chain rule for fractional calculus, Phys. Lett. A 376 (4) (2012) 257–259.

[21] X.-J. Yang, Local Fractional Functional Analysis and Its Applications, Asian Academic Publisher Limited, Hong Kong, 2011.

[22] X.-J. Yang, H.M. Srivastava, J.-H. He, D. Baleanu, Cantor-type cylindrical-coordinate method for differential equations with local fractional derivatives, Phys. Lett. A 377 (28) (2013) 1696–1700.

[23] X.-J. Yang, D. Baleanu, H.M. Srivastava, Local fractional similarity solution for the diffusion equation defined on Cantor sets, Appl. Math. Lett. 47 (2015) 54–60.

[24] E.C.D. Oliveira, J.A.T. Machado, A review of definitions for fractional derivatives and integral, Math. Probl. Eng. (2014) Article ID 238459, 6 pages.

[25] Y. Zhang, Solving initial-boundary value problems for local fractional differential equation by local fractional Fourier series method, Abstr. Appl. Anal. (2014) Article ID 912464, 5 pages.

[26] W. Chen, X.-D. Zhang, D. Korovsak, Investigation on fractional and fractal derivative relaxation-oscillation models, Int. J. Nonlinear Sci. Numer. Simul. 11 (1) (2010) 3–10.

[27] A.-M. Yang, Y.-Z. Zhang, Y. Long, The Yang-Fourier transforms to heat-conduction in a semi-infinite fractal bar, Therm. Sci. 17 (3) (2013) 707–713.

[28] S.G. Samko, A.A. Kilbas, O.I. Marichev, Fractional Integrals and Derivatives: Theory and Applications, Gordon and Breach, Yverdon, 1993.

[29] M.D. Ortigueira, Fractional central differences and derivatives, J. Vib. Control. 14 (9-10) (2008) 1255–1266.

[30] M.D. Ortigueira, J.J. Trujillo, Generalized Grünwald-Letnikov fractional derivative and its Laplace and Fourier transforms, J. Comput. Nonlinear Dyn. 6 (3) (2011) Article ID 034501.

[31] J.A.T. Machado, Fractional derivatives: probability interpretation and frequency response of rational approximations, Commun. Nonlinear Sci. Numer. Simul. 14 (9) (2009) 3492–3497.

[32] J.A.T. Machado, Fractional coins and fractional derivatives, Abstr. Appl. Anal. (2013) Article ID 205097, 5 pages.

[33] M.D. Ortigueira, F. Coito, From differences to derivatives, Fract. Calc. Appl. Anal. 7 (4) (2004) 459.

[34] A. Atangana, A. Secer, A note on fractional order derivatives and table of fractional derivatives of some special functions, Abstr. Appl. Anal. (2013) Article ID 279681, 8 pages.

[35] G. Jumarie, On the representation of fractional Brownian motion as an integral with respect to $(dt)^\alpha$, Appl. Math. Lett. 18 (7) (2005) 739–748.

[36] G. Jumarie, Modified Riemann-Liouville derivative and fractional Taylor series of nondifferentiable functions further results, Comput. Math. Appl. 51 (9) (2006) 1367–1376.

[37] G. Jumarie, Modeling fractional stochastic systems as non-random fractional dynamics driven by Brownian motions, Appl. Math. Model. 32 (5) (2008) 836–859.

[38] G. Jumarie, Table of some basic fractional calculus formulae derived from a modified Riemann-Liouville derivative for nondifferentiable functions, Appl. Math. Lett. 22 (3) (2009) 378–385.

[39] C.-P. Li, Z.-G. Zhao, Introduction to fractional integrability and differentiability, Eur. Phys. J. 193 (1) (2011) 5–26.

[40] R. Khalil, M. Al-Horani, A. Yousef, M. Sababheh, A new definition of fractional derivative, J. Comput. Appl. Math. 264 (2014) 65–70.

[41] T. Abdeljawad, On conformable fractional calculus, J. Comput. Appl. Math. 279 (2015) 57–66.

[42] F. Sabzikar, M.M. Meerschaert, J. Chen, Tempered fractional calculus, J. Comput. Phys. (2014) URL http://dx.doi.org/10.1016/j.jcp.2014.04.024.
[43] U.N. Katugampola, New approach to a generalized fractional integral, Appl. Math. Comput. 218 (3) (2011) 860–865.
[44] M. Caputo, M. Fabrizio, A new definition of fractional derivative without singular kernel, Prog. Fract. Differ. Appl. 1 (2) (2015) 73–85.
[45] J. Losada, J.J. Nieto, Properties of a new fractional derivative without singular kernel, Prog. Fract. Differ. Appl. 1 (2) (2015) 87–92.
[46] I. Podlubny, Fractional Differential Equations: An Introduction to Fractional Derivatives, Fractional Differential Equations to Methods of Their Solution and Some of Their Applications, vol. 198, Academic Press, New York, 1998.
[47] J.A.T. Machado, V. Kiryakova, F. Mainardi, A poster about the recent history of fractional calculus, Fract. Calc. Appl. Anal. 13 (3) (2010) 329–334.
[48] J.A.T. Machado, V. Kiryakova, F. Mainardi, Recent history of fractional calculus, Commun. Nonlinear Sci. Numer. Simul. 16 (3) (2011) 1140–1153.
[49] R.L. Magin, Fractional Calculus in Bioengineering, vol. 149, Begell House Publishers, Redding, 2006.
[50] V.S. Kiryakova, Generalized Fractional Calculus and Applications, vol. 301, Longman Scientific and Technical, Harlow (Essex), 1993.
[51] L. Debnath, A brief historical introduction to fractional calculus, Int. J. Math. Educ. Sci. Technol. 35 (4) (2004) 487–501.
[52] V.E. Tarasov, Fractional Dynamics: Applications of Fractional Calculus to Dynamics of Particles, Fields and Media, Springer Science and Business Media, New York, 2011.
[53] M.D. Ortigueira, Fractional Calculus for Scientists and Engineers, vol. 84, Springer Science and Business Media, New York, 2011.
[54] M.M. Meerschaert, A. Sikorskii, Stochastic Models for Fractional Calculus, vol. 43, Walter de Gruyter, New York, 2011.
[55] J.A.T. Machado, A.M. Galhano, J.J. Trujillo, On development of fractional calculus during the last fifty years, Scientometrics 98 (1) (2014) 577–582.
[56] C.A. Monje, Y. Chen, B.M. Vinagre, D. Xue, V. Feliu-Batlle, Fractional-Order Systems and Controls: Fundamentals and Applications, Springer Science and Business Media, New York, 2010.
[57] D. Baleanu, K. Diethelm, E. Scalas, J.J. Trujillo, Models and Numerical Methods, vol. 3, World Scientific, Singapore, 2012, 10–16 pp.
[58] R. Caponetto, Fractional Order Systems: Modeling and Control Applications, vol. 72, World Scientific, Singapore, 2010.
[59] G.A. Anastassiou, Fractional Differentiation Inequalities, Springer, New York, 2009.
[60] B. West, M. Bologna, P. Grigolini, Physics of Fractal Operators, Springer Science and Business Media, New York, 2003.
[61] G.M. Zaslavsky, Hamiltonian Chaos and Fractional Dynamics, Oxford University (Clarendon) Press, Oxford, London, New York, 2008.
[62] D. Baleanu, J.A.T. Machado, A.-C. Luo, Fractional Dynamics and Control, Springer Science and Business Media, New York, 2011.
[63] J. Klafter, S.-C. Lim, R. Metzler, Fractional Dynamics: Recent Advances, World Scientific, Singapore, 2012.
[64] K. Diethelm, The Analysis of Fractional Differential Equations: An Application-Oriented Exposition Using Differential Operators of Caputo Type, Springer Science and Business Media, New York, 2004.

[65] F. Mainardi, Fractional Calculus Waves in Linear Viscoelasticity: An Introduction to Mathematical Models, World Scientific, Singapore, 2010.

[66] A.B. Malinowska, D.F.M. Torres, Fractional Calculus of Variations, Imperial College Press, Singapore, 2012.

[67] S. Das, I. Pan, Fractional Order Signal Processing: Introductory Concepts and Applications, Springer Science and Business Media, New York, 2011.

[68] V.V. Uchaikin, Fractional Derivatives for Physicists and Engineers, Springer, Berlin, 2013.

[69] R. Herrmann, Fractional Calculus: An Introduction for Physicists, World Scientific, Singapore, 2014.

[70] G.-S. Chen, Generalizations of Hölder's and some related integral inequalities on fractal space, J. Funct. Spaces Appl. (2013) Article ID 198405, 9 pages.

[71] W. Wei, H.M. Srivastava, Y. Zhang, L. Wang, P. Shen, J. Zhang, A local fractional integral inequality on fractal space analogous to Anderson's inequality, Abstr. Appl. Anal. (2014) Article ID 797561, 7 pages.

[72] G.-S. Chen, H.M. Srivastava, P. Wang, W. Wie, Some further generalizations of Hölder's inequality and related results on fractal space, Abstr. Appl. Anal. (2014) Article ID 832802, 7 pages.

[73] Y. Zhang, D. Baleanu, X.-J. Yang, On a local fractional wave equation under fixed entropy arising in fractal hydrodynamics, Entropy 16 (12) (2014) 6254–6262.

[74] Y.-Y. Li, L.Y. Zhao, G.-N. Xie, D. Baleanu, X.-J. Yang, K. Zhao, Local fractional Poisson and Laplace equations with applications to electrostatics in fractal domain, Adv. Math. Phys. (2014) Article ID 590574, 5 pages.

[75] X.-J. Yang, D. Baleanu, Local fractional variational iteration method for Fokker-Planck equation on a Cantor set, Acta Univ. 23 (2) (2013) 3–8.

[76] X.-J. Yang, D. Baleanu, J.A.T. Machado, Mathematical aspects of the Heisenberg uncertainty principle within local fractional Fourier analysis, Bound. Value Probl. 1 (2013) 1–16.

[77] W.-H. Su, D. Baleanu, X.-J. Yang, H. Jafari, Damped wave equation and dissipative wave equation in fractal strings within the local fractional variational iteration method, Fixed Point Theory Appl. 1 (2013) 1–11.

[78] X.-J. Yang, D. Baleanu, J.A.T. Machado, Systems of Navier-Stokes equations on Cantor sets, Math. Probl. Eng. (2013) Article ID 769724, 8 pages.

[79] H.-Y. Liu, J.-H. He, Z.B. Li, Fractional calculus for nanoscale flow and heat transfer, Int. J. Numer. Methods Heat Fluid Flow 24 (6) (2014) 1227–1250.

[80] Y. Zhao, D. Baleanu, C. Cattani, D.-F. Cheng, X.-J. Yang, Maxwell's equations on Cantor sets: a local fractional approach, Adv. High Energy Phys. (2013) Article ID 686371, 6 pages.

[81] L.-F. Wang, X.-J. Yang, D. Baleanu, C. Cattani, Y. Zhao, Fractal dynamical model of vehicular traffic flow within the local fractional conservation laws, Abstr. Appl. Anal. (2014) Article ID 635760, 5 pages.

[82] A.-M. Yang, Y.-Z. Zhang, C. Cattani, G.-N. Xie, M.M. Rashidi, Y.-J. Zhou, X.-J. Yang, Application of local fractional series expansion method to solve Klein-Gordon equations on Cantor sets, Abstr. Appl. Anal. (2014) Article ID 372741, 6 pages.

[83] X.-J. Yang, D. Baleanu, J.-H. He, Transport equations in fractal porous media within fractional complex transform method, Proc. Rom. Acad. Series A 14 (4) (2013) 287–292.

[84] Y.-J. Hao, H.M. Srivastava, H. Jafari, X.-J. Yang, Helmholtz and diffusion equations associated with local fractional derivative operators involving the Cantorian and Cantor-type cylindrical coordinates, Adv. Math. Phys. (2013) Article ID 754248, 5 pages.

[85] X.-J. Yang, J. Hristov, H.M. Srivastava, B. Ahmad, Modelling fractal waves on shallow water surfaces via local fractional Korteweg-de Vries equation, Abstr. Appl. Anal. (2014) Article ID 278672, 10 pages.
[86] X.-J. Yang, J.A.T. Machado, J. Hristov, Nonlinear dynamics for local fractional Burgers' equation arising in fractal flow, Nonlinear Dyn. (2015) doi:10.1007/s11071-015-2085-2.
[87] X.-F. Niu, C.-L. Zhang, Z.-B. Li, Y. Zhao, Local fractional derivative boundary value problems for Tricomi equation arising in fractal transonic flow, Abstr. Appl. Anal. (2014) Article ID 872318, 5 pages.
[88] C. Cattani, H.M. Srivastava, X.-J. Yang, Fractional Dynamics, Emerging Science Publishers, Berlin, 2015.
[89] Y.-J. Yang, D. Baleanu, M.C. Baleanu, Observing diffusion problems defined on Cantor sets in different coordinate systems, Thermal Sci. (2015) doi:10.2298/TSCI141126065Y.
[90] H.F. Davis, Fourier Series and Orthogonal Functions, Dover, New York, 1963.
[91] H. Dym, H.P. McKean, Fourier Series and Integrals, Academic Press, New York, 1972.
[92] T.W. Körner, Fourier Analysis, Cambridge University Press, Cambridge, London, New York, 1988.
[93] N. Morrison, Introduction to Fourier Analysis, John Wiley and Sons, New York, 1994.
[94] M. Liao, X.-J. Yang, Q. Yan, A new viewpoint to Fourier analysis in fractal space, in: Advances in Applied Mathematics and Approximation Theory, Springer, New York, 2013 pp. 397–409.
[95] D. Baleanu, X.-J. Yang, Local fractional Fourier series with applications to representations of fractal signals, in: ASME 2013 International Design Engineering Technical Conferences and Computers and Information in Engineering Conference, American Society of Mechanical Engineers, 2013, pp. V07BT10A037–V07BT10A037.
[96] X.-J. Yang, D. Baleanu, J.A.T. Machado, Application of the local fractional Fourier series to fractal signals, in: Discontinuity and Complexity in Nonlinear Physical Systems, Springer International Publishing, 2014, pp. 63–89.
[97] Y. Zhao, D. Baleanu, M.C. Baleanu, D.-F. Cheng, X.-J. Yang, Mappings for special functions on Cantor sets and special integral transforms via local fractional operators, Abstr. Appl. Anal. (2013) Article ID 316978, 6 pages.
[98] Z.-Y. Chen, C. Cattani, W.-P. Zhong, Signal processing for nondifferentiable data defined on Cantor sets: a local fractional Fourier series approach, Adv. Math. Phys. (2014) Article ID 561434, 7 pages.
[99] X.-J. Yang, Y. Zhang, A.-M. Yang, 1-D heat conduction in a fractal medium: a solution by the local fractional Fourier series method, Therm. Sci. 17 (3) (2013) 953–956.
[100] X.-J. Yang, D. Baleanu, J.A.T. Machado, On analytical methods for differential equations with local fractional derivative operators, Chapter 4, in: R.A.Z. Daou, X. Moreau (Eds.), Fractional Calculus: Theory, Nova Science Publishers, New York, 2014, pp. 65–88.
[101] Y.-J. Yang, D. Baleanu, X.-J. Yang, Analysis of fractal wave equations by local fractional Fourier series method, Adv. Math. Phys. (2013) Article ID 632309, 6 pages.
[102] M.-S. Hu, R.P. Agarwal, X.-J. Yang, Local fractional Fourier series with application to wave equation in fractal vibrating string, Abstr. Appl. Anal. (2012) Article ID 567401, 15 pages.
[103] Y.-J. Yang, S.-Q. Wang, Local fractional Fourier series method for solving nonlinear equations with local fractional operators, Math. Probl. Eng. 2015 (2015) 1–9, Article ID 481905.
[104] S.-Q. Wang, Y.-J. Yang, H.K. Jassim, Local fractional function decomposition method for solving inhomogeneous wave equations with local fractional derivative, Abstr. Appl. Anal. (2014) Article ID 176395, 7 pages.

[105] W.-P. Zhong, F. Gao, X.-M. Shen, Applications of Yang-Fourier transform to local fractional equations with local fractional derivative and local fractional integral, Adv. Mater. Res. 461 (2012) 306–310.

[106] J.-H. He, Asymptotic methods for solitary solutions and compactions, Abstr. Appl. Anal. (2012) Article ID 916793, 130 pages.

[107] X.-J. Yang, M.-K. Liao, J.-W. Chen, A novel approach to processing fractal signals using the Yang-Fourier transforms, Proc. Eng. 29 (2012) 2950–2954.

[108] K. Liu, R.-J. Hu, C. Cattani, G.-N. Xie, X.-J. Yang, Y. Zhao, Local fractional Z-transforms with applications to signals on Cantor sets, Abstr. Appl. Anal. (2014) Article ID 638648, 6 pages.

[109] J.L. Schiff, The Laplace Transform: Theory and Applications, Springer Science and Business Media, New York, 1999.

[110] Y.-Z. Zhang, A.-M. Yang, Y. Long, Initial boundary value problem for fractal heat equation in the semi-infinite region by Yang-Laplace transform, Therm. Sci. 18 (2) (2014) 677–681.

[111] C.-F. Liu, S.-S. Kong, S.-J. Yuan, Reconstructive schemes for variational iteration method within Yang-Laplace transform with application to fractal heat conduction problem, Therm. Sci. 17 (3) (2013) 715–721.

[112] C.-G. Zhao, A.-M. Yang, H. Jafari, A. Haghbin, The Yang-Laplace transform for solving the IVPs with local fractional derivative, Abstr. Appl. Anal. (2014) Article ID 386459, 5 pages.

[113] Y. Li, L.-F. Wang, S.-D. Zeng, Y. Zhao, Local fractional Laplace variational iteration method for fractal vehicular traffic flow, Adv. Math. Phys. (2014) Article ID 649318, 7 pages.

[114] A.-M. Yang, J. Li, H.M. Srivastava, G.-N. Xie, X.-J. Yang, Local fractional Laplace variational iteration method for solving linear partial differential equations with local fractional derivative, Discrete Dyn Nat Soc (2014) Article ID 365981, 8 pages.

[115] S.-P. Yan, H. Jafari, H.K. Jassim, Local fractional Z-transforms with applications to signals on Cantor sets, Adv. Math. Phys. (2014) Article ID 161580, 7 pages.

[116] J.-H. He, Variational iteration method: a kind of non-linear analytical technique: some examples, Int. J. Nonlinear Mech. 34 (4) (1999) 699–708.

[117] J.-H. He, Variational iteration method: some recent results and new interpretations, J. Comput. Appl. Math. 207 (1) (2007) 3–17.

[118] X.-J. Yang, D. Baleanu, Fractal heat conduction problem solved by local fractional variation iteration method, Therm. Sci. 17 (2) (2013) 625–628.

[119] J.-H. He, Local fractional variational iteration method for fractal heat transfer in silk cocoon hierarchy, Nonlinear Sci. Lett. A 4 (1) (2013) 15–20.

[120] J.-H. He, A tutorial review on fractal spacetime and fractional calculus, Int. J. Theor. Phys. 53 (11) (2014) 3698–3718.

[121] X.-J. Yang, D. Baleanu, Y. Khan, S.T. Mohyud-Din, Local fractional variational iteration method for diffusion and wave equations on Cantor sets, Rom. J. Phys. 59 (1–2) (2014) 36–48.

[122] D. Baleanu, H.M. Srivastava, X.-J. Yang, Local fractional variational iteration algorithms for the parabolic Fokker-Planck equation defined on Cantor sets, Prog. Fract. Differ. Appl. 1 (1) (2015) 1–10.

[123] L. Chen, Y. Zhao, H. Jafari, J.A.T. Machado, X.-J. Yang, Local fractional variational iteration method for local fractional Poisson equations in two independent variables, Abstr. Appl. Anal. (2014) Article ID 484323, 7 pages.

[124] W.-H. Su, D. Baleanu, X.-J. Yang, H. Jafari, Damped wave equation and dissipative wave equation in fractal strings within the local fractional variational iteration method, Fixed Point Theory Appl. 1 (2013) 1–11.

[125] D. Baleanu, J.A.T. Machado, C. Cattani, M.C. Baleanu, X.-J. Yang, Local fractional variational iteration and decomposition methods for wave equation on Cantor sets within local fractional operators, Abstr. Appl. Anal. (2014) Article ID 535048, 6 pages.

[126] G. Adomian, Convergent series solution of nonlinear equations, J. Comput. Appl. Math. 11 (2) (1984) 225–230.

[127] G. Adomian, Solving Frontier Problems of Physics: The Decomposition Method, Kluwer Academic Publishers, Boston, 1994.

[128] X.-J. Yang, D. Baleanu, W.-P. Zhong, Approximate solutions for diffusion equations on Cantor space-time, Proc. Rom. Acad. Series A 14 (2) (2013) 127–133.

[129] A.-M. Yang, C. Cattani, H. Jafari, X.-J. Yang, Analytical solutions of the one-dimensional heat equations arising in fractal transient conduction with local fractional derivative, Abstr. Appl. Anal. (2013) Article ID 462535, 5 pages.

Index

Note: Page numbers followed by *f* indicate figures and *t* indicate tables.

D

E

Printed in the United States
By Bookmasters